Albrecht Schnabel, Vanessa Farr (Eds)

Back to the Roots: Security Sector Reform and Development

Geneva Centre for the
Democratic Control of Armed Forces
(DCAF)

LIT

Albrecht Schnabel, Vanessa Farr (Eds)

Back to the Roots: Security Sector Reform and Development

LIT

Gedruckt auf alterungsbeständigem Werkdruckpapier entsprechend
ANSI Z3948 DIN ISO 9706

Bibliographic information published by the Deutsche Nationalbibliothek
The Deutsche Nationalbibliothek lists this publication in the Deutsche Nationalbibliografie; detailed bibliographic data are available in the Internet at http://dnb.d-nb.de.

ISBN 978-3-643-80117-3

A catalogue record for this book is available from the British Library

Klosbachstr. 107
CH-8032 Zürich
Tel. +41 (0) 44-251 75 05
Fax +41 (0) 44-251 75 06
e-Mail: zuerich@lit-verlag.ch
http://www.lit-verlag.ch

LIT VERLAG Dr. W. Hopf
Berlin 2012
Fresnostr. 2
D-48159 Münster
Tel. +49 (0) 2 51-620 320
Fax +49 (0) 2 51-23 19 72
e-Mail: lit@lit-verlag.de
http://www.lit-verlag.de

Distribution:
In Germany: LIT Verlag Fresnostr. 2, D-48159 Münster
Tel. +49 (0) 2 51-620 32 22, Fax +49 (0) 2 51-922 60 99, e-mail: vertrieb@lit-verlag.de
In Austria: Medienlogistik Pichler-ÖBZ, e-mail: mlo@medien-logistik.at
In the UK: Global Book Marketing, e-mail: mo@centralbooks.com

Contents

Part IV: SSR, DDR and Development

Part V: Towards 'Developing' SSR Policy?

Part VI: Conclusion

Preface

The Geneva Centre for the Democratic Control of Armed Forces (DCAF) is an international foundation whose mission is to assist the international community in promoting good governance and reform of the security sector. Beyond a range of publications linked to specific activities, each year DCAF dedicates one book to a topic that is of particular relevance to its research and operational agenda. The first volume in the Yearly Book series, *Challenges of Security Sector Governance*, was published in 2003. Subsequent Yearly Books focused on *Reform and Reconstruction of the Security Sector* (2004), *Security Governance in Post-Conflict Peacebuilding* (2005), *Private Actors and Security Governance* (2006), *Intergovernmental Organisations and Security Sector Reform* (2007), *Local Ownership and Security Sector Reform* (2008), *Security Sector Reform in Challenging Environments* (2009) and, last year, *Security Sector Transformation in Africa* (2010). *Back to the Roots: Security Sector Reform and Development* is the title of the 2011 edition, which hopes to provide practitioners and academics within the larger development and security communities with lessons, suggestions and practical advice for approaching SSR as an instrument that serves both security and development objectives.

The intention of the ninth volume is to reawaken an important conversation about how SSR can work for the development community and return to its roots as a vehicle to help meet development objectives. In the field of operations, SSR and development communities have strayed from their shared origins and sometimes forget their common goals. These start with a commitment to promoting peace and justice via consultative and bottom-up programme planning. Both development and SSR actors are mandated to include everyone in the communities they serve, especially women and youth. Both need to embrace multi-stakeholder and whole-of-government approaches, honour their 'do-no-harm' commitments, focus on structural security and positive peace, and make operational their commitment to long-term processes for sustainable outcomes. Finally, they need to remember the complexity of the individuals and institutions with which they work, and deliver SSR and development programmes that contribute holistically to resolving a wide spectrum of post-conflict problems.

The debates collected in this book will, I hope, both encourage the development community to revisit some of its concerns about collaborating with the SSR community and remind SSR practitioners that they have much to learn from development insights. This will not happen unless SSR and development practitioners discuss how they can benefit from one another, explore more effective forms of collaboration and think again about how sustained investment in enhancing SSR's development dividends can be accomplished. The empirical research into development dividends and priorities within SSR programme planning and implementation presented in this volume can, I believe, assist both communities to move beyond simple rhetorical pronouncements and back into these important theoretical and operational discussions.

As this volume reminds us, the primary goal of good governance is to provide justice, security and development alike. I would therefore like to thank the editors for having produced a much-needed state-of-the art study that has practical utility; I invite its readers to take this conversation further through their engagement with the arguments it presents.

Ambassador Theodor H. Winkler
Director, DCAF

Acknowledgements

We would like to acknowledge the contributions of a number of individuals without whose cooperation and assistance this volume would not have been completed within the timeline of one year available for the completion of a DCAF Yearly Book.

We are grateful to Herbert Wulf, our principal reviewer, whose valuable comments have helped improve individual chapters while ensuring that the entire text, including the opening and concluding chapters, is coherent. A number of colleagues from inside and outside DCAF contributed with their own country expertise to the improvement of a number of chapters, or offered feedback during the development of the thematic focus and scope of this volume. Our thanks go to Alan Bryden, Cornelius Friesendorf, Heiner Hänggi, Markus Heiniger and Armin Rieser for providing valuable inputs and advice at different stages of the book project. We greatly appreciate the excellent contributions made by Cherry Ekins, who has copyedited and proofread the entire manuscript. Thanks to Yvonne Guo and Ulrike Franke for assisting us at various stages of the manuscript preparation. We thank Yury Korobovsky and his colleagues for assisting in meeting the publisher's formatting requirements and guiding the publication process. Thanks to the publisher, LIT Verlag, for the publication of this newest addition to the DCAF Yearly Book series. Special thanks go to Marc Krupanski, whose crucial role as editorial assistant has significantly shaped the book project from its inception, ranging from selecting chapter contributors to offering an 'outside' opinion on proposals and draft chapters, providing valuable and substantive comments and feedback during chapter revisions and copyediting draft chapters. He furthermore joined us in co-authoring the closing chapter of the book.

Finally, we thank the contributors for working under considerable time constraints and responding diligently to comments provided on several draft versions of their contributions. Agreement with contributors on the final approach and scope of their chapters included their willingness to address the key questions and puzzles which we wished to cover in the book. The final make-up of the chapters, the main sections and the organisation of the overall volume were the result of a dynamic process in close consultation with the respective authors and author teams.

The views expressed in this volume are those of the authors and do not necessarily reflect the opinions of the organisations they are affiliated with, including DCAF.

Albrecht Schnabel and Vanessa Farr
Geneva and Jerusalem, October 2011

Abbreviations

ANP	Afghan National Police
ASEAN	Association of Southeast Asian Nations
AusAID	Australian Agency for International Development
BRICS	Brazil, Russia, India, China, South Africa
BMZ	German Federal Ministry for Economic Cooperation and Development
CAVR	Commisão de Acolhimento, Verdade e Reconciliação de Timor-Leste
CFSP	common foreign and security policy (European Union)
CIDA	Canadian International Development Agency
CPA	Accra Comprehensive Peace Agreement (Liberia)
CPSG	community problem-solving group (Albania)
CSDP	common security and defence policy (European Union)
CSO	civil society organisation
DAC	OECD Development Assistance Committee
DANIDA	Danish International Development Agency
DDR	disarmament, demobilisation and reintegration
DFID	Department for International Development (UK)
DIAG	Disbandment of Illegal Armed Groups programme (Afghanistan)
DRC	Democratic Republic of the Congo
EC	European Community
ECOWAS	Economic Community of West African States
EU	European Union
Falintil	Forças Armadas de Libertação Nacional de Timor-Leste
FCS	fragile and conflict-affected situations
FDD	Focused District Development programme (Afghanistan)
F-FDTL	Falintil-Forças de Defesa de Timor-Leste
FRAP	Falintil Reinsertion Assistance Program (East Timor)
GAD	gender and development
GAM	Gerakan Aceh Merdeka (Free Aceh Movement)
GBV	gender-based violence
GDP	gross domestic product
HDR	*Human Development Report*
HLP	housing, land and property
ICG	International Crisis Group

IDDRS	UN Integrated Disarmament, Demobilization and Reintegration Standards
IMF	International Monetary Fund
IOM	International Organization for Migration
M&E	monitoring and evaluation
MDG	Millennium Development Goal
MEF	Malaita Eagle Force (Solomon Islands)
MOFA-ODA	Ministry of Foreign Affairs Section on Official Development Assistance (Japan)
NATO	North Atlantic Treaty Organization
NCDRR	National Commission for Demobilisation, Reinsertion and Reintegration (Burundi)
NGO	non-governmental organisation
ODA	official development assistance
OECD	Organisation for Economic Co-operation and Development
ONS	Office for National Security (Sierra Leone)
OSCE	Organization for Security and Co-operation in Europe
PEAP	Poverty Eradication Action Plan (Uganda)
PIF	Pacific Islands Forum
PNTL	Polícia Nacional de Timor-Leste
PRSP	poverty reduction strategy paper
PSC	private security company
RAMSI	Regional Assistance Mission to Solomon Islands
RESPECT	Recovery, Employment and Stability Programme for Ex-Combatants and Communities in East Timor
RRR	reform, restructuring and rebuilding
RSIPF	Royal Solomon Islands Police Force
SGBV	sexual and gender-based violence
SI	security institution
SIDA	Swedish International Development Cooperation Agency
SILSEP	Sierra Leone Security Sector Reform Programme
SSG	security sector governance
SSI	security sector institution
SSR	security sector reform
SSSU	security sector support unit
TCU	transnational crime unit
TNA	Tentara Neugara Aceh
UN	United Nations
UNAMSIL	UN Mission in Sierra Leone
UNDP	UN Development Programme

UNHCR	Office of the UN High Commissioner for Refugees
UNICEF	UN Children's Fund
UNIDIR	UN Institute for Disarmament Research
UNIFEM	UN Fund for Women
UNMIL	UN Mission in Liberia
UNMIT	UN Integrated Mission in Timor-Leste
UNODC	UN Office on Drugs and Crime
UNOTIL	UN Office in Timor-Leste
UNPOL	UN police
UNSCR	UN Security Council resolution
UNTAET	UN Transitional Administration in East Timor
UNW	UN Women
USAID	US Agency for International Development
VAW	violence against women
WACI	West Africa Coast Initiative
WDR	*World Development Report*
WID	women in development

PART I

INTRODUCTION

Chapter 1

Returning to the Development Roots of Security Sector Reform

Albrecht Schnabel and Vanessa Farr

> In an increasingly interconnected world, progress in the areas of development, security and human rights must go hand in hand. There will be no development without security and no security without development. And both development and security also depend on respect for human rights and the rule of law.
> *UN Secretary-General (2005)*[1]

> The Security Council recognizes that the establishment of an effective, professional and accountable security sector is one of the necessary elements for laying the foundations for peace and sustainable development.
> *President of the UN Security Council (2008)*[2]

Introduction

While the Arab Spring of 2011 swept across parts of North Africa and the Middle East, the World Bank issued its *World Development Report 2011: Conflict, Security and Development* (WDR).[3] The report strikes a number of chords with these defining events. It recognises what has for a long time been a truism among analysts, policy-makers and affected populations alike: repeated cycles of political and criminal violence cause human misery and disrupt development. Additionally, low levels of human development can contribute to instability and conflict. Affected populations 'are being left far behind, their economic growth compromised and their human indicators stagnant'.[4] As the WDR argues, 'strengthening legitimate institutions and governance to provide citizen security, justice, and jobs is crucial to break cycles of violence'.[5]

The people taking to the streets in the Arab Spring are calling for legitimate national institutions and more security, justice, employment and improved livelihoods for all. The Arab Spring was (and is) about human security and human development in addition to human rights. Populations across the region are willing to risk their lives and in certain instances take up arms to topple their governments or, at the very least, demand urgent action from their governments to provide authentic democratisation, development and good governance – and importantly, jobs to maintain a livelihood. The demands emphasise good governance of security sectors that have primarily secured the longevity of an otherwise illegitimate state and its elite, rather than the needs of the population.[6]

Since its inception, security sector reform (SSR) has been considered a critical component of establishing a security environment conducive to long-term sustainable development, especially in transitional societies. Despite the considerable scepticism that accompanied the first decade of serious debate on SSR and its governance and development linkages, major international organisations nonetheless assert the critical role played by SSR processes in the long-term pursuit of both security and development objectives. Serious engagement with SSR by the OECD (Organisation for Economic Co-operation and Development), the European Union, the World Bank and the United Nations, and regional organisations such as the African Union and ECOWAS (Economic Community of West African States), testifies to the perception, if not conviction, that SSR can and does make considerable contributions to a society's quest for both sustained security and sustainable development. Commitment to the human security concept, which, like SSR, is a product of the progressive understanding and new security thinking that emerged from the end of the Cold War, is a central component of people-centred approaches to both security and development; contested as it may be, human security has focused on the primacy of freedom from fear and want, and the provision of security from both direct and structural violence and threats.

In this volume we consider SSR as an expression, conceptually and practically, of the security-development nexus, as much as this nexus might be contested, misunderstood, poorly expressed or unclearly defined. Even if we are not able to define and explain the exact nature, scope and consequences of the nexus between security and development, there are – possibly mutual – links between the provision of security and resulting opportunities for development. If SSR assures that security providers are prepared to meet security threats and do not themselves compromise the security of the population or societies' ability to meet their own welfare and development needs, it makes a critical contribution to building and consolidating stable, just, inclusive, secure and well-to-do societies. But enough of assumptions and theory: the real questions remain. Have specific SSR programmes, as implemented on the ground, helped make this nexus more beneficial for populations and their diverse communities? How can SSR most effectively be used to promote long-term security, and how can it most effectively learn from inclusive development approaches? Can SSR contribute to more helpful security-development links, thus validating evolving human security approaches to the provision of security and development?

Our book addresses these questions by revisiting and advancing ongoing discussions on the theoretical and practical relevance of linking activities that connect the provision of security with development. While exploring the relevance of a broader security-development nexus, the specific focus of the book is on the roles SSR has played, is playing and has the potential to play as a major building block for sustainable human development.

In this opening chapter we introduce the underlying objectives, definitions and assumptions of the book as a whole. We then frame the main thematic focus of this volume and place it in the context of its theoretical, conceptual and practical bearing on current discussions of international affairs. We also demonstrate its relevance to academic debates, policy-making and practical work in both the security and development fields. We conclude by highlighting the main purpose, subjects and findings presented by each contribution.

About this book

The purpose of this book is to return to the roots of the SSR concept and debate: SSR is meant to be a set of activities ensuring that the provision of security does not stand in the way of – but rather supports – a society's sustainable development. Moreover, SSR is a people-centred activity that, at the end of the day, is able to advance human security and human development. We embarked on this book project with a number of key assumptions and questions, which we will revisit in the closing chapter.

While we are confident that a society's quests for security and development are both highly interrelated and interdependent, we acknowledge that it is difficult to apply such sweeping assumptions to theoretical analysis and policy and programming in highly diverse contexts.[7]

We consider SSR to be a potentially effective and powerful instrument in explaining and actualising the security-development nexus. SSR's contribution to security is well analysed and the subject of the majority of SSR literature. In order to understand and appreciate SSR's role in (and assumed contribution to) sustainable development, however, we need to ask about the potential and actual contributions that it makes to economic growth, human development and poverty reduction. To what degree is SSR able – and to what degree has it been used – to advance both security and development agendas simultaneously? Closely related is the question of whether the conceptual development of SSR, as well as experience gained in implementing the concept in the past decade, has sharpened or weakened confidence in its potential contribution to development. If we want to learn from past practice and improve the harmonisation of both broader security-development and narrower SSR agendas, we need to examine the major difficulties and drawbacks in doing so, analyse how these can be explained and overcome, and identify what remains to be accomplished to ensure that SSR – in theory and practice – is one of the main engines of equitable and sustainable development.

In contrast to the literature on the security-development nexus, very little analytical material of real substance and value so far exists on the development-SSR link. Thus, as surprising as it may be, the contributions in this book could not draw on much secondary literature on this subject, but rather had to evolve their arguments from a number of key puzzles identified by the editors. In the concluding chapter we will revisit and propose issues that require further study on the same and similar questions, from both conceptual and practical perspectives.

Beyond the intrinsic value of revisiting and contributing to the ongoing debate on the security-development nexus and the role SSR might play as a catalyst and expression of this nexus, we believe there are a number of practical benefits of doing so for the development and security/SSR communities. Joint debates on this topic have great potential to bring together the development and security epistemic and practitioner communities in order to bridge the theory and practice of security, development and SSR. SSR should not be seen as a separate 'third community', but a community made up of individuals and institutions that are actively engaged in both security and development provision. In such a way SSR can serve as a bridge between policy and research and between different epistemological communities. The analyses and suggestions presented throughout this volume remind development and security communities to take each other's concerns seriously and into account when planning, implementing and evaluating their own activities, both outside and within the context of SSR.

Moreover, the issue of governance features prominently in several chapters and has implications for the practical implementation of development and SSR work. We are thus reminded that the ideal of promoting good governance, a core development objective, is an incentive for the development community to engage in institutional change, as the WDR 2011 encourages; and it is a reminder for the SSR community not to compromise on the governance dimension of the security sector. Defining and understanding the range of meanings implicit in the term 'good governance' are thus matters of joint concern for both communities.

The discussions in this volume also remind us that there are serious concerns about the (sometimes real and sometimes wrongly assumed) potential for a securitisation of development. Such concerns need to be addressed. As Chapter 2 argues, securitisation as such does not need to mean militarisation. Indeed, if it is engaged in addressing comprehensive (both horizontal and vertical, direct and structural) security concerns, it stands for a return to human security provision. In this light, debating what is meant by good governance can achieve a third objective as well, that of supporting clear boundaries between security and development roles where these are needed. Good governance, in other words, can offer both the glue that holds development and security achievements together sustainably, and a brake preventing two mandates from overlapping in unhealthy ways that would result in development becoming more militarised. It should therefore be pursued by setting commonly agreed and joint objectives through which hierarchies of intentions and clear boundaries are set at the same time as areas of mutual benefit are identified.

Having laid out the central themes, objectives and assumptions in this book, we will move into a discussion of the theoretical, conceptual and practical relevance of this volume's central theme to current discussions of international affairs. We also demonstrate its relevance to academic debates, policy-making and practical work in both security and development fields.

From the end of the Cold War to the new millennium: Hoping for peace dividends

Since the end of the Cold War the peace dividends – and thus security and development dividends – that should result from a generally more peaceful and less volatile and violent world have yet to materialise. A more stable and peaceful world would have less reason to fear – and prepare accordingly for – violent internal and international war. Fewer wars, in turn, would lead to at least three positive outcomes. First, fewer and shorter wars would result in lower numbers of direct and indirect human casualties and less overall violence, among both combatants and civilians, as well as fewer internally displaced and international refugees. Second, as the likelihood and occurrence of war declined, fewer resources would need to be invested in preparing a society for defence against external and internal threats; the need for resource-intensive reconstruction would be similarly reduced. Resources spent on preparing for regular large-scale violence or dealing with its aftermath could thus be invested in serving immediate needs in the form of a broad range of public services, not least the reduction of poverty and the creation of livelihood options for societies, rich and poor alike. And third, less insecurity, war and destruction would promote and enhance development momentum, spreading and consolidating development gains in countries that have previously been torn apart by violence, instability, exclusion, underdevelopment or poor governance. As a result, a peace dividend of more and sustained security and economic well-being would be produced.

From 9/11 to the Arab Spring

The turn of the century provided a symbolic opportunity to learn from the past and take some forward-looking measures to set the growing (in terms of both population and the number of states) international community on a path towards greater peace, security and justice. The terrorist attacks of 11 September 2001 and the international responses triggered by these events crudely interrupted international commitment – and the momentum building around it – to consolidating and expanding human development, human security and security sector reform. The sudden and rude global awakening resulting from the attacks showed that the momentous opportunity and hope for peace would not be given a chance to materialise anytime soon. Instead, a new enemy – global terrorism – emerged, quashing hopes for more worldwide peace and stability. The focus on this new threat triggered a so-called 'global war on terror' that escalated, started or was used to justify highly destructive, costly and long-lasting wars (such as in Afghanistan and Iraq) without, in a sustainable manner, significantly reducing the threat they were meant to contain. The war on terror further antagonised and possibly escalated the schisms between the 'haves' and the 'have-nots' that intensified some of the popular support behind the extremist energy driving the creation, maintenance and perpetuation of international terrorist movements. The war on terror might also have further weakened the hollow, failing state structures that offered breeding, training and hiding grounds for the handful of terrorists who managed to hold the world in a decade-long grip of fear and counter-violence.

The results of the global war on terror have worn heavily on societies caught in turmoil and transition, as well as those which intervened with the intention of bringing peace, stability and political change in a matter of a few months or years. The recent popular uprisings throughout Northern Africa and the Middle East have reminded us that an appearance of regional and international stability (in the past at least partly assured by strongman despots throughout the region) comes at a very high price for the affected populations. Across the region, large parts of society have grown confident enough to express their disaffection with authoritarian states, slow development or underdevelopment – and security institutions that serve the interests of the ruling elites and state structures, instead of, and all too often at the expense of, the security and well-being of the population. In the case of the popular armed uprising in Libya the international community has come out in strong support of the armed rebellion and provided substantial material and rhetorical aid. Such assistance found expression in NATO air campaigns against government forces and for the protection of civilians. Official recognition of the rebels' transition government by many states and the United Nations sealed the international community's decision to side with those challenging the government and the state's monopoly of force. However, the uneven responses of many states to uprisings in other Arab countries have failed to establish a common, unified and human-security-centred response to citizens' desire for freedom. The implications of this lack of unity, for both national and regional development and security, remain to be seen.

Governing the security sector in transitional societies

Security governance dynamics at global, regional and national levels have been a central feature triggering as well as responding to old and new wars, conflicts and instability within and between societies. A society's security sector ideally provides justice and security to all persons living within a country's borders according to good governance principles such as participation, rule of law, transparency, responsiveness, consensus orientation, effectiveness and efficiency, and accountability.[8] Unfortunately, many security sectors do not operate under good governance premises. To the contrary, in authoritarian and non-democratic states they are often instruments of state suppression. It is this suppression – and oppressive security institutions' defence of illegitimate states and leaders – which perpetuates state-sanctioned organised violence. To return to the example of NATO military involvement in Libya, while in 1999 NATO assisted the Kosovo Liberation Army in its armed insurgency against Serb oppression, particularly in response to a presumed attempt at ethnic cleansing against the majority Albanian population, in Libya NATO came to the rescue of a population fighting against their own government's political oppression and tyranny. The fact that it helped topple a regime with which it had previously eagerly traded arms and oil, along with the apparent unwillingness to extend similar assistance to opposition movements in Syria, for instance, raises doubts as to the intentions behind NATO's behaviour. Even more concern was raised by events in the UN Security Council in October 2011, in which Russia and China – both of which have major oil concessions in Syria – vetoed an attempted resolution to stop the Syrian regime's crackdown on citizen-led protests. A unified stance on popular uprisings against poor governance and state-sanctioned violence, then, is clearly not much closer than it was in the days of the Cold War.

Once violence subsides, the existing regime has been overthrown and a new or transition government is in place, the political system and the security sector need to be put in the service of the people. In brief, this is what SSR is all about. Its intentions are to put in place a well-governed, well-tasked security sector which will provide the population with the type, extent and level of security – and with the requisite instruments at its disposal – to defend themselves against threats or impediments to their social, economic and political development. Security institutions will then be able to provide a public service for their clientele, the state and society.

However, if security provision turns once more into a security threat, what was intended to be a service for all, financed by the population through taxes or a society's resource wealth, turns – again – into a bulwark of a few and an existential threat against the many. In such a context development is not possible. In societies that are undergoing major political, economic and social transitions, SSR is intended to enable, support and facilitate both long-term development (thus triggering 'development of development') and state, community and personal security. A major goal is to manage and prevent the shifting dynamics of violence and relative peace that, especially in fragile states, often accompany political events such as electoral cycles.

One can fairly confidently argue that both 9/11 and the Arab Spring are, at least in part, a result of the legacy of failed decolonisation and subsequent democratisation. Despite high hopes, populations have not experienced the political freedoms, development and security that would have been required in order to prevent the type of upheaval we have been experiencing. What would have been a more responsive approach? Is a global focus on development producing more security? The evidence would suggest not: the anticipated peace dividend of development seems not to have been significant enough; and where it has occurred unhindered by war and conflict, its results have been modest. At the same time, a decade into the new millennium, it is also more apparent that political stability does not necessarily produce security and legitimacy of state institutions, particularly if it is built on the continued oppression of political challengers by ruling elites.

Nevertheless, the international community appears to be heeding lessons from the past. Short-term preservation of regional security and stability neither builds sustainable regional and international peace and stability nor affords people freedom from fear and want. This is the case especially when such short-term objectives are pursued at the expense of societies' desire for transformation, expressed through their quest for political freedom, human rights, gender equality, economic justice and the right to determine their own fate and well-being. Particularly after war and turmoil, and as a key contribution to political, economic and social transition, SSR processes are considered central to ensuring that security and development needs and provisions are synchronised with overall transition processes and a society's long-term national objectives and visions.

The need to rethink SSR in the context of the security-development nexus

SSR and development interventions in transitional societies are based not only on vague links and complementarities, but essential – and quite possibly existential – joint objectives. Therefore, it appears to be crucially important that SSR activities are closely envisioned, planned, designed and implemented with development activities, and that both are eventually monitored and evaluated for their ability to enhance each other's positive impact on society. While a commitment to such collaboration and partnership has been voiced in the past, it is not being actualised on the ground. A number of obstacles that stand in the way of close collaboration will be highlighted throughout the volume. By way of introduction, a selection of these challenges is discussed here.

Some of the more common obstacles include the prevalence of donors and development practitioners whose development and security initiatives remain stove-piped and divorced from each other – within their own organisational structures as well as in their interactions with other entities. One example is the unevenness with which the UN Development Programme (UNDP) applies a conflict-sensitive lens to its development interventions, even when these take place in fragile, conflict and post-conflict settings. This separation far too often reflects a dominant donor culture that has created and emphasised rigidly divided disciplinary approaches and organisational structures. Such divisions are often far removed from people's needs on the ground. The inability to reflect and project a sense of security and development that is primarily relevant to local populations rather than donor bodies can also lead to the prioritisation of objectives that do not match those aspired to by local communities, in all their diversity.

An additional challenge appears in the observation that neither our understanding of the security-development nexus nor our appreciation of SSR's role and contribution seems to have evolved much, despite considerable work in recent years on development, security, human security and SSR, often in the context of post-conflict peace-building strategies and all, presumably, contributing towards a better understanding of the connections between security, development and SSR. The need to promote such an understanding within the security and development communities has been re-emphasised by 9/11 and its consequences for security and development assistance across the world. While leading to increased calls for collaboration between the two communities, some of the experiences and struggles have also driven them further apart. However, the connection between security and development issues has recently been reignited by the Arab Spring revolutions of 2011. Although these events caught many donor communities off-guard, they would not have come as a surprise to those who have closely followed political developments in these countries or read the 2009 UNDP 'Arab Human Development Report' on human security in the region.[9] The failure to provide people-centred development and security created a vacuum on both counts – helping to make these countries prime candidates for the revolutions we have been witnessing since the spring of 2011.

An important lesson has also been that short-cutting SSR, particularly by failing to take the governance dimensions seriously and pursuing SSR in isolation from larger development and governance reform priorities, will do little good and may even further destabilise both state and society. The pursuit of SSR without a core and overarching objective of fostering security sector governance might be easier, quicker and politically less sensitive to accomplish, but in the long run it may turn out to be counterproductive not just to SSR aims, but to development and peace-building objectives as well. With the emergence of new political leaders and the willingness of local, national and international actors to invest in rebuilding the state, society and security sector of conflict-affected nations, opportunities arise to enable the reform of the security sector in line with people-centred, realistic development objectives. Opportunities also arise to pursue reforms in tandem with processes and institutions that improve democratic governance, accountability and the rule of law. Particularly, institutional reforms have been identified by the WDR 2011 as among the key ingredients in generating justice, security and jobs – and thus a safe and productive livelihood for these nations' populations. Therefore, we need to return to the conceptual roots of SSR, in the form of programmes that are intended to facilitate and support a nation's overall development objectives. A window of opportunity has now opened in Arab Spring societies that calls for reconsidering, re-evaluating and, in some instances, recalibrating recent applications of SSR to ensure that they are brought closer in line with these original intentions and objectives.

In the wake of the global war on terror, however, international engagement in peace, stability and development particularly in fragile states has in some cases (such as Afghanistan and Haiti) been militarised and thus, in a very narrow sense, securitised. By reducing poverty and decreasing state fragility, development was also seen as a pathway to eroding the support base for national and global terrorist networks. As in Afghanistan, in some instances development assistance was pursued in tandem with military action against militarised terrorist groups. When military troops become actively engaged in humanitarian or development assistance activities, local populations find it increasingly difficult to separate the aid community from fighting forces. Thus such assistance can be subsumed into a military campaign, while development can become militarised.

Very much the same can happen to SSR. When armed forces simultaneously wage a military campaign and train local military or police in the name of security sector reform, SSR is reduced to technical 'train and equip' exercises unrelated to authentic SSR strategies. Fundamental SSR ingredients get lost, or are not delivered through a holistic, comprehensive and broadly defined approach that includes many if not all security-providing institutions in the society. Nor is democratic, civilian oversight and management of the security institutions fostered. Of course, pursuing SSR in the midst of war is a considerable challenge; when military campaigns are under way between formal and informal armed groups, executive, legislative, judiciary and civil society institutions are unprepared (or not available or unwilling) to provide and establish good governance of the security sector. Still, at least preparatory SSR strategies can be pursued even under the most difficult circumstances.[10] Difficulties in implementing comprehensive SSR approaches should not, however, prevent the development of long-term SSR strategies that can be communicated and understood by those actors who are the subject of reforms as well as the broader society which will benefit from the reform of the security sector.

Returning to the roots of SSR: Renewed appreciation of the development-SSR link

SSR is a concept that has evolved from debates about the relevance of security and violence for development processes – exemplified most prominently by the efforts of the OECD to promote an understanding and appreciation of, and generate guidelines for, making the link between security and development policy operational. Most recently, the Geneva Declaration on Armed Violence and Development[11] and the WDR 2011 focused on the nexus between security and development, and its significance for meeting many of today's security and development challenges. Ideally, SSR serves as a linchpin to operationalise security and development objectives in transitional societies. By improving the delivery of security services to society, a reformed and thus appropriately sized and equipped, well-designed and well-governed security sector contributes to a positive and conducive environment for development processes and activities. Yet the contributions to this book show that, while these objectives are at the heart of SSR's purpose and goals, two issues in particular stand in the way of meeting these promises: a lack of evidence that SSR has been designed in tandem with development objectives, and the absence of mutual learning and sharing of experience between development and SSR communities.

The SSR-development link: By default or by design?

Rhetoric aside, there is very little empirical evidence that SSR programmes and strategies have in fact been designed specifically with development objectives in mind. This means in many instances that SSR agendas might have been planned in isolation from already existing development agendas – which of course is not conducive to a close and purposeful link between the two. In such cases SSR's contributions to development are at best coincidental, not intentional. Common-sense assumptions about the link between SSR and development, along the lines of common-sense statements about the mutually beneficial relationship between security and development, abound. Some might prove real due to an intuitive assessment of particular situations where security and development agendas meet and overlap. However, as these situations are highly context-specific, their relevance to other contexts is unknown.

Still, there will always be some intrinsic positive consequences of improved security for development processes – such as the mostly unchallenged fact that in the absence of open as well as low-level armed violence, development activities are easier to implement and recipient communities of assistance can be more readily reached. In addition, most development activities are undoubtedly conducive to the pursuit of long-term human, community and national security, as they provide an environment in which populations are better able to meet their basic livelihood and human needs and, as part of broader development agendas (as is the case with SSR agendas), good and democratic governance is promoted. At the same time, it has become common sense that development activities need to be conflict-sensitive and follow 'do-no-harm' principles to avoid an unravelling of existing levels of stability and security. Both security and development communities can do much better than hoping for some automatic, default benefits of their respective activities for the activities of the other. This, however, requires mutual understanding and respect of each one's approach and activities, early cooperation and proper planning and coordination of work.

The SSR-development link: Similar goals but different paths?

In addition to a lack of evidence of purposely designed, development-sensitive SSR and security-sensitive development programming, it becomes clear in this volume that both types of activities, although so closely linked and in pursuit of similar goals and objectives, invest little effort in learning from each other's experience, let alone utilising such understanding in their respective activities. Ideally, what we refer to as the 'SSR community' should include a conglomeration of thinkers, practitioners and policy- and decision-makers who understand themselves equally as development and security protagonists, espousing the ideals of an enlightened security community as much as those of an enlightened development community.

The contributions to this book show that a number of very useful insights, articulated in contemporary SSR literature, have been developed and applied in the development community. Those insights have the potential to contribute significantly to better SSR programming design and performance. In particular, they can help to achieve two important objectives: they can improve overall SSR performance and results, and they can help bring development and security communities closer together in the common pursuit of their collective goals. Indeed, the development community maintains a long, rich experience of pursuing highly political, locally sensitive and people-focused activities that help stabilise a population's development trajectory, build capacities for self-help and improve the framing conditions (e.g. the promotion of gender equality or support for improving democratic governance) for successful and sustainable development. Among those experiences is the feminist discourse that has characterised development work in the past years, leading to greater visibility of women as not just objects but actors in development processes and capable on matters of security. A similar focus has evolved on the roles of youth and children for both security and development.

Additionally, there have been significant contributions of international organisations to norms and standard setting in furthering peace, security and development. Over the past decade they have contributed considerably to establishing SSR as an integral component of managing and supporting change in transitional societies. Forward-looking discussions and commitments have found prominent expression in major policy statements (by the OECD, United Nations and European Union, with the WDR 2011 being only the latest such report), with a focus on immediate challenges (such as humanitarian aid) and the subsequent transition to longer-term activities and objectives. Experience also shows that communities' willingness to support and drive development and security provision, even if unconventional, is quite high. However, recurring experience shows that positive expressions of programme priorities and development processes tend to outpace equally necessary institutional developments within both formal and informal institutions.

The challenge lies in the fact that the long-running, sustainable approaches needed for successful SSR do not always gain the respect and support of short-sighted external actors who prefer to replicate templates and blueprints that have worked for them at home or in other seemingly similar contexts. Such a viewpoint also tends to favour a highly technical approach rather than engaging honestly with the historically contingent political nature of the context. The notion that cut-and-paste replication and a short-term vision will fail to produce sustainable results is well understood in the development community, learned through evolving practice over the past years and decades. SSR practitioners need to act on this learning.

There is evidence that the relevance of some of these lessons and linkages between development and SSR experience and practice is being recognised in SSR academic and policy discourses, although they are not being mainstreamed in SSR practice. As will be seen from the evidence presented in this book, there is a growing focus on issues of importance for socially inclusive development work: the gender dimensions in SSR; the roles of women, youth and children; the need to honour the primacy of local ownership, along with traditional justice and security providers; the need to employ inclusive and consultative programme planning and design; the significance of pursuing good governance principles in SSR and security sector governance; and the necessity to create an 'enabling environment' by building democratic and representative government institutions and investing in non-violent conflict management or job creation.[12] These themes, picked up in recent SSR work, attest to the growing realisation within the SSR community that it cannot achieve much in isolation from other crucial society-, state- and broader peace-building processes. However, as many chapters in this volume confirm, these links have yet to be translated adequately into practice. In addition, the links can be misused to allow for a significant range of activities that, as some authors in this volume argue, misread or even distort their real meaning or generate negative consequences for peace-building processes.

With this said, we will now turn to an overview of the remaining chapters in this volume. The authors have approached the question of the role of SSR within the security-development nexus and its relevance for development work from a variety of vantage points and backgrounds. Their views inform the findings already presented in this introduction, which will be returned to in the conclusion.

Chapter overviews

As much as possible, the different parts of the book build on each other, while sections within chapters are written to complement each other. Nevertheless, each chapter may also be read in its own right, for the contribution to knowledge it makes beyond the specific questions that are at the centre and within the scope of this book. In the conclusion we will highlight the common perspectives and cumulative knowledge gained from the individual chapters.[13]

The contributions to the volume are organised in six parts: the introductory part, which includes this chapter and a conceptual overview of the security-development nexus and SSR's place within it; 'Gender, security and development', which contains two chapters examining the consequences and meanings of gender differences within SSR and the security-development nexus; 'SSR and development – Regional perspectives', with two chapters that focus on West Africa and East Timor and the Solomon Islands; 'SSR, DDR and development', which contains two chapters that probe the challenges and links between DDR and SSR programmes and their potential development dividends; a section that asks if we are moving 'towards "developing" SSR policy', featuring three chapters that examine broader trends and practices, particularly focused on approaches taken by leading international donors; and a concluding chapter, which pulls together and compares the arguments and findings presented throughout the book and offers recommendations for a future research agenda and more immediate policy and programmatic changes in SSR and development work.

Following this opening chapter, in the second contribution Albrecht Schnabel traces the evolving discussions surrounding the security-development nexus debate. He discusses definitions of the terms security and development, the notion of a nexus between security and development as theoretical concepts, programmatic approaches and policy instruments, as well as concerns about securitising development as a result of the operationalisation of the nexus. The chapter then shifts to the role played by security sector reform in pursuing both security and development objectives. Is SSR an effective bridge between security and development, advancing both for mutual benefit? Arguing that this could in fact be the case, the author proceeds to explore the rationale behind the development community's role in triggering the original debate and practical implementation of the SSR discourse. As long as both security and development actors do not divert from their own principles, he argues, development and SSR objectives as well as approaches are highly complementary. Diverting from one's own principles, such as in the case of what Schnabel calls 'quasi-SSR' activities, runs counter to efforts to create the context for sustained security and sustainable human development. He concludes that SSR can be considered both an expression and an application of the security-development nexus in practice – in planning, implementing and evaluating both development and SSR activities.

Gender, security and development

The second part of the book is focused on the intersecting issues of gender, security and development. Both chapters in this section offer an examination of approaches to incorporating gender analysis and programming into SSR activities in order to advance gender equity in security and development. Although limited progress is noted, both chapters highlight the continued gender challenges of SSR.

In Chapter 3, Heidi Hudson provides a critical analysis of the security-development nexus and the challenges of gender mainstreaming in SSR practices. In addition to reviewing broad conceptual and programmatic approaches, Hudson conducts her examination through an analytical focus on sexual and gender-based violence (SGBV) in Liberia in particular. In her contribution, Hudson identifies and deconstructs conceptual contradictions and ambiguities within SSR and the security-development nexus more broadly, which in the end impede the potential positive impact of SSR programming. She concludes that the conceptual problems of the security-development nexus, and consequently SSR, are related to the broader neoliberal frameworks and ambitions for intervention and addressing SGBV. These frameworks and ambitions, she argues, focus merely on managing risks rather than addressing root causes of insecurity, underdevelopment and inequality. In addition, they rely on the use of rhetorical and anecdotal flourishes that naturalise an otherwise unproven security-development nexus in order to justify a humanitarian interventionist agenda that remains largely unaccountable and irrelevant to those local communities most directly impacted by insecurity, underdevelopment and inequality. Examining the gender implications of this framework, Hudson contends that SSR activities have been dominated by a liberal additive feminist approach that conflates gender and women, essentialises women's and men's roles in SSR, proposes a limited and tokenistic representation of women in security sectors rather than broader change and, lastly, results in a fixation on technical tools and bureaucratic processes. Instead, she urges an approach that employs an analytical lens of intersectionality and a practical approach that emphasises long-term advocacy centred on solidarity through care.

In Chapter 4, Rahel Kunz and Kristin Valasek detail the common shortfalls of SSR programmes' contributions to development objectives as a result of prevailing state-centrism and gender-blindness of SSR practice. Like Hudson's analysis, Kunz and Valasek identify and break down SSR's dominant conceptual and practical approaches to local ownership and civil society participation, which they consider to be greatly limited in both meaning and application, especially concerning gender issues. The authors argue that this weakness of SSR is partly due to an inadequate understanding of the meaning and purpose of development within SSR discourse and practice. For the most part, although SSR is supposed to advance development aims, within SSR design and practice the meaning of development is merely assumed and is treated as a pre-defined reality that will emerge once security is provided. In an effort to help confront this deficiency, Kunz and Valasek review lessons from the field of development, particularly on participation and gender, and identify ways in which these can be applied in the design and implementation of future SSR initiatives. The authors show that SSR programmes could more ably contribute to development objectives by adapting development practitioner approaches that emphasise participatory and gender-sensitive practices and fix definitions of development and security to local meanings. These approaches, they argue, could enable more authentic local ownership, legitimacy and civil society participation in SSR initiatives. As a result, a more accurate understanding of local security and development needs and goals can be developed and addressed.

SSR and development – Regional perspectives

The third part of this volume presents two chapters that analyse SSR practice within the security-development nexus through a focus on regional approaches and experiences. Through a detailed and close examination of recent SSR and SSR-type activities, Chapters 5 and 6 illustrate potential links, benefits and challenges of SSR practice in meeting both security and development objectives.

In Chapter 5, Tim Goudsmid, Andrea Mancini and Andrés Vanegas Canosa focus on the threat and impact of 'serious crime' in West Africa as a means to illustrate the relevance of the security-development nexus as an analytical framework and the importance of SSR to advance security and development objectives. Serious crime, they note, 'brings violence, distorts local economies, corrupts institutions and fuels conflict'. It poses a significant threat in fragile states and post-conflict situations and carries regional consequences. The authors examine the efforts of the ECOWAS West Africa Coast Initiative (WACI), a regional SSR-type initiative. They identify ways in which SSR can be implemented practically in order to contest serious crime and thus enhance security and development conditions. The authors identify specific innovations and features of WACI, such as transnational crime units and cooperation across security institutions at both intra- and inter-state levels, which may prove beneficial for other SSR initiatives. It is the regional approach of WACI that the authors identify as its central value. While SSR has generally been considered a national project, the authors argue that in order to address most usefully the security and development dimensions of cross-border threats such as serious crime, a regional and comprehensive approach to SSR is necessary. However, they note that a regional approach, while beneficial, is still challenging, as it tends to be a cumbersome and time-consuming endeavour. Nonetheless, the authors conclude that such an approach, when centred on good governance and local accountability, is the best means currently available to address these challenges.

In Chapter 6, Derek McDougall examines SSR activities in East Timor and the Solomon Islands for their impact on development. In each, the security institutions proved instrumental in the breakdown of order at different moments in their post-conflict periods, and SSR-related interventions have been significant and well documented. According to McDougall, the two locations provide valuable case studies for assessing the impact of SSR on development, although a truly comprehensive, full-scale SSR programme was not pursued in either country. After reviewing SSR-type activities that have occurred in each location, McDougall contends that SSR's development dividends in East Timor have been limited, as the activities have focused primarily on institutions rather than on the security sector as a whole. The failure to pursue comprehensive SSR activities limited their development potential. Development dividends have been larger in the Solomon Islands because SSR initiatives there have emphasised greater overall coordination. Still, as McDougall concludes, while in both locations researchers and practitioners have recognised the importance of SSR programmes for post-conflict reconstruction, these activities have generally failed to engage in an explicit focus on development objectives and impact and thus significantly limited SSR's potential positive effect. This critical omission retroactively illustrates the impact of security on development – both examples show that continuing instability emanating from unreformed security institutions may negatively affect development. In the end, McDougall claims that a closer and intentional incorporation of development demands within SSR programming, along with a broader focus on the security sector as a whole rather than on specific institutions, could have prevented some of the instability that continues to affect East Timor and the Solomon Islands.

SSR, DDR and development

The fourth part of this volume focuses on the impact and contribution of disarmament, demobilisation and reintegration (DDR) programmes on security, development and SSR. The authors of both chapters identify DDR as an activity that is closely related to SSR, if not itself a tool within a broader SSR package, in addition to constituting a central pillar of development activity in armed conflict recovery and prevention. Nonetheless, they recognise certain ruptures and challenges between DDR, SSR and development. In this way, a review and analysis of DDR approaches and practices provide a valuable opening to examine broader meanings, possibilities and problems of the nexus between security and development.

In Chapter 7, Alan Bryden probes the ways in which DDR and SSR activities are carried out in practice and contribute to development aims. Through an analysis of their conceptual approaches and case studies of Afghanistan and Burundi, Bryden posits that although the nexus between DDR and SSR is often recognised within policy literature, in practice the two activities continue to operate independently. In addition, despite their assumed pay-offs for development, both activities tend to maintain a narrow focus on hard security demands. As a result, potential and ideal collaborations that could generate more substantial security and development dividends are overlooked. In his analysis of conceptual approaches and their application in Afghanistan and Burundi, Bryden details three specific gaps in knowledge, as well as common practical approaches that account for this deficiency. Based on this analysis, Bryden proposes that positive collaboration could occur by addressing these critical knowledge gaps, utilising a human-security-driven approach and focusing on security sector governance as a basis for reforming security institutions. Not only will such collaboration assist in providing more immediate positive results for DDR and SSR activities, but it will also lay a solid foundation for longer-term development.

In Chapter 8, Henri Myrttinen examines the challenges of DDR and SSR contributions to development in Aceh (Indonesia) and East Timor. After outlining the recent conflict history in each case, as well as the nature and scope of SSR and DDR interventions in each location, Myrttinen identifies connections between SSR, DDR and development processes. He argues that these connections were not reflected explicitly in policy or practice due to the relatively *ad hoc* style of intervention. Instead, they materialised as the result of underlying security-development linkages.

Despite these links and their potential for positive results, Myrttinen locates several challenges that hinder the impact of SSR (through DDR) on development. Chief among these critiques, as Myrttinen argues, is the fact that the dominant approaches taken in East Timor and Aceh have been tolerant of some degree of corruption, nepotism and other illegal economic activities. They have prioritised efforts to coopt a small number of potential troublemakers and protagonists of the armed conflict despite the fact that they often maintained higher existing privileges and power in society. To overcome the problems this is now known to cause, Myrttinen calls for a broad and inclusive approach that would benefit wider segments of the affected community, especially those who were not active combatants, most often women and children, and who have the potential to benefit more sustainably from interventions, thus contributing to greater development results.

Towards 'developing' SSR policy?

The fifth part of the volume contains three chapters that explore broader conceptual and practical challenges of supporting development through SSR. The chapters focus on dominant approaches to SSR and the security-development nexus taken by leading international donors. While each chapter recognises the legitimacy of a security-development nexus and the shortcomings of international donors to advance both security and development goals, they offer varied opinions on how the nexus should be understood and achieved within donor activity.

In Chapter 9, Paul Jackson examines the lessons learned from SSR programmes within the context of post-conflict state-building. The dominant method of state-building, Jackson argues, has approached SSR as a technical process alone and has failed to recognise adequately the deeply political and historical processes integral to legitimate and lasting state-building. Additionally, he criticises leading donor approaches to state-building as being centred on their own security interests and the interests of local national elites, rather than promoting a more complex and, consequently, more time-consuming approach that prioritises the needs and interests of the local populations. Jackson concludes that this has resulted in a relative failure of state-building initiatives, especially in terms of supporting development agendas. To remedy these failures, he proposes a conceptual framework that places 'governance' rather than 'government' at its core and an approach that engages substate and regional networks. This framework and approach would better enable legitimate and accountable security provision, and thus better meet the security and development needs of the communities in post-conflict contexts.

In Chapter 10, Willem van Eekelen reviews the dominant approaches taken by international organisations to synthesise security and development analyses and objectives, principally through SSR strategies and activities. Van Eekelen begins his contribution by reviewing the Millennium Development Goals (MDGs) and highlighting both their conceptual and their practical shortcomings. Still, with the MDGs established as the prime development objectives, he then examines the ways in which leading international organisations and donors have approached SSR and the security-development nexus and, by extension, the MDGs. He focuses on the conceptual and practical approaches taken by the OECD, the United Nations, select state development agencies, the World Bank and the International Monetary Fund, NATO and the European Union. Van Eekelen's analysis centres on the approach taken by the European Union in the past decade, identifying a number of key challenges and offering recommendations to align security and development objectives better through SSR. He sees great potential for more development-sensitive SSR through stronger collaboration, particularly between NATO and the EU, both featuring comparative advantages that make them especially suitable for either security or development tasks within common SSR strategies.

In Chapter 11, Ann Fitz-Gerald argues that there is a separation between the conceptual treatment of SSR, which recognises and incorporates development objectives, and the practical implementation of SSR, which frequently fails to engage the development community. To substantiate this argument, Fitz-Gerald tracks the pathways through which SSR activities are conceptualised, created and ultimately delivered. She begins by reviewing the developments in SSR debate in terms of both policy and operation, and highlights gaps between the normative frameworks and delivery on the ground. Next, Fitz-Gerald introduces a review of national security strategies and national development processes in order to assess in what ways and to what extent these can be incorporated into SSR design, ultimately leading to better development impacts. The bulk of her chapter draws on case studies of Uganda, Sierra Leone and East Timor, which she utilises to examine whether or not local development strategies influence and are incorporated into donor-led SSR activities. Through this examination, Fitz-Gerald concludes that while there is a well-established normative framework for SSR, the biggest challenge remains in implementing this framework on the ground in the face of continued mutual disengagement by both SSR and development actors. In reference to the case studies Fitz-Gerald argues that national strategic development processes should better engage civil society actors and thus secure greater local legitimacy and accountability. Moreover, a combination of these strategies with national security strategies should be incorporated into SSR planning and practice in order to enhance the developmental impact of SSR.

In the concluding Chapter 12, Vanessa Farr, Albrecht Schnabel and Marc Krupanski highlight the main findings presented in the volume, focusing on common patterns and arguments, revisit initial assumptions and show how the SSR and development communities can benefit from closer collaboration and joint activities in societies in need of external development and SSR assistance. This closing chapter concludes with suggestions for future research and policy discussions on the links between security and development, and particularly the role of SSR in providing and supporting a society's security and development objectives.

The 2009 DCAF Yearly Book on *Security Sector Reform in Challenging Environments* has already identified a growing lack of commitment to the original intentions and approaches to SSR as a cause for poorly and incompletely implemented reforms, resulting in – at best – partially successful reform outcomes.[14] If SSR is reduced to reforms of the military or technical 'train and equip' programmes, little can be expected in terms of security-sector-wide reforms and improvements that will make a difference to the achievement of development aims. It is important not to lose sight of the original intentions of SSR, one of which was to ensure that SSR contributes substantially to the creation of a development-friendly environment. This can happen only if SSR is designed, perceived and implemented as a development project as much as a security project. In this regard SSR, as practised over the past decade, has not been very effective. The following chapters highlight this shortcoming, elaborate on the consequences and point to remedies that allow a return of SSR to its development roots.

Notes

[1] Report of the Secretary-General, 'In Larger Freedom: Towards Development, Security and Human Rights for All', General Assembly, Fifty-ninth Session, Agenda Items 44 and 55 – Integrated and coordinated implementation of and follow-up to the outcomes of the major United Nations conferences and summits in the economic, social and related fields, UN Doc. A/59/2005 (New York: United Nations, 21 March 2005).

[2] 'Statement by the President of the Security Council', UN Doc. S/PRST/2008/14 (New York: United Nations, 12 May 2008), available at www.securitycouncilreport.org/atf/cf/%7B65BFCF9B-6D27-4E9C-8CD3-CF6E4FF96FF9%7D/SSR%20S%20PRST%202008%2014.pdf.

[3] World Bank, *World Development Report 2011: Conflict, Security, and Development* (Washington, DC: World Bank, 2011).

[4] Ibid.: 1.

[5] Ibid.: 2.

[6] Albrecht Schnabel and Amin Saikal, eds, *Democratization in the Middle East: Experiences, Struggles, Challenges* (Tokyo: United Nations University Press, 2003): 26.

[7] Some main definitions are introduced in Chapter 2, although these definitions are not binding and do not constrain the authors of the subsequent chapters. Where the contributors' definitions diverge, the chapter-specific use of key terms will be explained. The reader should note that, overall, the book is written primarily from external and donor perspectives. It is written in a self-reflective attempt to contribute to donor debates on security, development and international engagement in assisting transitional and post-conflict societies with institutional reforms that ideally are locally and nationally initiated, and only supported by external actors at the explicit request of recipient societies. Thus, in our opinion, a similar study from those representing the societies and institutions receiving such reform activities would constitute an important and welcome companion to this volume. In fact, such a contribution would provide critical guidance and analysis to meet the challenges discussed here.

[8] For the discussion on good governance and good governance of the security sector the authors follow the definition of 'good governance' as provided by the UN Economic and Social Commission for Asia and the Pacific, available at www.unescap.org/pdd/prs/ProjectActivities/Ongoing/gg/governance.asp. For a helpful discussion of the World Bank's definition of good governance, see Carlos Santiso, 'Good Governance and Aid Effectiveness: The World Bank and Conditionality', *Georgetown Public Policy Review* 7, no. 1 (2001): 1–22.

[9] UN Development Programme, 'Arab Human Development Report 2009: Challenges to Human Security in the Arab Countries' (New York: UNDP Regional Bureau for Arab States, 2009).

[10] See Hans Born and Albrecht Schnabel, eds, *Security Sector Reform in Challenging Environments* (Münster: LIT Verlag, 2009).

[11] For further information on the Geneva Declaration on Armed Violence and Development, a diplomatic initiative that is aimed at addressing the interrelations between armed violence and development, see www.genevadeclaration.org/. The second Ministerial Review Conference, held from 31 October to 1 November 2011 in Geneva, was dedicated to the theme 'Reduce Armed Violence, Enable Development'.

[12] Please refer to the selected bibliography at the back of the book.

[13] The authors are grateful to Marc Krupanski for his substantial contribution in drafting the chapter summaries presented in this section, and for providing valuable feedback on first drafts of this chapter.

[14] Born and Schnabel, note 10 above.

Chapter 2

The Security-Development Discourse and the Role of SSR as a Development Instrument

Albrecht Schnabel

Introduction

Ten years after 9/11 and in the midst of mass uprisings across Northern Africa and the Middle East, we are reminded that it is not effective to pursue humanitarian assistance, development assistance and security sector reform as part of larger political and military campaigns to counter potential threats to national, regional or global security without focusing first on the development and security needs of the affected populations. Nor does an exclusively outward focus lead to locally supported and owned, and thus sustainable, results. Moreover, as the wave of uprisings known as the Arab Spring shows, and the recent *World Development Report 2011* (WDR) confirms, effective, democratically controlled and legitimate security and justice institutions are crucial for peace, stability and sustainable development.[1] Furthermore, development and security assistance must prioritise the needs and interests of the country's population, rather than those of the state or elites, in order to achieve locally owned and sustainable results.

The security-development nexus posits that there is an interaction between the security situation and development outcomes, between the development situation and security outcomes, and between performance and outcomes in security and development assistance. Security sector reform (SSR) contributes to making this interdependence mutually beneficial, and helps ensure that the security and development communities interact constructively, without compromising their respective mandates. Ideally, this interaction would be closely coordinated from the planning to the implementation and the evaluation phases.

Recent experiences with SSR in fragile, often post-conflict states – some of which are examined more closely in this volume – should remind those working on SSR of the necessity to commit to full-scale strategies instead of cutting short key objectives and principles by settling on light or quasi-SSR approaches that all too often renege on the governance dimension and cross-sectoral, holistic approaches needed for the hoped-for 'reform' aspects to take hold and flourish.[2] It is also necessary to (re)focus attention on the core contribution of SSR as a development assistance instrument. This requires honest efforts to reassert evidence of how development fares when SSR is conducted as a development exercise rather than a traditional defence reform or civil-military assistance project. SSR is an instrument to achieve both security and development – it is meant to do so by synchronising security with development objectives so that both support, not harm, each other. In this chapter I argue that the interdependence between development and SSR is significant and mutually beneficial; if SSR is pursued according to its original objectives, advantages of closer cooperation between security and development communities will outweigh possible disadvantages.

This volume focuses on examining three puzzles: the relationship between security and development, the role of security sector reform as a linchpin in the interaction between security and development, and the contribution made by SSR to development objectives and outcomes. These three sets of relationships also guide this conceptual chapter. It begins with a brief discussion of our current understanding of the terms security and development, focusing on why the provision of security and planning of development objectives should be jointly designed and implemented as a holistic 'project' or mission, rather than as separate activities. Advancing this argument, recent evolutions in the international donor community's approach to and practice of security and development assistance in fragile and transitional states are discussed. This is followed by an examination of how the concept of a security-development nexus explains, but sometimes also confuses, the connection between security and development assistance. As an extension of this point, the concept and practice of SSR as an instrument to advance security and development in transitional societies are explored.

SSR is examined as a hybrid activity that improves a society's capacity to pursue both security and development. Yet the security-development nexus does not mean conflating security and development goals, activities and timelines. SSR espouses the ideals of a security-development nexus in practice and follows, on one hand, a very simple logic: without the provision of a minimum of security, safety and stability, the pursuit of development objectives is a futile activity. On the other hand, however, is a much more difficult pursuit: how to ensure that inputs into security simultaneously advance development prospects. If the right balance is not found, long-term security and stability are threatened. I argue that there is no reason not to approach SSR as a win-win activity for both security and development communities. In the human security paradigm, both safety and economic well-being (freedom from fear and freedom from want) are key ingredients of a stable society with a promising future. SSR ensures that a nation's security institutions are effective, provide for the safety and security of the population and the state, and are overseen and controlled by civil society organisations and democratically elected representatives. The intention is to ensure that the security sector offers protection from external and internal threats without itself becoming a threat, resulting in an environment that is safer and less prone to violence and instability and thus encourages economic growth, poverty reduction and human development.

At least, this is the theory. In reality, and particularly in post-conflict societies, SSR's security mandate has been more pronounced than its development mandate. It is commonly considered to be an activity of the security community,[3] and sometimes seen as being little different from earlier interventions on civil-military relations instead of a genuinely new activity designed to meet both security and development objectives. The focus of this chapter – and the book overall – is therefore on the pronounced gap between assertions of SSR's contribution to the security-development nexus and its actual contribution to development on the ground. Is SSR an effective bridge between security and development, advancing both for mutual benefit? I argue that it is, at least conceptually and from a policy perspective, and in the way it has been embraced by national and international actors in their security and development policies for transitional states in need of security sector reforms. I also contend that even the often-mentioned fear felt by development actors anxious not to expose themselves to undue securitisation is not as pronounced as it might be. In reality, many development actors are quite engaged in supporting SSR activities, a practice that has been recognised and accelerated by the OECD's decision to make many aspects previously characterised as security support now qualify as official development assistance (ODA) activities. This has happened not to boost member states' ODA contributions artificially, but to recognise the importance of security-related contributions as a bedrock activity for development actors. Moreover, such initiatives by development actors significantly increase the chance for development-sensitive security interventions to take shape.

Security, development and the security-development nexus

SSR as a concept, process and practice has evolved from attempts within the development community to engage with security-related challenges that have become increasingly relevant to its activities in fragile and post-conflict societies.[4] We will come back to the dynamics of this conceptual and practical evolution of SSR later in this chapter. At this point, it should be noted that while the security-development nexus is of considerable relevance to SSR, it is not necessarily easy to grasp. As Stern and Öjendal point out, 'Understanding, responding to or enacting a security-development nexus promises to be a daunting project.'[5] This chapter assumes that there is in fact a security-development nexus, that SSR is an embodiment of this nexus serving security and development objectives both separately and concurrently, and that SSR can therefore be considered a development project as much as a security project.

The different relationships in the nexus are discussed here, beginning with a brief examination of the concepts of security and development – what each is or is not, and how each can be understood and utilised conceptually and practically. I then explore the notion of a nexus between security and development, before turning to the relationships between SSR and security and, the main subject of my discussion, between SSR and development.

Understanding security

The end of the Cold War, commencing in 1989 when the Berlin Wall came down and culminating in 1991 when the Soviet Union dissolved, and the emergence of 'new' security threats triggered a number of major changes in the security debate. First, we experienced an expansion of the concept of 'security', as the preoccupation with a bipolar world order between East and West, the arms race and a potential nuclear Armageddon gave way to a focus on intra-state conflicts. This shift was accompanied by increased public and official focus on ethnic and minority conflicts, and expanded later to include environmental and other (root) causes of armed violence. There was greater emphasis on the prevention of violent conflict as well as on something new and bold: attention was given to options for external efforts to prevent internal conflict. These efforts were in part spearheaded by the United

Nations. Prominent advocates of this new thinking were UN Secretary-General Boutros Boutros-Ghali, whose 1992 'Agenda for Peace' defined much of the subsequent policy and academic debate,[6] followed by Secretary-General Kofi Annan, who wanted the United Nations and the international community of states, as well as other actors such as the business community, to move from a 'culture of reaction to a culture of prevention'.[7]

A better understanding and appreciation emerged about the significance of structural violence, such as the inequalities and injustices between the rich and poor, North and South, men and women, in contributing to conflict.[8] These newly accentuated facets of insecurity brought about greater appreciation of how security is perceived and experienced by different sexes, ethnic groups, elites and individuals. It became more obvious that multiple forms of insecurity affect different people at different times and for diverse reasons. The idea that 'security' could be delivered in the same way, at the same time, but in different locations, began to be questioned. Such questioning of cause, effect and remedy was much better able to expose the real, interlinked origins of many violent conflicts. As a direct result the interconnected and multilayered nature of insecurity and underdevelopment was understood as both the roots and the triggers of social tensions, conflict escalation and armed violence. Broader, more comprehensive definitions and approaches to security began to emerge.

To be sure, there are drawbacks to this broadening. The simplicity of characterising the so-called 'long peace'[9] of the Cold War as a bipolar struggle between two easily discernible ideologically, geostrategically and geopolitically opposed camps has given way to a considerably more complex security concept. That also makes it more difficult to analyse security and insecurity and design complex strategies of security provision. The burgeoning number of international and national actors whose security inputs need coordination is also a new challenge.

However, the time has passed when threats were simply identified in terms of national security and insecurity – and when response strategies focused mainly on serving ideational and ideological security objectives defined by political elites, into which all other security dimensions and actors were simply subsumed. The time has also passed when human and group security needs at home and abroad were of secondary concern to national political security concerns.

In traditional thinking the security and rights of individuals and groups could be sacrificed for the sake of national security objectives, even in democracies, without much resistance from populations that trusted, or did not engage much with, the arguments of their political leaders. A similar dynamic developed in the aftermath of 9/11, when civil liberties were sacrificed for what were perceived to be and sold as larger national and global security interests. Likewise, in traditional approaches, resources for security provision focused on the political and military aspects of security: the defence of borders, investment in quality and quantity of military personnel, material and equipment, and the support of countries that belonged to the same ideological camp. Other needs – especially structural security – were serviced only when resources were available and populations claimed the right to argue for responses to different needs and entitlements through democratic decision-making processes. In countries with less wealth and political participation, structural insecurity was at best a distant secondary priority for their governments.

Such approaches are now clearly outmoded, and there has been a widespread change of mind and argument. New security debates helped us to refocus on the multidimensional nature of security, away from a politically and ideologically motivated oversimplification to the empirically and reality-driven complexity of security provision that we see today. Comparing, understanding, matching and merging security and development concerns and responses are the only logical consequences of such new thinking. Human security, as a holistic concept, has the simple goal of making people safer at the core of the organisations, institutions and processes that were created to meet our human needs and offer protection from threats to our survival and well-being.

In such 'new security' thinking, both horizontal and vertical dynamics are at play.[10] The horizontal encompasses different thematic dimensions of 'security'. The *militaristic* dimension refers to the role of armed forces, military doctrine, defence, deterrence, arms control, military alliances, demilitarisation, disarmament, demobilisation and reintegration (DDR), the wider scope of the security sector, SSR and security governance. The *political* dimension includes norms and values, democracy and the stability of the political system. The *economic* dimension refers to public finances, currency stability, trade balance and access to or dependence on resources. The *environmental* dimension relates to the depletion and use of natural resources, climate change, biological diversity, the greenhouse effect, global warming and access to water. The *social* dimension includes issues of culture, religion, identity, language, minorities, gender equality, human rights and health; and the *personal* dimension of security includes issues of crime, domestic violence and human trafficking.

The vertical dynamic of security alludes to the different dimensions of insecurity and our analytical responses to them. Firstly, there is global security, meaning those threats that are relevant across borders and require international, even global, responses. The globalisation of world trade exacerbates local vulnerability to fluctuating global economic dynamics; and financial or political instability in one country affects the wider region and the world. The United Nations was created in part to address such global threats. Secondly, regional security addresses conflicts that have cross-border and regional repercussions in terms of both causes and responses, and most regional organisations were created to enhance regional security by means of support to their members' national security and/or economic growth and development. Thirdly, national security is the main preoccupation of national decision-makers who are, at least in democracies, mainly accountable to their fellow citizens. The challenge with national security interests is that at times they are misunderstood or seen too narrowly, so that larger dynamics, such as regional and global perspectives, get lost. A narrow focus on national security can also cause local-level distortions when it prioritises the security and well-being of the state, the government and the ruling elite at the expense of the population. The human security concept, which works both at home and abroad, is a corrective to this problem. It articulates well with the assumption that when human security is provided for, national, regional and global security will also benefit.

Differentiated interpretations of whose security matters – or matters most – are not new. However, focusing on the individual and communities as the main referent objects of security – rather than the state – is novel and potentially very sensitive. It challenges state sovereignty and forces new questions, for example on the role the state occupies *vis-à-vis* its citizens. How can human security concerns be met when political authorities and elites prioritise their own interests and cannot or do not want to focus on the needs of the larger population? From the perspective of states, international organisations and many researchers, the answer is broadly uncontested: the state remains in its current central position but acknowledges its responsibility and accountability to the population and the international community of states. Thus new security thinking has paved the way for a new approach to state sovereignty, with human security as the essential ingredient. Concepts and emerging norms such as the 'responsibility to protect' are among the consequences of such new thinking.[11]

Human security is based on the assumption that threats to the basic human needs of individuals and communities cause human suffering, as well as social and communal deterioration. Ultimately, they can trigger direct and structural violence, possibly leading to armed violence. That, in turn, increases human frustration, feeding into a vicious, cyclical relationship. By contrast, if individuals and communities feel secure and protected from the existential threats that emerge from social, political and economic injustice, military violence, environmental disruptions or natural disasters – that is, if their human security is protected and guaranteed – then individual human suffering and also communal, regional and international conflict and violence can be significantly reduced. The concept of human security focuses not only on armed conflict and its consequences for civilians, but also on many non-traditional security threats, including disease and economic, environmental or inter-group security threats. Moreover, the human costs of non-traditional security threats – those not related to armed conflict, which reportedly has been declining[12] – are devastating, and it is now understood that such threats can escalate into armed violence and war.[13]

When the concept of human security was introduced in the 1994 UNDP *Human Development Report* (HDR), it was used as a comprehensive approach to encompass all threats to human rights, security and development experienced by individuals and communities.[14] Human security was meant to represent a key instrument, or rather an agenda, to fight poverty and improve human livelihoods – it was seen as providing both 'freedom from fear' and 'freedom from want'. Human security also for the first time introduced, from a development perspective, the security and development conundrum to a larger global community of practitioners, policy-makers and researchers. Later in this chapter I discuss the validity and nature of the nexus this thinking created.

Since its introduction, many governments and international organisations have acknowledged the concept of human security as an important item on their national and international security and development agendas. As Brzoska notes:

> the concept has given somewhat more intellectual depth to the development donors' idea of reducing military expenditure. Here was a concept that justified looking hard at the level of military expenditure, taking into account all threats to the survival and health of people. In fact, the 1994 UNDP Human Development Report unabashedly argued for deep cuts in military expenditure … On the other hand, by arguing that violence was but one threat among many to peoples' lives, it helped the development donor community take all threats – including those from violence – seriously. If development policy needed to address all threats to life and health, the development donors

> could also claim responsibility for all such policies, including those addressing protection from the threat of collective or individual violence.[15]

The concept has been promoted particularly by governments and non-governmental organisations that put less emphasis on traditional power politics, especially geostrategic politics. This includes countries such as Sweden, Norway, Japan, Switzerland, Canada and other members of the Human Security Network, an informal group of countries that is devoted to the promotion of the human security concept.[16] Former Canadian foreign minister Lloyd Axworthy's introduction of the concept in the UN Security Council during Canada's 1999–2000 presidency of the Council and Japan's initiative of the Commission on Human Security have given it prominence and worldwide recognition. The call for a return to a human security focus expressed in the recent WDR 2011 has revitalised the concept after some years of silence in academic and policy debates.

Human security has been at the forefront of the security-development nexus debate. The OECD acknowledged this shift in its 2001 guidelines, arguing that the new conceptualisation of security 'includes the responsibility, principally of the state, to ensure the well-being of people. As a consequence, discussion of security issues, "systems" and actors has become comprehensive and no longer refers to military systems only.'[17] Human security focuses on the individual and the population as the 'referent objects' of security – which is in the first instance about individuals, communities and populations. Two important dimensions define responses to human security threats. On the one hand measures need to be put in place to reduce or prevent them, as prevention has also been reintroduced as an important approach to security and development by the WDR 2011. On the other hand people's coping capacities need to be strengthened to adapt to ongoing human insecurity. As Mark Duffield so pointedly observes:

> In order to understand the nature and implications of the contemporary development-security nexus, development and underdevelopment are reconceived biopolitically. Rather than a labour of theory, however, this is more a question of drawing out how aid policy itself now attentively focuses on issues of life and community; on how life can be supported, maintained and enhanced; and within what limits and level of need people are required to live. In terms of development discourse, the emergence of concepts such as human development and human security are important. UNDP, for example, launched its annual *Human Development Report* in 1990, dedicating it to 'ending the mismeasure of human progress by economic growth alone' … Where human development marks the formal shift from an earlier economic

> paradigm to a 'people-centred' frame of development for the global south, human security effects a similar change in relation to security.[18]

Some protagonists of human security favour a narrow definition of the concept, one that is focused on *freedom from fear*. As it is politically more expedient and intellectually simpler, they focus on personal security, immediate threats of violent conflict and as a main objective the provision and maintenance of a *negative peace*. Under such conditions there would be no armed conflict, little or no violent crime – the streets would be safe – but people would continue to suffer from structural violence. They might not run the risk of getting killed by an armed group, government forces or thugs, but they might die or suffer from hunger or lack of medical services. Others have taken a different approach. The 1994 HDR and the Commission on Human Security's report 'Human Security Now' both argue that a broader range of threats must be addressed and existential threats of individuals must be dealt with regardless of their source.[19] The Commission on Human Security argues that 'Human security means protecting fundamental freedoms – freedoms that are the essence of life. It means protecting people from critical (severe) and pervasive (widespread) threats and situations. It means using processes that build on people's strengths and aspirations. It means creating political, social, environmental, economic, military and cultural systems that together give people the building blocks of survival, livelihood and dignity.'[20] This broad approach to human security overlaps heavily with the human development agenda and embodies the intersectional dynamics observed in the security-development nexus.

Understanding development

In order to grasp the role security and SSR play in development, we need to understand what development stands for and what the development community hopes to achieve through its assistance in countries that require support in meeting minimum human development standards. The latter term – human development – shows that development is not merely about economic growth and poverty reduction, but addresses people's humanity as well as their political, economic and social rights, as expressed in the Universal Declaration of Human Rights.[21]

In this subsection I first analyse a number of key definitions and approaches to development and recall some of the major indicators used to measure it. This gives an impression of the human condition that is to be affected and improved by development assistance. I also briefly revisit the definitions used by some donor nations' development agencies, particularly those which already embrace security and SSR-related issues in their approaches to development assistance.

Since its first publication in 1990, UNDP's annual HDR has significantly shaped development thinking globally and, with regional human development reports, in specific regional contexts. Its Human Development Index 'presents agenda-setting data and analysis and calls international attentions to issues and policy options that put people at the center of strategies to meet the challenges of development'.[22] UNDP's 1996 HDR, entitled *Economic Growth and Human Development*, for instance notes that 'Human development went far beyond income and growth to cover the full flourishing of all human capabilities. It emphasized the importance of putting people – their needs, their aspirations, their choices – at the centre of the development effort.' The report argues that 'human development can be expressed as a process of enlarging people's choices'. The 1997 HDR, *Human Development to Eradicate Poverty*, defines human development as 'widening people's choices and the level of well-being they achieve'. It explains that 'regardless of the level of development, the three essential choices for people are to lead a long and healthy life, to acquire knowledge and to have access to the resources needed for a decent standard of living. Human development does not end there, however. Other choices, highly valued by many people, range from political, economic and social freedom to opportunities for being creative and productive and enjoying self-respect and guaranteed human rights.'[23] The HDR's indicators of human development reflect a broad range of factors that measure the economic condition, livelihood and various aspects of structural security and insecurity.[24]

As a global initiative to focus attention worldwide on building a safer and more prosperous and equitable world, the Millennium Declaration[25] was endorsed by 189 world leaders at the United Nations in September 2000. The declaration was translated into the Millennium Development Goals (MDGs), consisting of eight time-bound and measurable goals that were to be reached by 2015. The first goal envisions the eradication of extreme poverty and hunger, including the reduction by half of the proportion of people whose income is less than US$1 a day and the proportion who suffer from hunger. The second goal targets the achievement of universal primary education, ensuring that all boys and girls complete a full course of primary schooling. The third goal focuses on the promotion of gender equality and empowerment of women, by eliminating gender disparity in primary and secondary education preferably by 2005, and in all levels of education no later than 2015. The fourth goal envisions the reduction of the mortality of children under five by two-thirds. The fifth goal focuses on improvements in maternal health, by reducing maternal mortality by three-quarters. The sixth goal is targeted at combating HIV/AIDS, malaria and other major diseases, by halting and reversing their spread. The seventh goal addresses the need to ensure environmental sustainability, by integrating principles of sustainable development into country policies and programmes, reversing the loss of environmental resources, halving the proportion of people without access to safe drinking water and basic sanitation, and improving the lives of at least 100 million slum dwellers by 2020. The eighth goal is dedicated to the development of a global partnership for development, by creating and expanding an open, rule-based, predictable, non-discriminatory trading and financial system; addressing special needs of the least developed countries, landlocked countries and small island developing states; dealing with developing countries' debt; cooperating with developing countries; developing and implementing strategies for decent work for youth; and, in cooperation with the private sector, making available the benefits of new technologies, especially information and communications.[26] Significantly, however, the MDGs do not mention traditional security threats, direct violence or their development impacts. Nevertheless, it would be difficult to imagine how many of these objectives can be reached outside an environment where the safety and security of the population and the state are guaranteed. Moreover, even where progress has been made, internal instability, organised crime, armed violence or even large-scale conflict reverses advances that may have been made towards meeting MDGs and more general human development and human security objectives. A supportive security sector, along with a stable, functioning and well-governed state, is a crucial element of creating an enabling environment for achieving and maintaining sustainable human development. Development agencies have recognised this fact, yet this recognition alone has not put to rest their critical and sceptical attitude towards collaborating closer with the security community in a common pursuit of presumably common goals.

As Brzoska noted in his influential 2003 study, development agencies showed different degrees of enthusiasm for SSR: 'The willingness of development donors to engage and work with the new concept of security sector reform has differed markedly from agency to agency in the years since it was first coined. The UK government, which took the lead, has found a number of followers in the Nordic countries, as well as in Belgium, the Netherlands, Germany, Switzerland and the United States.'[27] However, in recent years development ministries and agencies around the globe have incorporated security issues in their definitions of and criteria for development aid, including activities very specifically related to SSR, presumably following the OECD's decision to make SSR activities ODA-eligible, as discussed below. The following highlights a small selection of development donors' approaches to both development and SSR activities.

The Australian Agency for International Development (AusAID) focuses on accelerating economic growth, fostering functioning and effective states by investing in people and promoting regional stability and cooperation. Its aid categories include disability, disaster risk reduction, economic growth, education, environment, food security, gender equality, governance, health, human rights, infrastructure, the MDGs, mine action, regional stability, rural development and water and sanitation. Specific SSR projects include, among others, aid to strengthen Vanuatu's police services (2002), the Australia-East Timor Police Development Programme (2008–2010), and the Regional Assistance Mission to Solomon Islands since 2002.[28]

The Canadian International Development Agency (CIDA) invests in advancing food security, securing the future of children and youth, and stimulating sustainable economic growth. Among its specific goals it includes reducing the frequency and intensity of violent conflict and increasing civilian oversight, accountability and transparency of security systems. CIDA's specific SSR projects range from training and professional development of the Haitian National Police's managerial staff (2008–2015) to assistance in reforming the correctional system in Serbia through a grant to the Council of Europe.[29]

The Danish International Development Agency (DANIDA) offers development support to ensure people's freedom from poverty, fear, degradation, powerlessness and abuse, but also freedom to take charge of one's own destiny and responsibility for one's own life. DANIDA focuses on growth and development; freedom; democracy and human rights; gender equality; stability and fragility; and environment and climate issues. Within its programmatic area of conflict prevention activities, DANIDA contributes to nation-building and democratisation, both from the top down (involving state institutions and local authorities) and bottom up (involving civil society organisations and the private sector). It expects to achieve this through the promotion of and respect for human freedom and human rights, strengthening the rule of law, reform of the security sector, inclusive political processes and a responsible and more efficient state. DANIDA is currently engaged in setting up a whole-of-government stabilisation, security and justice sector development and peace-building programme in the East Africa/Horn of Africa/Yemen region.[30]

Germany's Federal Ministry for Economic Cooperation and Development (BMZ) promotes freedom and development for all, and assistance in securing a life without poverty, fear and environmental destruction. SSR is one of its key activities, designed to transform the state's entire security system through multi-stakeholder processes, the promotion of democratic norms, the enforcement of the state's monopoly of force and democratic control of the security sector. BMZ is involved in a wide array of SSR and SSR-related activities, ranging from justice sector reform training of police and ex-combatants to support for former child soldiers. German International Cooperation is involved in the promotion of civilian security and community policing, improving accountability and quality management in the judicial sector, the promotion of democratic control of security institutions and DDR programming. It is active in a number of countries, ranging from Cambodia to Afghanistan, Guatemala, Nicaragua, Morocco, Uganda and the Occupied Palestinian Territory.[31]

Japan's Ministry of Foreign Affairs Section on Official Development Assistance (MOFA-ODA) defines its objectives as contributing 'to the peace and development of the international community … thereby to help ensure Japan's own security and prosperity'. Few donors formally link their development activities so closely with national security and economic interests. Priority areas include poverty reduction, sustainable growth, addressing global issues and peace-building. MOFA-ODA recognises SSR as 'one of the critical foundations of a state and … an essential element for the return and resettlement of refugees and internally displaced persons, as well as for rebuilding the life of the local population'. Its activities focus on DDR in Afghanistan (especially of armed groups), military, police and justice reform in the Democratic Republic of the Congo (DRC), human resource and infrastructure development in East Timor and public sector reform in Mongolia. The Japan International Cooperation Agency is equally actively involved in SSR programmes, particularly in Afghanistan and Cambodia.[32]

Norway's Ministry of Foreign Affairs' International Development Programme (with the Norwegian Agency for Development Cooperation) focuses on fighting poverty and bringing about social justice. In its understanding of development, one of the key factors is 'a well-functioning state that safeguards peace, security and human rights, delivers basic services to the population, and ensures that there are good conditions for healthy economic activity and trade'. SSR cooperation programmes exist with Macedonia, Montenegro, Serbia and Indonesia. Activities include various SSR programmes in Bosnia-Herzegovina (2011), Liberia (2008), Afghanistan (2005), the DRC (2009), Ukraine (2007) and Sudan (2005–2010).[33]

The Swedish International Development Cooperation Agency (SIDA) aims to 'help create conditions that will enable poor people to improve the quality of their lives'. It asserts that 'by reducing injustices and poverty throughout the world, better opportunities are created for development, peace and security for all people and nations'. Among SIDA's five key areas for development, peace and security feature prominently, as it considers armed conflict and post-conflict situations as some of the main obstacles for development and poverty reduction in the world. It approaches SSR as one of the tools available to promote peace and security, and has been supporting SSR-related institutional reforms and capacity-building activities in South Africa, Serbia, Bosnia-Herzegovina, Rwanda, the DRC and Liberia, among others.[34]

The UK Department for International Development (DFID) has been instrumental in the development of the SSR concept. DFID focuses on, in the broadest sense, furthering sustainable development and improving the welfare of populations. The reduction of poverty, respect for human rights and other international obligations, and improving public financial management, promoting good governance and transparency and fighting corruption are key issues in reaching these goals. It defines security and justice sector reform as 'a people-centred approach to justice and security', with rule of law, accountability, transparency, accessibility and affordability as central components on this agenda. DFID highlights the establishment of democratic control of the security sector and capable, professional and accountable security services and justice systems, and a supportive culture for these reform objectives with the political, security and justice leadership. DFID is currently involved in over 20 SSR projects in Sudan, Kenya, Liberia, Sierra Leone and numerous other countries.[35]

Similar to Japan, the US Agency for International Development (USAID) openly links its development goals to its national interests. It defines development assistance as 'programs, projects, and activities carried out by USAID that improve the lives of the citizens of developing countries while furthering U.S. foreign policy interests in expanding democracy and promoting free market economic growth'. USAID has been active in SSR and SSR-related programmes in numerous countries spanning the entire globe.[36]

This brief review of several donors' self-proclaimed development objectives shows that not only are many of those objectives closely linked to the stabilisation and improvement of government services for the poor, but that security-related activities – including SSR – are considered to be key facilitators in providing development assistance and making the results of such assistance last. Nevertheless, worries abound about reaching too far beyond one's core development activities, as the following subsection will show.

Securitising development and/or developmentalising security?

Here I further explore the reasons for closer cooperation between security and development actors, particularly in fragile post-conflict contexts where security and development agendas and requirements have been increasingly difficult to separate and a formerly antagonistic relationship has now evolved into mutually supportive coexistence to achieve cooperation. This is hardly surprising, considering the close proximity of even traditional security issues with many of the factors that drive and characterise human development. Development actors' fairly active engagement with security-related issues, as seen in the overview above, attests to the centrality of justice, security, violence and conflict management for development work. Moreover, considering the broadened security agenda which has evolved not only in the academic debate but also, since the end of the Cold War, within policy debates, issues such as epidemic diseases, access to drinking water, gender-based violence, hunger and joblessness are taken seriously as (root) causes of structural, direct and, eventually, armed violence.

Yet both security and development actors are wary of closer proximity to the work of the other community, although unsure about the impact, possibly through joint activities, that it has on their own preparedness and capacity to address their core business. On the one hand, there is a tendency among policy-makers to merge political-military and humanitarian-development activities with a broader trend towards the politicisation of aid. Humanitarian and development action is treated as a political instrument in violent conflicts or a substitute for political action in regions that are peripheral to national strategic interests. Military support of such assistance activities might be seen as a legitimate instrument in the toolbox of conflict management. In some cases, however, humanitarian or development 'labels' are abused to justify political or military action – a development characteristic of international actors' post-9/11 campaign in Afghanistan.[37]

The early academic debate on 'securitisation' in the 1990s features a similar phenomenon: a horizontal broadening of a plethora of security issues, from poverty to health and the environment, combined with efforts to establish the direct or indirect links of such issues to a potential escalation to armed conflict. This approach has elevated some of these threats, which were previously not at the centre of traditional security thinking, to the level of serious national security concerns. Addressing them would in turn require and possibly trigger responses equal to those that meet traditional major national security threats, such as nuclear arms proliferation. Securitising a 'non-traditional' security threat would thus attach enough significance and urgency to elevate it to a top national security concern. Critics could not help but notice an inherent danger in such reasoning. In cases where a particular threat – such as poverty or HIV/AIDS, for example – could not be convincingly linked to an eventual outbreak of violent armed conflict, its significance as a key issue for immediate preventive action may in fact decrease. It was feared that threats not obviously correlated with potential armed conflict would fall through the cracks of national and international security and conflict management, thus drawing less, not more, attention to a number of the new threats identified in new security thinking.

On the other hand, and on a more practical level, there is great fear that the collaboration of security and development communities in the design and implementation of joint projects, including jointly planned and administered SSR programmes, may lead to a 'takeover' by one agenda and one set of actors. While for the purposes of expediency and efficiency, in field activities a lead actor often takes the initiative to provide guidance, such leadership can skew the nature of the joint activity. There is a real danger that, due to its organisationally more rigorous structure and culture, financial capacities, access and territorial reach, members of the security community will dominate and possibly take over. More research is needed to generate empirical knowledge on this tension and address the extent to which it is actually happening, to either confirm or disprove the validity of fears in the development community that development assistance is becoming securitised and militarised. Strong evidence for the likely domination of the security community over the development community in shared missions could put an end to joint initiatives, re-create silos, add parallel tracks in the pursuit of similar objectives and create considerable duplication and confusion in the implementation of joint programmes for the common good.

The security community also has its concerns. In modern multidimensional and complex peace operations, the military and civilian components are increasingly tasked with delivering humanitarian assistance and aid.[38] Security actors are not necessarily keen on closer cooperation with development actors: military forces in particular fear the developmentalisation of their missions. They fear that merging their tasks and operations with development (and humanitarian aid) activities will put pressure on them to move beyond their initial mandates to provide public security and protect the personnel of civilian humanitarian and relief missions. In complex emergencies this might complicate or even compromise their military missions, particularly in early stabilisation phases after the formal conclusion of an armed conflict. Moreover, military troops are not always fully trained for and sensitised to the needs of humanitarian and development activities, increasing the risk of their unintentionally engaging in inappropriate behaviour towards civilians in these activities and triggering public relations disasters in the mission and at home.[39]

Still, the debate on securitisation needs to be more even-handed. Depending on what one considers to be the hallmarks of the 'security' concept, securitisation is not necessarily synonymous with militarisation. If we see it as encompassing both structural and direct security, particularly in the context of the broader security (sector) community, securitisation may be as much about improved justice provision and reducing government corruption as it is about providing military assistance to stabilise post-conflict situations.

Security and development – as well as their 'offspring' SSR and human security – are concepts and activities that do not have to compromise each other's objectives as long as they are not pursued as part of highly politicised or misappropriated agendas. If the latter is the case, and unfortunately the ongoing military engagements in both Afghanistan and Iraq are strong cases of misappropriated SSR, the common bases for otherwise constructive and fruitful joint objectives and collaboration, supported by complementary modes of operation, are jeopardised. If development activities become militarised and military operations become developmentalised, both SSR and human security objectives will be eroded and lose their legitimacy and utility for both donor and recipient communities.

Understanding – or envisioning – a security-development nexus[40]

In a recent volume on the linkages between security and development, its editors argue that while 'The call for greater convergence between security and development policies emerged in response to the complex and interlocking humanitarian, human rights, security, and development crises that confronted international policymakers in the immediate aftermath of the Cold War … interestingly, academic researchers initially had little to offer to the international policy debates and were slow in removing the blinders of their particular disciplines so as to better examine the linkages between security and development.'[41] All the same, since the 1990s there has been a continuous 'stream of policy documents by international institutions and bilateral and multilateral donors … [calling] for concerted international action to address these complex and multidimensional challenges'.[42]

As readers of this volume will realise, despite much innovative research, not much progress has been made to date. In fact, as Tschirgi et al. note, 'policymakers frequently become frustrated while trying to make sense of competing interpretations of the complex and pressing problems in [poor and unstable countries, which] may explain their frequent resort to ready-made policy formulas such as the security-development nexus, which has come to mean many things to many people'.[43]

The terms 'linkage', 'interdependence', 'connection' and 'relationship' crop up often in debates on development and security. However, the term 'nexus' is also increasingly used in academic debate. What is a nexus – particularly in the context of the linkage of multifaceted concepts? The lead authors of a recent special issue of *Security Dialogue* on 'The Security-Development Nexus Revisited' defined the term as 'a network of connections between disparate ideas, processes or objects; alluding to a nexus implies an infinite number of possible linkages and relations'.[44] This definition does not simplify the concept – and in their article the authors unsurprisingly struggle with the challenge to make sense of what the security-development nexus could possibly entail, explain or suggest. They point to an emerging literature on issues ranging from peace-building to complex emergencies, post-conflict reconstruction, human security and intervention – with this book comfortably fitting into the line-up – which reflects 'a seeming consensus that "security" and "development" are interconnected, and that their relationship is growing in significance given the evolving global political-economic landscape'.[45]

The security-development nexus seems to explain the inexplicable, the assumed and the incomprehensible – yet at the same time is seen as common sense, dictating a seemingly undeniable linkage. Security and development are linked, somehow, through relations, dependencies, interdependencies, causal links, claims, perceptions, convenience, sensations and similar assumptions about cooperation between and among communities and organisational structures. Stern and Öjendal make sense of this confusing, necessary and potentially – if used to inform security and development policy – life-saving relationship in the following way:

> The notion of a 'nexus' seems to provide a possible framework for acutely needed progressive policies designed to address the complex policy problems and challenges of today. Furthermore, and perhaps most importantly, an ever-growing amount of economic resources and political will is being poured into the 'security-development nexus' and the attendant revamping of national and multilateral institutions and actions designed to address it. Hence, 'the nexus' matters.[46]

Heidi Hudson (Chapter 3) describes 'intersectionality', a very similar phenomenon. Sometimes complex relationships cannot be simplified without losing the – challenging and sometimes frustrating – richness and depth of their very meaning and utility. However, it is often difficult to grasp and understand complex concepts and relationships in practice – a critique that is also levied against the concepts of human security and SSR. The programmatic agendas and activities of these concepts need to be demystified in order to generate confidence in their practical relevance.

Recognising a security-development nexus and policy responses to it might help stop unstable, poor or war-torn societies from descending into chaos. It might allow the merging of security and development agendas to achieve common goals in bringing peace and stability to otherwise fragile societies. Does the recognition of this nexus – and acting on this recognition – help make development assistance and security provision more sustainable? While the nexus could take the form of a social contract for those to whom security and development support is provided, it may also provide a sense of accountability and responsibility for those providing such support. As is the case with any external intervention, at some point initiatives are handed over to national and local actors, while international actors for the most part retreat. Activities that are based on the existence of a security-development nexus, such as SSR, will have to be continued by governments that enjoy a minimum of trust and legitimacy, yet are 'entrusted' with coordinating both development and security activities.

The goals, objectives and benchmarks for security and development activities need to be negotiated and determined ahead of launching externally supported programmes and before handing them over to national actors. The security-development nexus is a conceptual puzzle, an empirically questionable reality and a policy agenda. However, to turn the nexus into a set of specific points that can reliably inform the design and implementation of policy priorities which straddle both security and development objectives, such as SSR, we require more than a vague realisation that, by default and possibly under most circumstances, investments in security benefit development and vice versa.

I close this discussion on the security-development nexus with an assessment of the nature and utility of the nexus referred to at the beginning of this subsection. As Stern and Öjendal conclude:

> First, [we draw] attention to the claims that there is an empirically real and growing 'nexus', which is reflected in the increased usage of the term 'development-security nexus'. Although timely, we aver that this borders on the banal: 'the nexus', however conceived, reflects a reality that resonates in the experiences and imaginations of many, it is being used to 'describe' a growing realm. Second, and perhaps more intriguingly, the 'content' or form of 'the nexus' is not clear. It is therefore open for all kinds of (illicit) use under the guise of progressive and ethically palatable politics. We believe that … different discourses imbue 'the nexus' with different meanings. Third, as 'the nexus' is being and can be used as a 'recognizable' and seemingly comprehensible narrative, various processes can be pursued in the name of (more or less) in/compatible combinations of security-development.[47]

Put in somewhat simpler terms, the authors confirm that the exact nature of the security-development nexus is hard to grasp and might thus be considered more of a condition than a set of easily visible interconnecting factors and processes. Nevertheless, this does not make the security-development nexus any less important or less crucial for workable, sensible and effective security and development assistance activities. Comprehensive and holistic, security- and development-sensitive SSR occasionally engenders similar confusions and uncertainties about its empirical value and practical utility. Yet such a sense of complexity and 'intersectionality' does not make it any less important – or any less workable.

Security sector reform and the security-development nexus

Conceptual as well as practical debates on SSR suffer from a sometimes bewildering and counterproductive diversity of definitions of the institutions and actors that make up a security sector or specific tasks and activities that define the process of reforming the security sector.[48] In contrast, the 2008 report on SSR by the UN Secretary-General offers a solid framework for a common, comprehensive and coherent approach by the United Nations and its member states, reflecting shared principles, objectives and guidelines for the development and implementation of SSR.[49] The report notes that:

> It is generally accepted that the security sector includes defence, law enforcement, corrections, intelligence services and institutions responsible for border management, customs and civil emergencies. Elements of the judicial sector responsible for the adjudication of cases of alleged criminal conduct and misuse of force are, in many instances, also included. Furthermore, the security sector includes actors that play a role in managing and overseeing the design and implementation of security, such as ministries, legislative bodies and civil society groups. Other non-State actors that could be considered part of the security sector include customary or informal authorities and private security services.[50]

Moreover, in the words of the report, 'Security sector reform describes a process of assessment, review and implementation as well as monitoring and evaluation led by national authorities that has as its goal the enhancement of effective and accountable security for the State and its peoples without discrimination and with full respect for human rights and the rule of law.'[51]

As is characteristic for UN reports of this kind, the UN Secretary-General's definitions represent the result of extensive and broad consultation processes that generate broadly supported UN norms and guidelines for its member states. Although reflecting the result of a similarly careful and inclusive consultation process, the definition of SSR provided by the OECD Development Assistance Committee (DAC) is slightly more comprehensive and demanding in terms of its coverage of actors, processes and principles. The OECD/DAC's *Handbook on Security System Reform*, a much-referred-to standard elaboration on the concept of SSR, calls for a holistic approach to the security 'system' and offers helpful elaborations on the roles and tasks of all state and non-state institutions and actors that contribute to the provision of security for the state and its people. These actors include the following:

- *Core security actors:* the armed forces; police service; gendarmeries; paramilitary forces; presidential guards; intelligence and security services (both military and civilian); coastguards; border guards; customs authorities; and reserve and local security units (civil defence forces, national guards and militias).
- *Management and oversight bodies:* the executive, national security advisory bodies, legislative and select committees; ministries of defence, internal affairs and foreign affairs; customary and traditional

authorities; financial management bodies (finance ministries, budget officers and financial audit and planning units); and civil society organisations (civilian review boards and public complaints commissions).

- *Justice and the rule of law:* the judiciary and justice ministries; prisons; criminal investigation and prosecution services; human rights commissions; ombudspersons; and customary and traditional justice systems.
- *Non-statutory security forces:* liberation armies; guerrilla armies; private security and military companies; and political party militias.[52]

In addition, although not specifically mentioned in greater detail beyond their inclusion in the group of management and oversight bodies but usually thought to have considerable influence, there are civil society actors such as professional groups, the media, research organisations, advocacy groups, religious bodies, non-governmental organisations and community groups.[53]

Objectives of SSR

The main objectives of security sector reform are, first, to develop an effective, affordable and efficient security sector, for example by restructuring or building human and material capacity; and, second, to ensure democratic and civilian control of the security sector, for example through strengthening the management and oversight capacities of government ministries, parliament and civil society organisations.

In operational terms SSR covers a wide range of activities within five broad categories:[54]

- *Overarching activities*, such as security sector reviews and their development, needs assessments and development of SSR strategies and national security policies.*Activities related to security- and justice-providing institutions*, such as restructuring and reforming national defence, police and other law enforcement agencies as well as judicial and prison systems.
- *Activities related to civilian management and democratic oversight* of security and justice institutions, including executive management and control, parliamentary oversight, judicial review, oversight by independent bodies and civil society oversight.

- *Activities related to SSR in post-conflict environments*, such as DDR, control of small arms and light weapons, mine action and transitional justice.
- *Activities related to cross-cutting concerns*, such as gender issues and child protection.

In addition, SSR's contribution to peace-building has specific political, economic, social and institutional dimensions. The political dimension entails the promotion and facilitation of civil control over security institutions; the economic dimension ensures appropriate consumption and allocation of society's resources for the security sector; the social dimension holds that the provision of the population's physical security should in all cases be guaranteed, and not additionally threatened, by the assistance of the security sector; and, directly related, the institutional dimension focuses on the professionalisation of all actors in the security sector.[55]

In addition to these technical objectives of SSR efforts, the academic and practitioner literature as well as official statements and operational and institutional statements such as the OECD/DAC guidelines and the UN Secretary-General's report argue that SSR should embrace the following principles:

- SSR should be people-centred, locally owned and based on democratic norms, human rights principles and the rule of law, so that it can provide freedom from fear and measurable reductions in armed violence and crime. This principle must be upheld in both the design and implementation of SSR programmes. It should not simply remain at the level of proclamation and intention.[56]
- SSR must be seen as a framework to structure thinking about how to address diverse security challenges facing states and their populations, through more integrated development and security policies and greater civilian involvement and oversight. National, broad and public consultation processes as well as a national security strategy are thus inherent requirements of feasible SSR strategies.
- SSR activities should form part of multisectoral strategies, based on broad assessments of the range of security and justice needs of the people and the state. They have to respond to the needs of all stakeholders.

- SSR must be developed in adherence to basic governance principles, such as transparency, accountability and other principles of good governance.
- SSR must be implemented through clear processes and policies that enhance institutional and human capacities to ensure that security policy can function effectively and justice can be delivered equitably.[57]

How does one know if a security sector is in need of reform? Put simply, if the sector is not inclusive, is partial and corrupt, unresponsive, incoherent, ineffective and inefficient and/or unaccountable to the public, then it (or any of its affected institutions) is in need of reform. The term 'reform' describes an institutional transformation that leads to the improved overall performance of a legitimate, credible, well-functioning and well-governed security sector, which serves society in providing internal and external, direct and structural security and justice as public services.

The extent of the reform required depends on how much is needed to make the sector fulfil its roles accountably, and rarely means a total overhaul. Certain components and aspects of a nation's security sector might be functioning admirably well, while others might be in need of extensive improvements. Thus identifying where, how and when individual components must be (re)built, restructured, changed and/or fine-tuned is an important step and requires a solid assessment of the sector's roles, tasks and requirements in light of national and local assessments of society's security and development needs. SSR processes therefore vary from country to country, with each SSR context being different and unique.

The fallacy of 'SSR-light'

The full range of tasks and options ideally covered in SSR processes is comprehensive and demanding, but certainly not undoable if planned, prepared and implemented in collaboration with all relevant actors in a sensible, sequenced, phased and context-responsive strategy. SSR is a long-term exercise that does not lend itself to quick-fix approaches, even though there are some pieces that can be completed fairly swiftly. For example, an apparently short-term activity, such as a 'train and equip' programme for armed forces, police or border guards, or some other technical measure to address immediate security and stabilisation needs, will only succeed fully if it is seen as part of a longer-term reform approach. Development actors should judge the seriousness and genuineness of an SSR activity in light of how it contributes to longitudinal change.

Taking an 'SSR-light' approach may be tempting, as it might ensure quick approval by national actors who stand to lose influence, power and privileges as a result of full-fledged SSR programmes. Yet approaches that search for easy ways out are counterproductive to the improvement of stability, peace, security and development. If SSR is true to its own goals, approaches and principles, it strongly matches the objectives and approaches preferred by development actors. Neither should disagree about what needs to be done – or how – to support a society's transition process comprehensively. If only partial and quasi-SSR activities are being offered, development actors should stay away. They would take a big risk of being instrumentalised to implement a set of activities that cannot deliver what they claim. The same applies to SSR actors – they should accept all offers to engage with development donors and match and implement joint or complementary strategies and programmes to further their joint objectives. If development actors do not comply with generally respected standards of development assistance, for instance by benefiting one particular group over another or advocating a particular ideology or donor nation's strategic aims, collaboration should be avoided.

It is important to ensure that only genuine SSR is implemented as a companion to development assistance – meaning that it is pursued as a long-term project, designed in a participatory and inclusive manner in collaboration with state and non-state actors, and makes a strong commitment to local ownership and good governance, among other key principles. Quasi-SSR activities that do not meet those qualifications will do more harm than good.[58]

SSR as a security and development 'project'

After this brief discussion of definitions and approaches of SSR, I turn my focus to SSR as both an embodiment and a driving force for the interrelationship, the nexus, between security and development. Being a development 'project', yet working primarily with security institutions, what can SSR do for development?

The popularisation of SSR has been attributed to the former UK secretary of state for international development, Clare Short, who argued that 'A security sector that is well tasked and managed serves the interests of all, by providing security and stability – against both external and internal security threats. And obviously security is an essential prerequisite for sustainable development and poverty reduction.' Moreover, she insisted that 'a security sector of appropriate size, properly tasked and managed, is a key issue. We are therefore entering this new area of security sector reform in order to strengthen our contribution to development.'[59] These statements reflect DFID's commitment to engage in SSR to facilitate poverty reduction through development assistance. Short created the momentum for this development with a speech she gave at the Royal College of Defence Studies in London in May 1998, where she called for 'a partnership between the development community and the military' in an effort to address the 'inter-related issues of security, development and conflict prevention'.[60]

Herbert Wulf argues that:

> One criterion for using the term security sector reform is that this assistance is integrated into an overall strategy of development and democratisation of the society. This implies that security sector reform can never be implemented as a stand-alone programme but has to be embedded in a general peace-building and development programme. The military assistance programmes, implemented during the Cold War, which were essentially ideologically motivated, did not as a rule comply with the concept of security sector reform in use today, since they aimed merely to strengthen or modernise the armed forces in question and consolidate the influence of the donor countries. But they did not seek to help establish a democratically controlled security sector that would be conducive to development.[61]

The next subsection explores in more detail the evolution of SSR as a joint security and development 'project'. SSR is a highly intersectional concept – defining SSR priorities depends on effective collaboration and linking of security and development needs assessments, conducted and implemented by actors from both communities. It is necessary to reflect on past experiences and improve opportunities for learning how to do SSR right.

The origins of SSR in the development discourse

> It would seem obvious that there was a need to find a new term for a plethora of phenomena and activities related to reform of the sector of society charged with the provision of security.[62]

In tracing the evolution of the debate on SSR's relevance for the development community, I refer extensively to a study written for DCAF by

Michael Brzoska in 2003. Although somewhat dated, his analysis still offers one of the best examinations of the role SSR began to play for development actors. Little has been written on the subject since then – and it makes sense to develop further debates from Brzoska's observations. As he notes, in the early 1990s 'Security sector reform has its roots in the development donor debate, an on-going discussion among various groups of practitioners and theoreticians on how best to target and implement development assistance … previously, the donor community had largely refrained from discussing security-related issues. Many actors in the donor community have had, and continue to have, a strong bias against working with security sector players, particularly with the military.' Brzoska notes that with the lifting of the political and ideological constraints of the Cold War in the early 1990s, the development donor discourse began to shift to embrace security-related issues, while expectations grew that development actors would engage with issues such as conflict prevention, post-conflict peace-building and, after 9/11, anti-terrorism. Thus, in his words, 'Security sector reform can be understood as an attempt to connect, in one concept, the opportunities of expanding development assistance into security-related fields and the challenges of new demands on development donors, and to provide both with a common vision.'[63]

In addition to considerable scepticism about the 'wider adoption of security sector reform as an element of development donor programmes' by development agencies, who saw themselves in conflict with legal regimes that would limit such new activities. Brzoska argues that other ministries feared that development ministries would encroach upon their traditional overseas assistance and peace-building work. He notes:

> What is more, the ministries' primary local partners in the developing countries themselves may vary, and may sometimes even be in conflict with each other, thus reducing the coherence of the assistance offered. Whereas development ministries may well be perceived by the so-called 'power ministries' as being politically weak and full of 'do-gooders', there is often an aversion in development assistance circles to the 'command approaches' to problems with which such 'power ministries' are identified.[64]

Triggers of development communities' security commitments

Brzoska identifies the roots of the evolving SSR discourse within development circles as debates on military expenditure, conflict prevention, post-conflict reconstruction and public sector governance.

Military expenditure in development donor policy. As donor countries decreased their military spending in the 1990s, they felt morally justified to ask developing countries to do the same and thus generate a peace dividend that could be invested in development activities. 'The concept of security sector reform came in quite handy for development donors to keep the concern with "overspending" alive, [while] at the same time it relieved their policies of a possible "neo-colonialist" taint.'[65]

Post-conflict peacemaking and conflict prevention. A number of changes moved security issues up the development agenda, including the tremendous cost of wars and post-conflict rebuilding activities. The growing number of international peacekeeping missions, 'along with a wider spectrum of activities by development donors in post-war situations, led to new challenges that brought development donors into contact with uniformed forces, eg in demobilisation, demining, small arms control and policing'. In the aftermath of armed violence, all security actors, including the armed forces, non-state armed groups, police, the justice system and other actors within the security sector, need to be downsized, reformed and put back into the service of the entire population. Moreover, 'Wars also regularly leave a legacy of surplus weapons which can prove to be an impediment to development. Without de-mining, areas may remain inaccessible or unusable for productive activities such as agriculture. Widespread illegal use of small arms, in criminal acts and personal violence, reduces economic growth and development.'[66]

In the complex peace operations of the 1990s many of these tasks were addressed by peacekeepers, whose short-term mandates were not created to carry out long-term peace-building tasks. After their departure, development actors seemed to be their natural successors in contributing international support, if required. However, as Brzoska notes, 'While in theory, there is a "peacekeeping-to-development" continuum in security-related activities, similar to the "relief-to-development" continuum on the humanitarian side, in practice a gap has opened up in many cases between activities begun (or not begun) by peacekeepers and continued (or not continued) by development donors.'[67] While some development actors became involved in post-war peace-building activities, particularly DDR (as in the World Bank) or police reform (as in the case of UNDP), as Brzoska notes, 'it soon became clear that more coordination, more cooperation and a certain degree of conceptual clarity were needed … Again, the concept of security sector reform came in handy to describe a range of activities about which peacekeepers, UN administrations and development donors needed to talk.' As a result, 'Slowly, if reluctantly, at least some development donors expanded their envelope of activities to include those with security relevance, generally from judicial reform issues to police forces and, at least in a few cases, the control of military forces.' In that way SSR was also 'well suited to describe both the content and the objectives of security-related activities in conflict prevention'.[68]

Governance and public sector reform. As governance has emerged as a major concern of development policy since the early 1990s, reforming the provision of public services has become a major instrument of development policy. Improving the effectiveness and efficiency of public services includes the provision and governance of security services, which would thus quite naturally become a matter of concern for development actors.[69]

Formal recognition of SSR as a development instrument

Although the merits of SSR have been increasingly appreciated by development actors, legal constraints as well as political and institutional resistance limited the extent to which SSR activities could be included in – and funded by – development donor programmes. Yet scholars and practitioners have been calling for donors to make resources available to support SSR programmes and incorporate SSR activities in their own poverty reduction and public expenditure work. Since the mid-2000s considerable progress has been made in that direction.[70]

ODA eligibility of SSR programmes

A significant step towards formalising the development community's foray into the traditionally problematic area of security politics was taken by the OECD. As Wulf notes, in 2001 it 'published a Conceptual Framework with six broad categories of recommendations for members of the Development Assistance Committee to develop security sector reform policies and more integrated approaches to security and development'.[71] The OECD suggested recognising the developmental importance of security issues; conceptualising a comprehensive security system reform that outlines the appropriate roles for actors; identifying the required capacity and institutional reforms in donor countries; developing an effective division of labour among development and other relevant international actors; working towards the integration of security system concerns in overall foreign and trade policy; and providing assistance to enhance domestic ownership of and commitment to reform processes.[72]

Giving official blessing to an area of activity that has long been part of many development actors' work, particularly in post-war societies, the OECD widened 'the extent to which donor countries should be permitted to report as official development assistance (ODA) their spending in areas where development and security issues converge'.[73] At the DAC High Level Meeting of Ministers and Heads of Aid Agencies on 3 March 2005, following 18 months of deliberations, a number of activities were accepted as ODA-relevant. Consensus was reached on technical cooperation and civilian support for six items: management of security expenditure through improved civilian oversight and democratic control of budgeting, management, accountability and auditing of security expenditure; enhancing civil society's role in the security system to help ensure that it is managed in accordance with democratic norms and principles of accountability, transparency and good governance; supporting legislation for preventing the recruitment of child soldiers; security system reform to improve democratic governance and civilian control; civilian activities for peace-building, conflict prevention and conflict resolution; and controlling, preventing and reducing the proliferation of small arms and light weapons.[74]

The impact of the OECD/DAC's initiative is still felt and recognised years later. The World Bank, along with UNDP, another international 'trendsetter' in the debate and practice of development assistance, had for the most part remained relatively silent.

The World Development Report 2011

With the WDR 2011, the World Bank published an impressive elaboration on the linkages and mutual significance of conflict, security and development.[75] The report is based on the realisation that 'threats to development gains from organized violence, conflict, and fragility cannot be resolved by short-term or partial solutions in the absence of legitimate institutions that provide all citizens equitable access to security, justice, and jobs. Thus, international engagement in countries facing fragility, conflict, and violence must be early and rapid to build confidence, yet sustained over longer periods, and supportive of endogenous efforts and institution building.'[76] The Bank admits that:

> The 20 years of working to support institutions in post-transition countries (e.g., in Africa and Eastern Europe) and a decade of efforts to rebuild the state in high-profile environments (in particular Iraq, Afghanistan, and post-earthquake Haiti), have yielded uneven results. This discomforting realization

> is reinforced by new pressures for political transition in the Middle East, themselves a reflection of the need to review the accepted principles of institutional performance. It has become increasingly urgent for the WBG [World Bank Group] to position countries facing fragility, conflict, and violence at the core of its development mandate and to significantly adjust its operations model.[77]

Yet the informed reader cannot help but notice that very little in the report is new. Much of what is said, for instance with regard to security, development and SSR, could be found in even more detail in OECD/DAC documents almost a decade ago. Nor was the report designed to present new insights to the world community. Its significance lies in the fact that an organisation as weighty as the World Bank, which has traditionally been reluctant to address issues of conflict and security head on, has now decided to focus on the interlinkages between conflict, crime, security and development by picking up on established debates that have emanated from research and practitioner communities. In fact, the WDR was preceded by an extensive research, fact-finding and 'debate-finding' exercise, in an attempt to elevate these debates and arguments to a level at which policy communities in particular could not avoid engaging with them. Concepts and issues that seemed to lose significance in international policy debates – such as human security or conflict prevention – have been given new impetus by the report. Moreover, it shows that progress is possible – and has in fact been made – in lowering the number and impact of conflicts and increasing security and development options for even the poorest societies.[78]

At least as interesting as the WDR itself are the World Bank's plans to implement its findings and recommendations. For this purpose the WDR 2011 team drafted a report entitled 'Operationalizing the 2011 World Development Report: Conflict, Security, and Development'.[79] What does this report say – directly or indirectly – about SSR's significance for development? The short answer is that it makes no explicit mention of SSR. It emphasises the importance of focusing on fragile and conflict-affected situations (FCS), creating jobs in these states, forging links with external organisations, convincing donors to provide consistent funding and redefining risk tolerance, risk management and expected results. However, considerable emphasis is placed on institution-building as the key to enabling development, which is a critical component of SSR. The report's premise is that 'violence and other challenges plaguing FCS cannot be resolved by short-term or partial solutions in the absence of institutions that provide people with security, justice, and jobs'.[80] It notes that:

> today's realities engage development agencies in protracted periods of sustained violence or transition – and require an approach to restoring confidence and building institutions that is adapted to the local political context. Broadly, the main development challenge in countries facing fragility, conflict, and violence is a mismatch between the development community's current business models and the realities in these situations.[81]

The report emphasises the importance of building institutions that are both legitimate and functional, and calls upon development agencies to reform their strategies, behaviours and results metrics in countries facing fragility and risks of violence.[82] This is significant, as it presumes that development agencies need to change their traditional approach to providing assistance to FCS. The same applies to the Bank: by placing strengthening 'institutional capacity, inclusion, accountability and legitimacy'[83] among its main priorities and positioning fragility, conflict, and violence at the core of its development mandate, the Bank seeks to 'significantly adjust its operations model while remaining within its established mandate and focusing on development and poverty reduction'.[84]

A number of innovations can be observed. For instance, there is the call for long-term financial and political commitment, as 'it takes a long time to build legitimate and capable institutions (it commonly takes a generation or more for a fragile national institution to achieve reasonable functionality and legitimacy)'.[85] The report also recognised that blueprints based on the experience of stable, prosperous, developed countries may not work everywhere, as 'many of the most appropriate approaches for countries facing fragility, conflict, and violence are found in the experiences and expertise of other practitioners with experience in similar contexts, rather than in the "best practices" of more developed, more complex economies'.[86] The World Bank recognises and builds on the original initiative taken by the OECD in creating the momentum for the actualisation of much closer cooperation between security and development communities – a taboo issue not long ago – within less than a decade. It highlights its close cooperation with the OECD's International Network on Conflict and Fragility, and it co-chairs the network's Task Team on State-building.[87]

The report refers to the WDR's recognition that, particularly in the context of peace- and state-building, 'improved security and justice establish a context of credible exchange that can encourage markets, allow human development to proceed, and provide space for innovation'.[88] A concrete action that is directly relevant to both SSR and development objectives is to 'integrate the role of security actors to fully inform Bank strategies and operations in FCS and in countries faced with violent criminal networks'.[89]

Development agencies' activities on SSR

Similar to the World Bank, despite perceptions and sometimes assurances to the contrary, many development ministries and agencies feature a significant record of security and SSR-related work. Documents such as the OECD/DAC guidelines and handbooks or the WDR 2011 will not necessarily serve as the impetus or driving force for more security-directed work of the development donor community, but, perhaps at least as importantly, will make such work acceptable internally and externally and highlight its significance in supporting core development activities. Moreover, they draw attention to the fact that conflict and security challenges, along with associated risks and relevant responses, are part and parcel of the development discourse and practice. They cannot be disassociated and left to be dealt with by others – unless they do so in collaboration and partnership with development actors.

Earlier I offered a brief review of the security and SSR-related work of a number of development ministries and agencies. The degree to which security and SSR issues have been embedded in these organisations' work is impressive. It proves that many, although not all, such agencies and ministries have evolved in the five to ten years since Michael Brzoska conducted his research on the early engagement of the development donor community with the evolving SSR agenda and after the OECD/DAC officially recognised that many security-related tasks – including SSR – are part and parcel of development assistance.

Moving towards development-sensitive SSR: The security-development nexus in action

This chapter's examination of the role of SSR within the security-development nexus as – originally – a development concept triggers a number of suggestions that may be of interest to the development, security and overlapping SSR policy communities.

The development community should more openly stand by its original ownership of the SSR concept. In doing so it would be worth returning to the OECD/DAC's initial groundbreaking work on SSR. It will also be advantageous to engage fully with the WDR 2011 and join the World Bank's various initiatives in implementing the lessons and suggestions that emanated from this report. Among many messages, renewed emphasis on conflict prevention, human security and SSR has the potential to serve as a major impetus for a return to some critical debates that, after the diversions caused by 9/11 and subsequent global responses and preoccupations, have become side-tracked. Re-engaging with these debates and concepts, and learning lessons from experiences in implementing these concepts so far, will help the development donor community strengthen its relationship with the security community and facilitate joint ownership, perhaps in the spirit of 3D (defence, diplomacy and development), 3C (coherent, coordinated and complementary) and whole-of-government approaches.

Nevertheless, some caution is warranted. Future cooperation on SSR within a larger context of security-development activities should be reflective of a broader security approach, spanning a breadth of themes and actors when it comes to defining what security means and for whom, who should be involved in providing security and what role it plays for human development and vice versa. Such preparedness to look beyond one's own professional horizon will be required from all participating actors. Moreover, SSR needs to be respected for its most fundamental principles – these include commitment to democratic governance, accountability, the rule of law, human rights, inclusive approaches and adherence to other good governance principles. Defaulting on comprehensive SSR in favour of quick-fix, politically opportunist approaches to do 'something' with 'someone' will not win the trust of either the development or the security community. It will not lead to a serious long-term, and thus sustainable, venture to assure eventual good governance of an effective and accountable security sector that is capable of creating and safeguarding the best possible environment for sustainable human development.

Glitches, problems, inconsistencies, turf wars, seemingly irreconcilable organisational cultures and modes of operation, historical misgivings, fears and perceptions, and other factors that stand in the way of effective and efficient cooperation between security and development communities should be accepted for what they are – dynamics that need to be taken seriously and worked out cooperatively from case to case. SSR covers a range of activities that were previously pursued in isolation, bringing together activities and actors to increase the effectiveness, efficiency and accountability of the provision of security and development – for the state, the community and the individual. It is not necessarily a complex undertaking, but certainly one that requires all involved parties to think beyond usual patterns of action and interaction. With new objectives in mind, one needs to do things differently. Working out how to do business sustainably is the task of researchers, decision-makers and, most importantly and with guidance from the former two, those implementing the bits and pieces that all add up to reforming, operating and governing a new type of security sector that sees its main role as improving and sustaining both security and development for the state and society, within and beyond national borders.

Before closing this chapter with a few concluding thoughts, I will draw the reader's attention to two areas in particular that emerge from this discussion as potentially useful departures for future research priorities.

Moving beyond assumptions: The need to assess SSR's security and development impact

As has been shown in this chapter, and will be reflected in the remainder of this volume, SSR activities are rarely explicitly geared towards meeting specific development objectives. In addition, close linkages to national security policies, which are often still limited to traditional state security concerns, are often neglected. While the contributions of SSR to security and development objectives are assumed, their precise extent remains generally unknown. In addition, despite intentions to the contrary, donor-initiated SSR programmes tend to be primarily donor-driven, often with little input from beneficiaries at planning and implementation phases. Commitment to inclusive, representative and sustainable approaches in project planning and consultation practices tends to be weak. Also, nationally initiated programmes inadequately focus on the impact on local beneficiaries and the satisfaction of their security and development needs.

The insufficient impact of SSR in terms of security and development dividends can be traced back to poor planning and, more so, poor implementation. There are no mechanisms for assuring mutual accountability in SSR processes. For the most part, beneficiary populations have no recourse to hold donors accountable to their stated commitments to provide sustainable and effective development-sensitive SSR support; while donors cannot hold national state authorities and beneficiary populations accountable for ensuring that reforms are effectively implemented and security and development objectives are met. Further work on the security-development nexus and the role of SSR in development should focus on developing mechanisms to assure mutual accountability in synchronising

SSR programmes with security and development objectives, building specifically on the findings of our present book.

Such work could translate into and inform the creation of a *global compact for mutual accountability in supporting security and development through security sector reform* and its subsequent implementation and maintenance. Member-donors of the global compact would be responsible for adhering to and implementing norms and guidelines they have produced and agreed to. They would be held accountable in fulfilling the obligations and commitments they express in statements and documents accompanying and underlying the global compact. Similar commitments would be required from the beneficiary community. Donors and beneficiaries of SSR would be obliged to live up to their respective promises. Both sides would be careful not to start undertakings they are unable to complete, or raise expectations they are not able to meet. Moreover, beneficiaries would play an oversight role in monitoring and checking the accountability of donors' own assurances.

The objective of such work would be to achieve more effective, meaningful, impact-oriented and measurable provision of security and development through SSR. This implies that overall security and development objectives benefit rather than suffer from SSR; and that the security and development needs and expectations of a broad spectrum of society are solicited and well understood before SSR programmes are designed and implemented. The result of this approach would be changed and improved policy, programming (design and implementation), training and impact, supported by sustainable and inclusive security and development-responsive SSR programming.

A second and related priority should be a focus *on tracking and analysing the development-related roots, objectives and impacts of SSR programmes*. While SSR is expected to make significant contributions to improve both security and development in transition societies, thus far the main focus of programmes – in both design and implementation – appears to have been primarily on security dividends, while development dividends remain unspecified or vaguely defined as implicit and immeasurable outcomes of improved security conditions. This characteristic mirrors the broader work on the security-development nexus, which asserts (without much empirical basis) that increased security and stability are favourable conditions for economic growth, poverty reduction and human development – and vice versa. As discussed earlier in this chapter, the presumed correlations between security, development and SSR seem common sense and convincing. However, such rhetorical assertions constitute a weak basis from which to develop convincing, empirically based conclusions about the symbiotic relationship between security provision and development, and more specifically the role and impact of SSR in enhancing and supporting this relationship.

It is important to investigate and substantiate this assumed relationship. Without full recognition of SSR's design and capacity to support development, development actors often find it difficult to embrace SSR as a tool that is both necessary and worthy of their engagement. Similar scepticism – or mere lack of knowledge – about SSR's professed development mandate prevents SSR planners and practitioners from explicitly incorporating and engaging long-term development needs and objectives into their efforts. It would be helpful, from a policy planning and programme implementation perspective, to analyse the extent to which SSR activities have been designed, implemented and evaluated in terms of their development contributions. This will require systematic surveying and mapping of representative samples of major stakeholders involved in delivering SSR activities, either those conducted as holistic and comprehensive cross-sectoral activities that encompass several security and oversight/governance institutions, or quasi- and partial SSR activities that address only select components of wider SSR reform agendas.

Conclusion

SSR can be considered both an expression and an application of the security-development nexus in practice – in planning, implementing and evaluating both development and SSR activities. The fact that it often is not perceived as such cannot be blamed on faulty design or a lack of commitment from those dedicated to materialising the intentions in SSR concepts and policy, but results from shortcomings in translating these into programme designs and implementation. As is reflected in the other contributions to this volume, failing to tie very specific development objectives and priorities into SSR programming deprives SSR of its opportunity to live up to its potential and help transitional societies meet both their security and their development objectives. Simply assuming that, in one way or another, SSR is good for security and development – similar to the assumptions that more security is in general good for development – is not enough to establish specific expectations and objectives, set goals and design programmes, implement these and, finally, assess them for their effective contributions.

The lessons we can learn from a decade of SSR activities and their still mostly unknown, vague and merely assumed impact on development, as well as the still ambivalent interactions between security and development actors when it comes to joint contributions to SSR, match similar discussions and experiences in assessing and understanding the assumed and asserted security-development nexus. Clarifying these relationships would help researchers and practitioners get a better grip on designing development-enhancing SSR programmes and activities. Structures and processes need to be put in place to ensure that SSR serves the overall security and development goals of transitional societies and their human security and human development objectives – initially with the support of external actors, but in the long run by empowered and committed local and national actors.

Notes

[1] World Bank, *World Development Report 2011: Conflict, Security, and Development* (Washington, DC: World Bank, 2011).

[2] Hans Born and Albrecht Schnabel, eds, *Security Sector Reform in Challenging Environments* (Münster: LIT Verlag, 2009).

[3] In this chapter the term 'security community' refers to an epistemic community of organisations and individuals working on security issues. It is thus different from the concept of the security community developed in Karl W. Deutsch, Sidney A. Burrell, Robert A. Kann, Maurice Lee Jr, Martin Lichterman, Raymond E. Lindgren, Francis L. Loewenheim and Richard W. Van Wagenen, *Political Community and the North Atlantic Area* (Princeton, NJ: Princeton University Press, 1957).

[4] See Michael Brzoska, 'Development Donors and the Concept of Security Sector Reform', DCAF Occasional Paper no. 4 (Geneva: DCAF, November 2003).

[5] Maria Stern and Joakim Öjendal, 'Mapping the Security-Development Nexus: Conflict, Complexity, Cacophony, Convergence?', *Security Dialogue* 41, no. 5 (2010): 6.

[6] Boutros Boutros-Ghali, 'An Agenda for Peace: Preventive Diplomacy, Peacemaking and Peacekeeping', report pursuant to the statement adopted by the Summit Meeting of the Security Council on 31 January 1992, UN Doc. A/47/277–S/24111 (New York: United Nations, 17 June 1992).

[7] United Nations, 'Report of the Secretary-General on the Work of the Organization', Supplement no. 1 (A/54/1) (New York: United Nations, 31 August 1999), available at www.un.org/documents/ga/docs/54/plenary/a54-1.pdf: para. 61.

[8] Albrecht Schnabel, 'The Human Security Approach to Direct and Structural Violence', in *SIPRI Yearbook 2008: Armaments, Disarmament and International Security* (Oxford: Oxford University Press, 2008): 87–96.

[9] John Lewis Gaddis, 'The Long Peace: Elements of Stability in the Postwar International System', *International Security* 10, no. 4 (1986): 99–142.

[10] See Barry Buzan, *People, States and Fear: An Agenda for International Security Studies in the Post-Cold War Era*, 2nd edn (Boulder, CO: Lynne Rienner, 1991); Barry Buzan, 'New Patterns of Global Security in the Twenty-first Century', *International Affairs* 67, no. 3 (1991): 431–451.

[11] International Commission on Intervention and State Sovereignty, *The Responsibility to Protect* (Ottawa: International Development Research Center, 2001).

[12] Human Security Centre, *Human Security Report 2005: War and Peace in the 21st Century* (New York and Oxford: Oxford University Press, 2005), available at www.humansecurityreport.info.

[13] The Geneva Declaration on Armed Violence and Development is a diplomatic initiative aimed at addressing the interrelations between armed violence and development, as armed violence undermines development and aid effectiveness and hinders the achievement of the Millennium Development Goals. A core group of states and affiliated organisations advocate effective measures for the implementation of the Geneva Declaration, and so far more than 100 countries have signed it. For further information see www.genevadeclaration.org/.

[14] UNDP, *Human Development Report 1994: New Dimensions of Human Security* (New York and Oxford: Oxford University Press, 1994): 24.

[15] Brzoska, note 4 above: 20.

[16] The members of the Human Security Network are Austria, Chile, Costa Rica, Greece, Ireland, Jordan, Mali, Norway, Slovenia, Thailand and Switzerland, with South Africa as an observer. For further information see www.eda.admin.ch/eda/en/home/topics/intorg/un/missny/mnyhsn.html.

[17] Organisation for Economic Co-operation and Development/Development Assistance Committee (OECD/DAC), *DAC Guidelines on Helping Prevent Violent Conflict* (Paris: OECD, 2001): 37. Cited in Brzoska, note 4 above: 19.

[18] Mark Duffield, 'The Liberal Way of Development and the Development-Security Impasse: Exploring the Global Life-Chance Divide', *Security Dialogue* 41, no. 1 (2010): 53.

[19] UNDP, note 14 above; Commission on Human Security, 'Human Security Now' (New York: CHS, 2003), available at www.resdal.org/ultimos-documentos/com-seg-hum.pdf: 4; Albrecht Schnabel and Heinz Krummenacher, 'Towards a Human Security-Based Early Warning and Response System', in *Facing Global Environmental Change: Environmental, Human, Energy, Food, Health and Water Security Concepts*, eds Hans Günter Brauch, Úrsula Oswald Spring, John Grin, Czeslaw Mesjasz, Patricia Kameri-Mbote, Navnita Chadha Behera, Béchir Chourou and Heinz Krummenacher (Berlin, Heidelberg and New York: Springer, 2009): 1253–1264.

[20] Commission on Human Security, ibid.: 4.

[21] For a full text of the Universal Declaration of Human Rights, see www.un.org/en/documents/udhr/index.shtml.

[22] See '20 Years of Global Human Development Reports', available on UNDP's website at http://hdr.undp.org/en/reports/.

[23] UNDP, *Human Development Report 1996: Economic Growth and Human Development* (New York: Oxford University Press, 1996); UNDP, *Human Development Report 1997: Human Development to Eradicate Poverty* (New York: Oxford University Press, 1997). Cited in Sabina Alkire, 'Human Development: Definitions, Critiques, and Related Concepts', Human Development Research Paper 2010/01 (New York: UNDP, June 2010), available at http://hdr.undp.org/en/reports/global/hdr2010/papers/HDRP_2010_01.pdf: 7–8. These concepts in the HDR were very much influenced by Amartya Sen's concept of entitlements. See Amartya Sen, *Poverty and Famines: An Essay on Entitlement and Deprivation* (Oxford: Oxford University Press, 1982).

[24] UNDP's human development indicators include the following factors: adjusted net savings; adult literacy rate (both sexes); adolescent fertility rate (women aged 15–19 years); carbon dioxide emissions per capita; combined gross enrolment ratio in education (both sexes); expected years of schooling (of children); expenditure on education (% of gross domestic product [GDP]); expenditure on health, public (% of GDP); GDP per capita (2008 US$ PPP); gross national income (GNI) per capita (2008 US$ PPP); gender inequality index (updated); gender inequality index, value; HDI value; headcount (k greater than or equal to 3), population in poverty; homicide rate; household final consumption expenditure per capita PPP (constant 2005 international US$); inequality-adjusted education index; inequality-adjusted HDI value; inequality-adjusted income index; inequality-adjusted life expectancy index; income Gini coefficient; intensity of deprivation; internet users; life expectancy at birth; labour force participation rate, female/male ratio; mean years of schooling (of adults, female/male ratio); maternal mortality ratio; maternal mortality ratio (new estimates); multidimensional poverty index (k greater than or equal to 3); population affected by natural disasters (per million

inhabitants); population living below $1.25 PPP per day; population with at least secondary education, female/male ratio; prevalence of undernourishment in total population; protected area; refugees by country of origin (thousands); robbery rate; shares in parliament, female-male ratio; under-five mortality; and unemployment rate, total (%). For a complete description of the human development indicators see http://hdrstats.undp.org/en/indicators/default.html. The World Bank, by comparison, uses the following key indicators of development: population, average annual population growth rate; population density; population age composition, ages 0–14; gross national income (GNI); GNI per capita; purchasing power parity (PPP) GNI; PPP GNI per capita; gross domestic product (GDP) per capita growth; life expectancy at birth; and adult literacy rate. See World Bank, note 1 above: 357.

25 UN General Assembly, 'United Nations Millennium Declaration', UN Doc. A/RES/55/2 (New York: United Nations, 18 September 2000), available at www.un.org/millennium/declaration/ares552e.pdf.

26 UNDP, 'Fast Facts: Millennium Development Goals' (New York: UNDP, 3 March 2011), available at www.beta.undp.org/undp/en/home/librarypage/results/millennium-development-goals.html.

27 Brzoska, note 4 above: 4.

28 The information is drawn from the AusAID website, available at www.ausaid.gov.au/.

29 The information is drawn from the CIDA website, available at www.acdi-cida.gc.ca/home.

30 The information is drawn from the DANIDA website, available at http://um.dk/en/danida-en/.

31 The information is drawn from the BMZ website, available at www.bmz.de/en/; and the German International Cooperation website, available at www.giz.de/en/home.html.

32 The information is drawn from the MOFA-ODA website, available at www.mofa.go.jp/policy/oda/.

33 The information is drawn from the website of the Norwegian Ministry of Foreign Affairs International Development Programme, available at www.regjeringen.no/en/dep/ud/selected-topics/development_cooperation.html?id=1159; and the website of the Norwegian Agency for Development Cooperation, available at www.norad.no/en/.

34 The information is drawn from the SIDA website, available at www.sida.se/English/.

35 The information is drawn from the DFID website, available at www.dfid.gov.uk/.

36 The information is drawn from the USAID website, available at www.usaid.gov/. See also Nicole Ball, 'Promoting Security Sector Reform in Fragile States', PPC Issue Paper 11 (Washington, DC: USAID, 2005).

37 Albrecht Schnabel, Marc Krupanski and Ina Amann, 'Military Protection for Humanitarian Assistance Operations – Roles, Experiences, Challenges and Opportunities', report for Directorate for Security and Defence Policy of Swiss Federal Department of Defence, Civil Protection and Sports (Geneva: DCAF, May 2010): 40. See also Albrecht Schnabel, Marc Krupanski and Ina Amann, 'Military Protection for Humanitarian Assistance Operations? Guidelines, Experiences, Challenges and Options', paper presented at 16th International Humanitarian Conference: Humanitarian Space, organised by Webster University Geneva with the Office of the UN High Commissioner for Refugees and the International Committee of the Red Cross, Geneva, 27–28 January 2011.

38 Schnabel et al. (2010), ibid.: 40.

[39] Ibid.: 39.

[40] For literature on the security-development 'nexus', linkages and connection see, among others, Mark Duffield, *Global Governance and the New Wars: The Merging of Development and Security* (London: Zed Books, 2001); Mark Duffield, *Development, Security and Unending War: Governing the World of Peoples* (Cambridge: Polity, 2007); Neclâ Tschirgi, Michael S. Lund and Francesco Mancini, eds, *Security and Development: Searching for Critical Connections* (Boulder, CO: Lynne Rienner, 2010).

[41] Tschirgi et al., ibid.: 5.

[42] Ibid.

[43] Ibid.: 6.

[44] Stern and Öjendal, note 5 above: 11.

[45] Ibid.: 6.

[46] Ibid.

[47] Ibid., 24-25.

[48] This section draws on Albrecht Schnabel, 'Ideal Requirements versus Real Environments in Security Sector Reform', in *Security Sector Reform in Challenging Environments*, eds Hans Born and Albrecht Schnabel (Münster: LIT Verlag, 2009): 7–11. The section is provided particularly for readers who are not yet acquainted with the concept of SSR.

[49] Heiner Hänggi and Vincenza Scherrer, 'Recent Experience of UN Integrated Missions in Security Sector Reform', in *Security Sector Reform and Integrated Missions*, eds Heiner Hänggi and Vincenza Scherrer (Münster: LIT Verlag, 2008): 3–4.

[50] United Nations, 'Securing Peace and Development: The Role of the United Nations in Supporting Security Sector Reform', Report of the Secretary-General, UN Doc. A/62/659–S/2008/392 (New York: United Nations, 3 January 2008): para. 14.

[51] Ibid.: para. 17.

[52] OECD/DAC, *Handbook on Security System Reform: Supporting Security and Justice* (Paris: OECD, 2007); see also UNDP, *Human Development Report 2002* (New York: UNDP, 2002): 87.

[53] UNDP, ibid. For this expanded definition the UNDP was referring to Nicole Ball, Tsjeard Bouta and Luc van de Goor, 'Enhancing Democratic Governance of the Security Sector: An Institutional Assessment Framework' (The Hague: Clingendael Institute for the Netherlands Ministry of Foreign Affairs, 2003), available at www.clingendael.nl/publications/2003/20030800_cru_paper_ball.pdf: 32–33.

[54] These definitions have been elaborated by Hänggi and Scherrer, note 49 above: 15.

[55] Herbert Wulf, 'Security Sector Reform in Developing and Transitional Countries' (Berlin: Berghof Research Center, July 2004), available at www.berghof-handbook.net/documents/publications/dialogue2_wulf.pdf: 5.

[56] For excellent discussions of the dynamics of local ownership in SSR see Laurie Nathan, *No Ownership, No Commitment: A Guide to Local Ownership of Security Sector Reform* (Birmingham: GFN-SSR, University of Birmingham, October 2007); Timothy Donais, ed., *Local Ownership and Security Sector Reform* (Münster: LIT Verlag, 2008).

[57] United Nations, note 50 above. See also United Nations, 'Maintenance of International Peace and Security: Role of the Security Council in Supporting Security Sector Reform', concept paper prepared by the Slovak Republic for the Security Council open debate, UN Doc. S/2007/72, 9 February 2007; European Commission, 'A Concept for European Community Support for Security Sector Reform', communication from the Commission to the Council and the European Parliament, SEC (2006); Council of the European Union,

'EU Concept for ESDP Support to Security Sector Reform (SSR)', EU Doc. 12566/4/05; Heiner Hänggi, 'Security Sector Reform', in *Post-Conflict Peacebuilding – A Lexicon*, ed. Vincent Chetail (Oxford: Oxford University Press, 2009): 337–349.

58 Organisation for Economic Co-operation and Development, 'Security System Reform and Governance', DAC Guidelines and Reference Series (Paris: OECD, 2005), available at: www.oecd.org/dac: 16.

59 Clare Short, 'Security Sector Reform and the Elimination of Poverty', speech at Centre for Defence Studies, King's College (London, 9 March 1999), available at www.clareshort.co.uk/speeches/DFID/9%20March%201999.pdf.

60 Clare Short, 'Security, Development and Conflict Prevention', speech at Royal College of Defence Studies (London, 13 May 1998). Cited in Nicole Ball and Dylan Hendrickson, 'Trends in Security Sector Reform (SSR): Policy, Practice and Research', paper presented at workshop on 'New Directions in Security Sector Reform', Peace, Conflict and Development Program Initiative, Ottawa, 3–4 November 2005, available at www.idrc.ca/en/ev-83412-201-1-DO_TOPIC.html.

61 Wulf, note 55 above: 3.

62 Brzoska, note 4 above: 1.

63 Ibid.: 2–4.

64 Ibid.: 4.

65 Ibid.: 9.

66 Ibid.: 9–10.

67 Ibid.: 11.

68 Ibid.: 13–14.

69 Ibid.: 13.

70 See, for instance, Nicole Ball, 'Transforming Security Sectors: The IMF and World Bank Approaches', *Journal of Conflict, Security and Development* 1, no. 1 (2001): 45–66.

71 Wulf, note 55 above: 9.

72 As summarised in ibid.

73 Tillmann Elliesen, 'Security or Development Efforts?', *D+C: Development and Cooperation* 48, no. 5 (2007), available at www.inwent.org/ez/articles/054114/index.en.shtml: 206.

74 See OECD/DAC, 'Conflict Prevention and Peacebuilding: What Counts as ODA?' (Paris: OECD/DAC, 3 March 2005), available at www.oecd.org/dataoecd/32/32/34535173.pdf. See also 'Annex 1: Extracts from DAC Statistical Reporting Directives', DCD/DAC(2007)34, in OECD/DAC, *ODA Casebook on Conflict, Peace and Security Activities*, DCD/DAC(2007)20/REV1 (Paris: OECD, 13 September 2007), available at www.oecd.org/dataoecd/27/21/39967978.pdf: 62–63.

75 World Bank, note 1 above.

76 World Bank, 'Operationalizing the 2011 World Development Report: Conflict, Security, and Development' (Washington, DC: World Bank, 2011): 1.

77 Ibid.: 9.

78 Ibid.: 2.

79 Ibid.

80 Ibid.: iii.

81 Ibid.: 1.

82 Ibid.

83 Ibid.: 3.

[84] Ibid.: iii.
[85] Ibid.: 4.
[86] Ibid.
[87] Ibid.: paras 27, 31 and 45.
[88] Ibid.: 5.
[89] Ibid.: 8.

PART II

GENDER, SECURITY AND DEVELOPMENT

Chapter 3

A Bridge Too Far? The Gender Consequences of Linking Security and Development in SSR Discourse and Practice

Heidi Hudson

Introduction

This chapter focuses on the gender challenges faced by security sector reform (SSR), with specific reference to the developmental and security-related consequences of sexual and gender-based violence (SGBV). The thriving SSR industry reflects the growing recognition of the need to link security and development, particularly in post-conflict states. Gender roles play a key part in promoting or hindering the efficiency and professionalism of SSR.[1] Especially in the African context, the challenges of bringing the security and development concerns of both men and women into the equation are many. Attempts to bridge the gap between gender-sensitive security and development policy and practice have to contend with, among other things, a hostile environment marked by lawlessness, weak justice systems and high levels of SGBV. It is this implementation gap which explains why the linking of security and development in the SSR context is fraught with conceptual contradictions and ambiguities. The latter invariably lead to practical problems, such as the difficulty of using SSR to address SGBV meaningfully as both a security and a development issue.

The first contradiction relates to the fact that SSR is trapped in a neoliberal governance model which seeks to manage risks rather than address root causes of insecurity, underdevelopment and inequality. On paper SSR appears to alleviate root causes through its focus on reforming the security sector – itself often viewed as a root cause of conflict and structural violence. However, application in specific contexts becomes problematic.

For instance, in the case of SGBV protection and empowerment are assumed to go together, because security and development are assumed to be linked. I, however, argue that the conventional link between security and development should be revisited, because it is often based more on anecdotal than solid empirical evidence. With understanding of the causal mechanisms remaining rather fuzzy, there is room to contend that the assertion of these connections ultimately serves a political purpose, namely to justify a humanitarian interventionist agenda as part of the 'good' governance framework, considered to be in the best interest of 'all'.[2] Decision-makers and practitioners work with the assumption that what they do reflects the will of the international community and is therefore inherently 'good'. Collaboration among internal and external partners therefore contributes to the 'global good'.[3] Alternatively these conceptual conflations could serve to mask – through rhetorical commitments – much less benign global policy-making (e.g. by the United Nations, the World Bank and strong states).

Second, the assumption that there is a nexus, or that this nexus is assumed to exist and therefore need not be proven, has ramifications not only for how SSR is conducted in general, but particularly how the integration of gender is approached within SSR. Ironically, scholars have often used an emphasis on civil society or local ownership, specifically women's organisations, as the bridge to strengthen the common-sense connection between *human* security and *human* development.[4] The multilevel and multidimensional character of women's insecurity fits in well with security and development discourses, which prioritise protection and empowerment strategies respectively. But does this emphasis on the 'human' or 'women' signify a shift away from narrow security conceptualisations? Dare we argue that these labels could be misused to silence women or gloss over failures to address high levels of violence against women (VAW), due to complacency with a so-called all-encompassing and therefore morally justified concept that puts 'people' first?

In the context of the neoliberal framework that SSR subscribes to, a liberal additive feminist approach guides gender initiatives within the security sector. Based on the assumption that gender inequality inhibits development and potentially triggers conflict, gender equality is elevated as *the* route to emancipation.[5] This adherence to promoting gender equality has several implications. First, it leads to a conflation of gender and women in SSR discourse and subsequently of SGBV and VAW – in spite of growing evidence of sexual violence against men and boys.[6] Second, the gender equality perspective leads to the essentialising of women's and men's roles in SSR.[7] Third, this perspective further emphasises women's tokenistic representation at the expense of solutions that consider multiple gender identities and the implications of other overlapping identity constructions, such as class, ethnicity and sexual orientation. Lastly, the overemphasis on gender equality has implications for the choice of tools. In the absence of analyses of gender and power, planners become fixated on technical and bureaucratic processes.

I therefore contend that the flaws in the security-development nexus become the flaws of SSR, which in turn impact on how gender is integrated into SSR and how SGBV is addressed. Although recognising that SSR is about more than just police or defence reform (as it includes aspects of penal and justice reform, disarmament, demobilisation and reintegration (DDR) and border management, among others), this chapter is limited to a focus on the former.

Police and military are traditional strongholds of masculinity and violence. So far increased representation of women in key military positions has neither challenged the culture of militarism nor shifted dominant notions of masculinity and femininity.[8] In SSR cases which are generally regarded as successful, such as Liberia, Sierra Leone and Timor-Leste, women are still largely underrepresented, mainly in leadership positions. Furthermore, VAW and discrimination in terms of budgeting and access to resources and professional opportunities are still perpetrated and condoned by members of the armed forces.[9] In the case of Sierra Leone, for instance, both UNAMSIL (UN Mission in Sierra Leone) and International Alert reported images of militarised masculinity and high numbers of sexual offences persisting even in the absence of armed conflict.[10] Similarly, fluctuating gender identities, roles and perceptions are also closely associated with some of the security challenges faced by DDR and SSR processes in Timor-Leste.[11] Several cases of sexual assault and rape have been reported in the East Timorese police force, while female members of the armed forces have complained about widespread gender discrimination.[12] There is also insufficient understanding of the links between the public and private dimensions of SGBV as both a tool of war and a crime during peacetime, coupled with a lack of trust in the security forces and justice system. Consequently, while VAW is understood to be a pervasive form of insecurity with far-reaching socio-developmental implications, it is approached in an ahistorical and decontextualised manner. Recent research on sexual violence in the Democratic Republic of the Congo (DRC) has underscored the need to cast the net wider and not just explain such violence as a weapon of war. Additional underlying structural factors such as poverty, weak governance structures, ethnic identities and changing gender identities, norms and roles (accelerated by war) should also be considered.[13] The inability of a liberal feminist approach to interpret unequal gender relations in terms of power structures may thus perpetuate (and even exacerbate) the contradictions which emanate from the security-development nexus. This approach has also not helped to bring about a significant change in widely held beliefs about what constitutes security.

I conclude that SSR, as one of many security-development 'projects', needs to become more sensitive to its own 'externally imposed' and neoliberal underpinnings and how these shape the security-development nexus driving SSR. Although comparatively speaking SSR practitioners have been quite busy integrating gender into SSR, the results on the whole have been less than spectacular. Critical self-reflection about the narrow interpretations of gender equality can help to harmonise SSR initiatives and reinstall strategic direction. In other words, using a critical feminist perspective to address the tensions between SGBV and VAW could not only provide some coherence in SSR work, but would also offer an alternative way of looking at the poorly grasped nexus between security and development.

The aim of this contribution is therefore to examine the implications of linking security and development for gendering SSR, through a focus on SGBV. The chapter opens with an analysis of the security-development nexus, and its treatment of human security in the context of SSR. In the next section I link the conceptual problems of the nexus to broader neoliberal frameworks of global governance and strategic policy-making. This discussion prepares the way for an analysis of how gender has been integrated into SSR discourse and praxis, with specific reference to the problems of addressing SGBV and VAW. The latter provides a framework for an analysis of gender, SGBV and SSR in Liberia, illustrating the dilemma of how to achieve the objectives of efficiency as well as normativity. In the final section I make a case for not trying to mend the existing perforated links between security and development, but rather developing different kinds of links. I propose the active pursuit of an ethic of care combined with the use of an intersectional lens to investigate the overlapping interface of gender, ethnicity, race, class and sexuality.[14] Given the reality of backlash when tackling entrenched identities within the security sector, attention to different dominant and subordinated masculinities and femininities as well as other identities can be more effective in changing unequal relations. Of necessity this demands a long-term development approach. Such an approach is also more reflective of social reality than a blanket targeting of patriarchy. Providing culturally sensitive integration and support services through an ethic of care is ambitious, but offers a more relational alternative to current individualist rationalist approaches.

SSR and the flawed security-development nexus

The linking of development with security is currently widely acknowledged. For instance, Abrahamsen and Williams declare that 'there is little doubt that insecurity is a key concern for poor people and a significant obstacle to development and prosperity'.[15] According to Brzoska, SSR is a superior example of linking security and development, as it adopts a holistic approach to the provision of security and concentrates all reform activities on the promotion of development goals, especially poverty reduction.[16]

Lately, however, critical voices have questioned the nexus. SSR policy as well as security-development nexus literature generally make intuitive common-sense assumptions (correlations) about the link.[17] Despite strong statistical correlations between conflict and poverty, empirical evidence remains anecdotal, and theoretically the causal mechanisms that are necessary to explain observed and assumed correlations and the directions of causation remain vague.[18] There is also a tendency to make an analytical jump from the specific conflict-poverty correlations to the fuzzy and contested areas of the security-development nexus. To declare that 'without security there is no development and without development there is no security' is not helpful in explaining how developmental factors contribute to conflict.[19] The problem is that the vagueness of the mechanisms does not allow one to know where to start the analysis. The connection therefore relies on rhetorical claims rather than on careful empirical analysis.[20]

Two main lines of critique arise from the above-mentioned problem. Firstly, the link between conflict/insecurity and development is presented as a circular one, with the result that it tends to obscure rather than clarify the interface. Security leads to development; development leads to security. Conversely, poverty leads to insecurity; insecurity leads to underdevelopment.[21] The following DFID statement is indicative of the intuitive correlations which are prevalent in this discourse: 'Sub-Saharan Africa has experienced more conflict … than any other continent … it is no coincidence that Africa lags behind the rest of the world in progress towards the Millennium Development Goals.'[22] While security and development in the developing world cannot be divorced from security in the developed world and international security, the literature misses the point that poverty is but one of many causes of conflict.

Secondly, human security and human development coexist in a symbiotic relationship in which the concerns are with the basic freedoms people enjoy, but adding 'human' as a prefix does not necessarily mean that the discourse has become less state-centric. Viewed as a bridging concept between development, security and human rights, Krause and Jütersonke argue that the lens of human security offers a means of shining 'a spotlight on the links between violence and insecurity … and underdevelopment and poverty'.[23] Once again a common-sense connection is made between both 'freedom from fear' and development and 'freedom from want' and development. The critical point about understanding security comprehensively and holistically in terms of the real-life, everyday experiences of human beings and their complex social and economic relations as these are embedded within global structures gets lost amid the rhetoric of people-centredness.[24]

The very same 'intrinsic truths' which drive the logic of the security-development nexus in general appear in slightly different language within SSR discourse. The OECD/DAC handbook on SSR therefore makes or reinforces the following assumptions:

- A democratically run, accountable and efficient security system helps reduce the risk of conflict, thus creating an enabling environment for development to occur.
- Security is a prerequisite for the achievement of the Millennium Development Goals (MDGs), ensuring people's livelihoods and reducing poverty.
- Security creates an environment in which the vulnerable have access to social services.
- Poverty reduction takes place through enhanced service delivery.[25]

The human insecurity cycle of vulnerable groups, such as women, the elderly and children, underlines the need for SSR. When the system fails them through bad policing, weak or non-existent justice and penal systems or corrupt armed forces, they suffer disproportionately.[26] Dominant discourses aiming to overcome poverty and discrimination therefore tend to underline the fact that women's vulnerability puts them at risk of SGBV and human trafficking. Due to societal and cultural practices that circumscribe women's position as a result of divorce or widowhood and the consequent loss of access to assets such as land, they suffer disproportionately. (See in this regard the research on the obstacles Acholi women face in claiming land rights in the patrilineal cultural setting in northern Uganda after the conflict.[27]) However, one has to be careful not to lose sight of the fact that so-called less vulnerable groups, like men, are not homogeneous and many of them may suffer similar vulnerabilities.

Analysis and policy-making are made more complicated by the fact that 'security' and 'development' are fuzzy, contested and ideologically loaded concepts.[28] Both fields are also severely under-theorised. In practice this has led to *ad hoc* decision-making and gaps between policy and reality, i.e. an implementation gap. SSR literature therefore often calls for greater coherence and policy coordination across the security and development fields, greater attention to context, alignment of short- and long-term timeframes, civil society participation and sensitivity to differential access of particular social groups.[29]

The myth of poor coordination: It's 'our' security that depends on 'their' development, stupid!

I contend that the debate about lack of coordination/coherence is in fact a red herring – distracting from the fact that a liberal peace or governance problem lies at the root of the implementation gap. In the post-9/11 period the liberal peace thesis rests on two pillars, namely that effective liberal states are a bulwark against international instability, and failing or conflict-prone states represent a threat to international security. Against this background peace-building and particularly state-building have become part of the security agenda.

With regard to the security-development nexus some scholars contend that this ambiguity of interpretation (and perhaps also the lack of theoretical and conceptual clarity) is a deliberate strategic choice.[30] Keeping it vague allows scope for actors to pursue their own interests, namely to maintain international stability and prosperity. Duffield reminds us that the unproven idea that conflict is bad for development is sufficiently entrenched to justify in a strategic sense the need for intervention in the name of the liberal peace, effectively securitising underdevelopment.[31] The liberal governance model does not locate underdevelopment or fragile states in an unjust global system or a particular historical context of imperialism. Instead, leading states and international organisations promote a consensus position which internalises the causes of conflict and political instability (the now familiar 'us' versus 'them' bifurcation) – with the problem coming wholly from the inside and the solution provided from the outside. According to this thesis, the 'answer' is not the transformation of the global system but rather of individual societies.[32]

SSR forms part of this neoliberal peace-building framework, consisting of a range of seemingly benign and politically correct assumptions related to the rule of law, multilateralism, free markets, human rights, democracy, development and the importance of context and local ownership. But together these make up, according to Mac Ginty, a 'one-size-fits-all' IKEA-style model, with a formal and low-intensity peace as the outcome.[33] The point is that there is a gap between SSR intentions to honour context and the realities of implementation as well as donors' reading of policy advice. In this regard Donais states that while local ownership may be accepted in principle by the donor community (see OECD/DAC documents), the literature also displays uneasiness about transferring 'full' ownership to the locals.[34] Locals with some degree of autonomy can 'pick and choose' which elements of SSR to implement.[35] This state of affairs has two implications. First, it enforces limited local ownership (e.g. via local elite cooption or buy-in). Post-conflict states will therefore remain accountable to donors and other international organisations, which will perpetuate the legitimacy crisis of SSR and affect its efficacy. Second, it exposes the difficulty of implementing widespread calls and intentions within SSR circles to apply SSR in a context-sensitive manner. Ultimately the cultural context remains secondary, and is 'merely superimposed upon a core being, which is the liberal rational self'.[36]

Ultimately, then, the problem is not the intentions but the flawed neoliberal assumptions which drive the political solutions of current SSR. Since SSR is mainly donor-driven, its agenda has tended to reflect donor understandings and priorities. Countries such as Canada, the Netherlands and Switzerland, as well as Scandinavian countries, have used their sizeable overseas economic aid budgets to project their own political value systems.[37] Mac Ginty sums it up as follows:

> This [peace-building] is not necessarily a vast Machiavellian conspiracy through which the agents conspire to deliver a poor quality peace yet maintain the verbiage of liberalism. Rather the liberal peace results from a combination of the pursuit of rational self-interest by core elements of the international community, the promotion of peace guided by liberal optimism and a genuine belief that democracy and open markets provide the best route to its achievement.[38]

It would therefore be a gross overstatement to argue that the neoliberal consensus has given rise to a single-minded interventionist agenda by Western states in non-Western states, which is, anyway, not conducive to the promotion of mediated and more localised solutions. But it could be argued that we are seeing the other extreme. Chandler contends in this respect that 'Rather than a framework of coherent intervention, we are witnessing a framework of *ad hoc* intervention … and the disavowal of external or international responsibilities.'[39] What started off as an implementation gap caused by 'well-meaning' neoliberal assumptions has become – when confronted with the complexity of security issues in post-conflict contexts – an abdication of political purpose and strategic direction altogether. Linked to that, Chandler proposes that the conflation of security and development has implied the prioritisation of security over development.[40] Previously poverty reduction through broad-based economic development was pursued as a goal in itself, and not as a guarantee to prevent conflict. However, the current collapsing of development strategy into conflict resolution 'privileges security over development on the basis that this support of the status quo, rather than fundamental change, is the desire of the people in poor countries'. The World Bank's Voices of the Poor project is a case in point, where a predetermined research focus on well-being evaded the question of poor people's preferences and demands in policy priorities. In this way good governance, state capacity-building, anti-corruption and transparency took precedence over development.[41] As a result, pre-existing policy frameworks in the field of development have lost a clear political purpose. What remains is sophisticated policy rhetoric by major Western political actors in the place of strategic political implementation. Specific policy issues have made room for broader and more declaratory projects, such as 'saving Africa', 'preventing state failure' or 'eradicating poverty'.[42] These two explanations of the gap between policy and practice (intervention and disengagement) are not opposites, but rather interconnected. Both versions display an inability to deal with larger questions of transformation, and opt instead for technocratic governance strategies.

Because of the place of SSR early in the overall state-building process, it becomes the logical entry point for development donors with a gender/human security agenda. The treatment of SSR as the bridge between security and development as well as the positive climate created by UN Security Council Resolution (UNSCR) 1325 on Women, Peace and Security (2000) have served as important catalysts for women's agencies to enter the mainstream strategic debate.[43] According to Natalie Hudson, the UN Development Fund for Women (UNIFEM) – traditionally a development agency – has strategically used this comprehensive security framework for action to position itself as a legitimate SSR actor.[44] Gender equality advocates therefore also become part of the growing neoliberal (governance) consensus that the creation of a democratic and efficient security sector rests partially on the *inclusion* of women's and gender issues. With a dramatically enhanced budget (and possibly better coordination of gender and women's issues) the new UN Entity for Gender Equality and the Empowerment of Women (UN Women) also reflects this nexus.

But does gendering SSR make a substantial difference if the underpinnings of the system are questionable in terms of their value in reaching the twin objectives of security and development? In the next section I will show how this conflation of security and development has impacted on the gender dimensions of SSR. I apply Chandler's argument to the gender dimensions of SSR, and concur that by focusing only on 'the immediate conditions of marginalised groups', instead of their position as well, the gender project of SSR is diluted and reduced to problem-solving tinkering with a system that fundamentally fails to deliver to the vulnerable.[45] Developmental interventions which acknowledge the material nature of security are important, but do not exist in isolation from global processes. These interventions should therefore be conceptualised in tandem with an understanding that local conditions of women, for instance, may be linked to deeply entrenched masculinist attitudes within global politics.

From flawed to more flawed: SSR and gender through the lens of SGBV/VAW

There have indeed been some advances globally in terms of women's representation, partly due to the proliferation of international legislative instruments, e.g. the Convention on the Elimination of All Forms of Discrimination against Women (1979), the Declaration on the Elimination of Violence against Women (1993) and the Beijing Declaration and Platform for Action (1995). UNSCR 1325 has raised awareness about the differential impact of conflict on men and women as well as the important role that women can play in peace-building. Gender-based violence is now 'outlawed' as a crime against humanity through UNSCR 1820 (2008).[46] The list of SSR initiatives with a gender perspective has also grown to include the OECD Handbook on Security System Reform (2009 edition), and documents by the UN International Research and Training Institute for the Advancement of Women, the OSCE Office for Democratic Institutions and

Human Rights and the Geneva Centre for the Democratic Control of Armed Forces (DCAF). The latter produced a comprehensive toolkit for gender and SSR.[47] Nonetheless, there is an implementation gap because these policies are not translating into real changes for women on the ground.

Indeed, VAW has reached almost epidemic proportions, both during and after conflict.[48] In Africa for example, despite some success stories such as in Rwanda, the gains women have made after conflict are marred by the continued prevalence of patriarchal cultural norms, high levels of SGBV and a culture of lawlessness which threaten the security of both men and women.[49] In Liberia, SGBV persists despite the presence of large contingents of female UN peacekeepers and gender advisers. A similar situation prevails in the DRC, particularly in the eastern part of the country, as recent reports of mass rapes have indicated: since 2009 about 1,100 rapes have been reported each month.[50] In Sierra Leone, since the peace in 2003, the government has struggled to implement innovative laws to protect women's rights and outlaw sexual violence. In Somalia and Côte d'Ivoire rapes and other sexual offences were committed by both government and rebel forces.[51] African security institutions remain largely dominated by men, with a masculinist culture squarely in place.[52] There is increased (UN) awareness that violence in the post-war phase is often the result of the availability of weapons, trauma of males in the family, frustration related to poverty and a backlash against women.[53] For instance, tensions between women workers and their husbands increased in Timor-Leste in the post-conflict period.[54] However, there is insufficient understanding of the links between the public and private dimensions of SGBV as both a tool of war and a crime during peacetime, coupled with a lack of trust in the security forces and justice system. These combine to create a cycle of impunity and violence which is difficult to break.[55]

While organisations such as the United Nations often recognise the above-mentioned multi-causal explanations of SGBV, their discursive assumptions remain largely intact. This is because justifications by international institutions for incorporating a gender perspective into SSR draw on the cyclical connectedness of security and development, and make this the basis for elevating gender equality as *the* route to emancipation. Gender inequality is considered to be detrimental to development. At the same time VAW is understood to be a manifestation of insecurity with widespread negative effects for development in terms of its socio-economic and psychological costs, among others.[56] In short, women cannot work in the fields, access water points or go to the market and girls cannot go to school for fear of being raped.[57]

Paradoxically, then, because VAW/SGBV have such a huge impact on women and serve as a barrier to their participation in SSR and development, and because security forces are headed and predominantly staffed by men, are often the source of SGBV and discriminate against their female members, SSR is viewed as a key element in the long-term prevention of SGBV.[58] By including a gender perspective, SSR can – it is maintained – meet both normative and efficiency goals of making security forces more accountable, professional and respectful of human rights, and thereby help to decrease levels of violence and enhance productivity. The inclusion of women thus automatically means a strengthening of local ownership and oversight of the security sector.[59]

The argument has once again come full circle: women need security and development, and the security-development nexus needs women! However, justification for such policies is drawn mostly from anecdotal evidence – the assumption that women in security institutions not only change the cultural environment of the organisation but also make them more democratic.[60] What these linkages often lack is the recognition that SSR in the post-conflict context actually implies a fundamental redistribution of power and resources. A preoccupation with gender equality (extending individual freedom and liberal rights to women) is a typical liberal feminist notion which concentrates on 'amending gender discrimination' while refusing to connect 'gender issues with larger forms of oppression'.[61] The UN gender discourse on peace-building is rooted within such a liberal gender mainstreaming approach.[62] A liberal feminist approach, however, does not challenge the underlying institutions that reproduce gender hierarchies.[63] Instead, gender is grafted on to existing power structures, as the later analysis of UN discourse on SGBV related to armed conflict will show. What this argument misses is the fact that gender equality is not a panacea for an already dysfunctional system. Inserting women into such a system using the gender equality tool will not change skewed relationships and structures. This logic ignores the fact that gender is a construction and therefore the discourse becomes essentialist. This means that theorising and policy-making are guided by bifurcated notions of women as victims, peaceful and to be protected, as opposed to men as aggressors, warlike and being protectors.[64] The consequences of this ideological preference are twofold. On the one hand rationalist modes of masculinity and femininity are produced which effectively silence all 'other' versions of masculinity and femininity.[65] This legitimises (military) men's experiences as the norm and normalises the notion of women as lacking in agency. On the other hand, labels such as 'human' development, 'human' rights and 'human' security could become 'catch-all phrases'. For instance, in the SSR context, where human security is viewed as the means to bridge security and development, the possible collapsing of femininity and masculinity into the term 'human' could conceal the gendered underpinnings of security or power practices. Through an emphasis on gender equality in both development and security one can therefore argue that 'it has been possible ... to improve the *condition* of women without hurting the condition of men or challenging the *position* of men'.[66]

The 'womenandgender' issue

From this critique stems a number of specific contradictions or implications related to how SGBV is addressed through SSR. Firstly, an exclusive focus on gender equality can lead to an easy slippage between gender and women in both development and security discourses, partly because women are usually the ones drawing attention to gender issues. In a liberal feminist analysis the complexities of the link between gender and power are negated, and SGBV and VAW are often conflated. See for instance, how the proceedings of the UNHCR 2001 Inter-Agency Lessons Learned Conference declared that 'gender-based violence is predominantly men's violence towards women and children'.[67] The neglect of gender violence against men also becomes noticeable in the background paper for the Inter-Agency Standing Committee (1999). The paper claims to discuss the differential impact of conflict on men and women, but fails to relate it to men and boys, and only devotes one section to VAW.[68] This kind of gender approach offers very little theorising around gender identities and/or power relations. In Timor-Leste it appears as if SGBV was prioritised from the outset, but in the form of a focus on offences against women.[69]

The UN Declaration on the Elimination of Violence against Women (1993) quite rightly recognises VAW as 'a manifestation of historically unequal power relations between men and women, which has led to domination over and discrimination against women by men and ... is one of the crucial social mechanisms by which women are forced into a subordinate position compared to men'.[70] But herein lies the dilemma: greater awareness but weak understanding of SGBV have led to it being aligned mainly with women and victimhood. This understanding not only undermines women's and girls' agency but also negates men's experiences of sexual violence. In contrast, Nayak and Suchland's definition of gender violence underscores two pertinent issues. First, it highlights 'systematic, institutionalized and/or programmatic violence ... that operates through the constructs of gender and *often at the intersection of sexuality, race and national identity*'. This characterisation takes SGBV beyond gender. Second, the definition draws gender violence away from an exclusive focus on women towards 'the acts and practices that systematically target a person, group or community in order to dictate what "men" and "women" are supposed to be'.[71] And if intersectionality is taken seriously, SGBV may even lose its usefulness as an analytical construct. It could become more meaningful to talk about an analysis of how oppression or multiple identity constructions operate through the use of violence in different contexts.

The essence of essentialism

The gender equality perspective also leads to the essentialising of women's and men's roles in SSR, especially in the context of the argument around efficiency. There is a general acceptance that women bring a specific skill-set to peacekeeping in terms of searching and interrogating women; possessing often better communication skills; and building trust in the system more effectively (e.g. community policing). It is also argued that women are more likely to report incidents of SGBV to female officers, as is seen in evidence from the DRC, India and Sierra Leone. For these reasons family violence has been prioritised in SSR in Afghanistan, Kosovo, Liberia and Sierra Leone through the establishment of so-called family police units.[72] In this logic women, due to innate qualities, are considered to be better equipped to deal with SGBV.[73] For example, in Rwanda women were included in peace-building because they were considered to be less corrupt and more peaceful than men.[74] Women's organisations (e.g. the Liberian Women in Peacebuilding Network) also had to step in and provide services on behalf of UNMIL (UN Mission in Liberia) during the chaotic situation in Liberian DDR camps. Yet women had to struggle to find a voice at the peace table.[75]

The abuse of agency is therefore manifest in two broad areas. On the one hand there is very little or no interrogation of hypermasculinist and militarist culture in the security sector, or of the possibility that through integration in the military women may adopt those very same traits. On the other hand, agency may become a buzzword masking very specific one-dimensional roles. While the recognition of women's operational roles in SSR can be seen as a tacit acknowledgement of the need for changed gender relations, their role as actors does not necessarily mean they have agency. Women cannot be regarded as agents simply because they provide key security services – for instance, providing shelter to male and female victims of rape and torture; their expertise regarding security-related programming such as human trafficking; their skills in training on human rights and gender issues; and their greater access to the community and thus enhanced intelligence-gathering capacities (i.e. bridge-builders between local and security policy communities).[76] Women's local ownership is largely viewed as an issue of operational necessity, with the normative imperative being sidelined.[77] More often than not, women's key civil society watchdog role (i.e. to challenge structural or institutional barriers) makes governments uncomfortable. Women's agency is therefore selectively applied, and the recognition of their service delivery role is treated as a smokescreen for channelling donor assistance, rather than 'encouraging the development of independent policy interlocutors' in the area of security decision-making.[78] The international community can help to address this selective application through critical self-reflection. Some introspection about the kind of language used by externals to frame women's roles will go a long way in creating awareness that operational roles for women do not automatically translate into a change of mindset. Women's advocacy groups in SSR, such as Women in the Security Sector in Sierra Leone, should therefore be supported by the international community in a way that makes substantive equality (rather than just numerical parity) a reality.

So while women's roles in SSR are narrowly circumscribed, men's changing gender identities are completely overlooked. As Dolan points out, in a context of severe poverty, impunity and endemic violence, men who fail to live up to stereotypical gender expectations, e.g. as breadwinners, also suffer stigmatisation, and maybe even more than women because they occupy a higher status in their communities. Changing gender relations become a source of tension, particularly as the change comes from within, through women/wives who are asserting their 'rights' at the expense of men 'who have become like women'.[79]

Where are the women?

Thirdly, the gender equality perspective emphasises women's representation, protection and empowerment at the expense of solutions that consider both genders and the implications of other overlapping identity constructions. Gender-sensitive reform of the security sector requires, according to conventional wisdom, more women in decision-making and in uniform, gender-sensitive budgets and a gender-sensitive code of conduct. The pursuit of quotas often does not prevent women from being relegated to administrative sectors within the security institutions. For instance, in Sierra

Leone female police officers often only acted as cooks for male officers.[80] Many of the female officers recruited also worked in the family support units. On the one hand this reflected a community-based approach, but it also signalled a separation between women's and men's jobs in the police force, and the lack of general sensitisation of the police regarding gender issues and SGBV.[81] In Timor-Leste local and international collaboration opened space for women's representation, but their role in decision-making remains limited.[82] Before 2006 the Policía Nacional de Timor-Leste had a higher percentage of female staff than most police forces, but they were doing mostly office work or stationed in the vulnerable persons units, once again signalling that women's work enjoyed a lower status.[83] The liberal answer to such challenges is usually one-dimensional – change the leadership, and provide more training and gender mainstreaming.

The gender mainstreaming literature is split between two main approaches, namely integrative (liberal) and a more radical model.[84] The integrative approach follows a strategy of gender balancing focused on increasing the recruitment, retention and advancement of women in security sector institutions.[85] It starts off with a statistical audit of 'where are the women' in policies and programmes, and then proceeds to develop organisational strategies to fill the gaps. In Sierra Leone, in a purely quantitative sense, the strategy was reasonably successful in that 1,550 female police officers were recruited by the end of 2007, but this approach runs the risk of marginalising those females if quotas are pursued in isolation from other more transformative measures to address social barriers.[86] The integrative model is narrowly women-focused. To compensate for this, the United Nations proposes a dual approach of presenting gender inclusion in terms of both efficiency and rights-based discourses.[87] However, the latter is somewhat self-defeating as it does not challenge universalist liberal assumptions, but instead buys into the liberal project. The second approach, the radical model, concentrates more on the long-term strategic analysis and transformation of gendered power relations placed within the broader transformational context of societal change.[88] The reality is that this alternative will require a different kind of leadership, and will be at odds with the short- to medium-term timeframe of most SSR interventions. It takes time to change attitudes, especially in an institution that is notoriously conservative.

Despite these ideological differences between the two approaches, it needs to be stated that all practitioners seek to change the agenda in one way or another. In the case of the former approach, however, there is a danger that it might *only* achieve integration and a 'quick fix' in terms of representation. Electoral reforms using legislative quotas ensured that Rwanda became the world leader in terms of women's representation in parliament, with 48.8 per cent of parliamentarians being women.[89] However, in the current increasingly authoritarian political climate women have resorted to pragmatic mobilisation around 'soft' issues of widowhood, rather than issues related to SGBV. Since the Rwandan nation-building project has prohibited reference to ethnic differences, the gender project (as it overlaps with other identities) is also harmed and women's ability to influence policy-making has decreased.[90] The successes in gender mainstreaming are therefore threatened by the latent conflict potential of new emergent divisions, such as along language (Rwanda now has four official languages) and racial (Bantu versus Hamite) lines in the Great Lakes area. Ultimately reaching a conservative SSR audience with 'soft' gender language becomes more important than dealing with root causes of gendered manifestations of inequality, discrimination and violence. Viewing the integrative approach as a stepping-stone towards a more transformative agenda can only work if the liberal underpinnings of the integrative approach are not treated as an end in themselves.

Problem-solving tools of the trade

This point brings us to the final implication of a gender equality perspective – the kinds of tools and instruments chosen for implementation. Since planners cater to donors' short-term, measurable demands rather than invest in real, long-term social change, they become fixated on technical and bureaucratic processes complete with workshops, handbooks and toolkits. In practice we therefore witness a paradoxical effect of the marginalisation of gender issues in spite of a thriving gender mainstreaming industry.[91]

Interestingly, the MDGs do not mention security, and UNSCR 1325 does not mention development.[92] However, the liberal peace agenda has sought to bridge this gap via SSR. Since policy documents on gender in SSR generally draw on UNSCR 1325, it is imperative to highlight some of its flaws and silences, as these may be perpetuated in the tools and plans devised for implementation. I concur with the NGO Working Group on Women, Peace and Security, who argue that the essence of gender mainstreaming dissipates when such plans draw only on women's experiences as a resource in peace-building and lose sight of the importance of gender as an analytical tool for rethinking peace-building policies.[93] Although UNSCR 1325 does not mention SSR specifically, it includes calls for adherence to broad normative principles of equal participation in the area of peace and security. The guiding principle for UNSCR 1325 is the neoliberal assumption that rights-based discourses of equality help sustain liberal democracy and a free market economy as the only rational alternative to war and underdevelopment. The resolution further makes a simplistic connection between women's presence and positive change, founded on the assumption that women as a group are good peace-builders.[94] Masculine identities are left uninterrogated, and there is also a silence on cultures of violence which lie at the root of armed conflicts and are carried over into the post-conflict period. Due to the preoccupation with women's violent experiences, the impact of global structures of violence on men as both victims and perpetrators of violence is negated.

At one level the adoption of UNSCR 1820 (2008), to fill the gaps of UNSCR 1325 by declaring conflict-related sexual violence a crime against humanity and a matter of international security, is a positive development. It has broken the silence about crimes such as rape and transcended the public/private divide. Sexual violence has now officially become a matter of high politics, but the robust international discourse has both shaped and overshadowed the national responses of non-governmental organisations and governments. UN advocacy in this respect has hinged on two pillars, namely exposing the crime, urging governments to condemn sexual violence, and demanding and offering justice for victims. As a result, the failure to acknowledge the incidence of SGBV (breaking the silence) is treated as the main cause of its prevalence.[95] This is indicative of a problem-solving approach whereby the contextual dimensions of a complex problem are being overlooked. Sexual violence is therefore viewed as a stand-alone issue with little connection to pre-existing gender and other power relations and the culture of violence that often permeates the social fabric of post-conflict countries.[96] Margot Wallström's statement that 'Sexual violence in conflict is neither cultural nor sexual. It is criminal' is indicative of the oversimplified approach adopted by the United Nations in this regard.[97] In contrast, Myrttinen argues that in Timorese society violent masculinities are legitimated by both men and women, 'by acquiescence and active support' in a complex mix of culture, power, patronage and gender.[98] The work by Dolan and International Alert on sexual violence in eastern DRC also testifies to the need to develop more comprehensive international-local solutions.[99]

Barrow's critique also rings true. She argues that UNSCR 1820 dilutes UNSCR 1325. The latter tried to establish women's agency in conflict prevention, resolution and peace-building processes, whereas provisions contained within the former may reinforce women's victimhood.[100] UNSCR

1820 is yet to be fully implemented (see UNSCR 1888 to speed up implementation). The resolution asks for inclusion of specific textual provisions related to SGBV in DDR and SSR processes. The logic for this is not to address root causes, but quite simply that peacekeepers, law enforcement agencies and the justice system are more likely to act if they have a mandate. Silence on SGBV is also logical, as often both government and rebel groups are guilty of committing sexual and gender violence and therefore prefer during peace negotiations to exclude it from consideration. There is a link between the absence of women at peace negotiations and this silence, yet the 2002 Sun City peace accord for the DRC showed that provisions against sexual violence do not necessarily end SGBV.[101] The principles and tools which underlie SSR are not sophisticated enough to deal with such complexities.

While I am generally critical of the neoliberal bias in UN-inspired documents and instruments related to women, peace and security, there are some positives. DCAF publications on gender and SSR try to be even-handed in the treatment of gender as an issue of both women and men, but the overarching theme is still one of 'adding women':[102] for example, Verwijk laments the absence of women in the police in Afghanistan.[103] In contrast, in the *Operational Guide to the Integrated Disarmament, Demobilization and Reintegration Standards* (2006), Section 5.10 on 'Women, Gender and DDR' covers topics not only dealing with gender mainstreaming but also SGBV, gender roles and relations, and gender identities: 'Gender roles and relations are defined by cultural, geographic and communal contexts.'[104] The training manual deals explicitly with violent masculinity, male victims of SGBV and men's traditional roles (e.g. as breadwinners).[105] The section on planning and programming further makes an explicit distinction between gender-responsive and female-specific DDR. This kind of treatment of the topic is encouraging as it offers a more nuanced and sophisticated tool of analysis.

But the dilemma remains – how to reach conservative audiences and achieve normative goals of emancipation at the same time? SSR tools may meet the needs of policy-makers and practitioners, but may do little to transform the military culture. Note for instance that the initial concept which preceded the SSR gender toolkit was more academic, including discussion papers and case studies, but with a non-academic and non-gender audience in mind, 'gender jargon and theoretical/abstract arguments' and terms such as 'militarised masculinities' were deliberately avoided.[106] It is easier to change policy than behaviour, but to ignore questions of how femininities are transformed when brought into the security sector and challenging societal/cultural stereotypes (of 'loose women and lesbians' in the security sector) also has long-term ramifications. The SSR gender toolkit has been around for almost three years, and should be fairly solidly established as a broad concept in the minds of SSR practitioners. That being said, it may be time to move on and produce a bolder revised edition.

Liberal frameworks of security, development and gender equality offer easy answers, but alternatives do not come easily. Ulrich, however, reminds us that to deal with 'GBV', states must adopt a discourse that 'questions the fundamental roots of such violence; those roots, however, extend deep into the international legal scheme and ironically into the instruments that specifically address violence against women'.[107] While resolutions and toolkits cannot enforce compliance, the textual language of international commitments is important in that it represents and frames dominant discourses and ideological meanings of, for instance, what counts as security or SGBV. A discourse analysis of such documents exposes the way in which the dominant neoliberal discourses on peace and security allow certain subjects to speak while forcing others to remain silent.

The implications of a liberal additive approach to gender in SSR are most tellingly revealed through the case of gendering SSR in Liberia. Although well documented, it offers useful illustrations of how a gender equality approach privileges representation over protection and empowerment; and how inadequate understandings of SGBV in its local context lead external donors to develop inappropriate responses.

SSR in Liberia – Security-development work in progress?

The Liberian reconstruction project serves as a 'laboratory' for the coming together of the agendas of collective security and development assistance. Gender forms one of the main pillars of Liberia's development strategy and its peace and security agenda. The UN Development Assistance Framework clearly states that the security agenda must be consistent with human rights obligations, in particular regarding VAW.[108] In this section the aim is not to provide a comprehensive overview of challenges and successes, but rather to highlight the predominance of liberal gender mainstreaming processes within SSR and their contradictions in relation to SGBV. External actors (e.g. the United States, UNMIL, donors and private military companies such as DynCorp International) have set the agenda for Liberian reconstruction, but it came at a price.

In 2003 the Accra Comprehensive Peace Agreement (CPA) ended the second civil war, which started in 1999. Since then great strides have been made in the areas of post-conflict reconstruction and development. However, given the limited capacity of both the army and the police and the lack of trust in the new police structures, the security and political situation remains fragile. Human insecurity, extreme levels of poverty, SGBV, setbacks with regard to reconciliation (see the bungled Truth and Reconciliation Commission report of 2009), lack of trust in the governance of the justice system, ethnic and religious tensions between Christians and Muslims, corruption and a flawed DDR and SSR process constitute threats to stability.[109]

In 2003 the Liberia's women's movement embarked on a 'Mass Action for Peace' campaign. Women activists from Guinea, Sierra Leone and Liberia (the Mano River Union Women Peace Network) not only pushed leaders of the three countries to enter into negotiations, but were also instrumental in Liberian women subsequently being given observer status in the 2003 Accra talks.[110] At the end of 2006 Liberian women's organisations called for an increase in the role of women in SSR.[111] It is noteworthy that these organisations accepted that their participation would 'transform public perception of the military and police'.[112] In early 2007 they addressed donors to Liberia at the World Bank and emphasised the interconnectedness of economic development, health, education and security, calling for a more holistic treatment of SSR processes by focusing on human security, particularly SGBV. The contribution of women towards development, gender equality and sustainable peace has been a central message of the Sirleaf administration, to the extent that Liberia was selected by Denmark as a model country for the implementation of MDG 3.[113]

These developments and 'progress' must be understood in light of the content of the CPA. Women's minimal input in the peace agreement helps to explain silences on key gender-related issues. While an estimated 40 per cent of the population were affected by sexual violence during the conflict, it was not mentioned in the CPA.[114] The peace accord did not call for gender reforms in relation to the security sector; instead it focused on specific technical aspects of restructuring related to issues of new command and enlistment of recruits. The CPA also fixed the mission of the new army in rather narrow state-centric terms, thus undermining women's empowerment and a people-centred understanding of security. Strategies used to integrate gender into SSR revolved largely around gender mainstreaming, gender equality and specific women-focused projects (to include women in decision-making within the security sector), but no real work in terms of challenging cultural and systemic barriers had been undertaken. SSR efforts concentrated on police reform at the expense of the military.[115] The process also lacked a clear strategy of gender mainstreaming, and efforts such as setting a target of 20 per cent for the inclusion of women in the police and armed forces, a code of conduct and the introduction of gender modules into the training curriculum of security institutions (e.g. prisons) lacked coherence.[116]

In Liberian SSR, representation is privileged over protection and empowerment, but as Jacob points out, 'Respecting women-as-actors within the security forces did not mean that society and security forces questioned or modified traditional roles.'[117] Sirleaf's election in 2005 placed Liberia on the map as a so-called 'pioneering feminist government' in the eyes of the international community, but it has also revealed a number of contradictions. First, statistics on women's representation in decision-making structures are selectively used to justify progress, yet these figures tell a somewhat less positive tale when viewed comparatively.[118] Although women fill 22 per cent of cabinet positions, their political representation – with about 15 per cent women in the legislature – lags behind other African countries such as Uganda, Rwanda and South Africa. So despite evidence that women have gained from the opportunities which opened up after the conflict, men still dominate the government, civil service and academia.[119] Second, the high emphasis on representation overshadows gaps in human security and the continuation of impunity. Gender violence persists despite the presence of a large contingent of female UNMIL peacekeepers and gender advisers, and both the UN envoy and deputy UN envoy being women. The high visibility of having a female president, a female UN special representative of the Secretary-General, a woman leading the police and the initial increase in female applications following the presence of the first all-female unit of Indian police officers are all positive developments, creating a favourable environment for the reduction of SGBV.[120] However, paradoxically, if a gender project is seen as being successful on the surface, less attention may be paid to the real problems. For instance, UNICEF reports that SGBV remains high, with most rape cases involving children.[121]

Ironically, the international community considers Liberia as one of the success stories in implementing UNSCR 1325 and addressing SGBV. The Rape Amendment Act (with stricter penalties) was passed in 2006 and Liberia is one of a few African countries to have adopted a national action plan for the implementation of UNSCRs 1325 and 1820.[122] The plan was the culmination of an extensive consultation process across civil society. For example, in December 2003 the UN Office of the Special Adviser on Gender

Issues and the Advancement of Women prepared a gender checklist for Liberia, but this could be creating a false picture of local ownership. The gender checklist, among others, has been found to be quite extensive and therefore difficult to implement.[123] So while civil society participation is to be commended, the danger of imposing (ambitious) external viewpoints is real, as is evident from the dominance of liberal equality language in the national action plan document. The Liberian document is openly pro-women and girls, and conflates women and gender. The inclusion of support systems for men's economic empowerment and psychosocial counselling appears to be merely rhetorical.[124] With regard to SSR the focus is on women's inclusion (recruitment, retention and promotion), participation, training and capacity-building – the usual liberal equality elements. The reasons for sensitising the population on the benefits of women's participation in security sector institutions are not explained.[125] A great deal of attention is paid to presenting women's and girls' roles and needs in a holistic manner, but it is diluted by essentialist representations of women as 'good female role models' in the security sector without offering any clarification.[126] In a positive sense the national action plan adopts a broad definition of violence. This is reinforced by efforts to change attitudes and behaviour, targeting parents, religious and traditional leaders and community elders, but this point is couched in soft, almost neutral, language. When it comes to the role of women's groups, the plan paradoxically not only devotes attention to the role of these groups as implementers and watchdogs of the process, but also stresses women's role as facilitators of funding from donors. The suitability of women's organisations is assumed, without considering the possible negative implications of their role as multipliers of international discourse. The plan thus superficially acknowledges the importance of local partnerships but neglects a deeper feminist exploration of power. For all its technical sophistication, the plan leaves many of the important rules and discursive practices of the international peace and security institutions in place and does not question the choice of a gender mainstreaming approach. The gender question remains peripheral – despite, or rather because of, the women-focused language. In the words of Liberian Minister of Gender and Development Vabbah Gayflor, 'while gender is a very good concept ... our own critical challenges now are in the area of women and children … because these are the two groups that ... were not handled appropriately'.[127]

External misconceptions about what constitutes protection and what counts as SGBV further complicate the gendering of SSR. The international community (e.g. UNIFEM) responded to women's advocacy by establishing a taskforce on gender violence comprising the Liberian National Police, the

UN Police, other UN agencies and local NGOs.[128] The Women and Children Protection Section of the Liberian National Police has officers trained in dealing with cases of SGBV. Some may hail this as an example of best practice, while others, such as Nagelhus Schia and De Carvalho, regard the establishment of women and children protection sections next to police stations as a quick fix by the international community to address SGBV without taking cognisance of the broader framework of the reconstruction of all rule-of-law institutions.[129] These sections are in all the major capitals but not in rural areas, thereby making these services inaccessible to the majority of people. Consequently victims must travel to the cities to report a crime; yet rape case evidence must be collected within 72 hours. Rape also often happens within families, and people therefore prefer to use the traditional system where the chiefs mediate. It is considered more practical and less stigmatising.[130]

This case reveals firstly that donors often overlook the complex interplay between statutory and customary systems of justice. They consequently offer simplistic solutions to very complex problems which require a holistic engagement with the contradictions of the local context and the hybridity of its institutions. For instance, donors pursue their own narrow agenda, wanting to supplant traditional systems with something formal or assuming that there are no institutions in place. Nagelhus Schia and De Carvalho contend that the United Nations does not understand Liberian traditional customs or its penal code.[131] They argue that rape dominates in terms of the focus of intervention, and female genital mutilation is neglected because of its heavy cultural baggage. The result is a fragmented attempt to address symptoms rather than causes. Secondly, this case shows how the dominant neoliberal discourses decide what counts as gender violence. It is no wonder then that such interventions are unable to address the root causes of SGBV (e.g. entrenched patriarchal/cultural attitudes, changing gender relations, backlash against women's increased empowerment, spillover from violent practices during the war), as they fail to 'speak to' the context-specific political, economic and cultural characteristics of gender violence.[132]

While the international community is certainly concerned about the prevalence of this form of violence, and various tools have been developed, much more needs to be done to address the tensions between representation, protection and empowerment.

Towards difficult answers: To mend the bridge or build a road around it?

Having pointed out the limitations of current ways of linking security and development and how this conflation harms the process of gendering SSR, the question remains – do we tinker with the SSR system or do we change course? How does one break through this barrier of conventional donor wisdom that women need security/development, and security/development needs women – a dispensation in which rights, morals and empowerment cannot be separated from measurable outcomes and a 'product' that offers value for money? And how can a normative argument be translated into workable policies and plans?

Do we take small incremental steps on the road to larger social change? For instance, do we accept the liberal feminist emphasis on representation in SSR but also push for deeper transformation? The argument in this chapter is a qualified response to this question since – as mentioned earlier – the evidence points to the pursuit of a neoliberal order as an end in itself (and in fact becoming both means and ends). As long as liberal democracy (with all its bells and whistles) is presented as the ultimate prize, the system remains closed off from the possibility of other alternatives. And this is where a critical (or post-colonial) feminist alternative comes in – to push for a re-envisioned understanding of the link between means and ends. Conventional views posit security as the prerequisite for peace, whereas in a critical feminist sense it is often the process or method that is more important than the goal.

It is only if the end result is not couched in exclusively liberal terms, but left open-ended, that complexities of consequential and pragmatic decision-making by SSR practitioners begin to make sense. For instance, while the tension between what policy-makers want to hear and the reality of the situation is a dilemma not exclusively reserved for the gender domain, it is exacerbated in the context of security. It therefore follows that toning down radical feminist language to reach conservative SSR audiences seems plausible from a pragmatic point of view. In this context lofty arguments about social justice and positive peace cannot compete with the rhetoric of a rules-based approach to governance. This approach claims to be motivated by an adherence to procedural rules to make SSR structures operate more efficiently, equitably and democratically. In practice, shying away from sensitive issues such as hypermasculinity[133] and women's and men's 'condoning' or internalisation of violence as a result of years of conflict seems to be more a case of consequential decision-making, i.e. whether the action produced a good or a bad result in a particular context.[134] This kind of decision-making is further reinforced by complex political contestation at the local level that often drives the SSR process. SSR practitioners may justify striking bargains with warlords (the new political elites) as an attempt to respect local culture, but this may lead to an entrenchment of patriarchal values. At the same time, locals manipulate the process to get the best of both worlds, as in the case of Sierra Leone where local and international decentralisation agendas are at odds.[135] This chapter has been quite critical of the Liberian reconstruction process, but others who work with a liberal framework would hail it as a huge success. I argue that it is more a case of having won the battle but losing the war in the end, because questions such as 'whose battle' and 'whose war' are negated.

So what is the way out of these dilemmas? Bringing the critical feminist view into the SSR discourse is not too much to ask from policy-makers, although such an approach will take longer. In this respect I argue for a shift in gaze to two feminist 'tools' by means of which root causes of all unequal (not just gender) relations and biases can be addressed in some incremental way. These tools are meant to be open-ended and do not foreground a specific ideological agenda. Their intrinsic multiplicity and open-endedness make them feasible to implement. In other words, it is not about adding yet another project to an already impossible 'to do' list of SSR practitioners on the ground. What is required is a mind shift away from a fixation on achieving holistic SSR. The fallacy of this ideal is that it is perceived to be coherent, stable and harmonious, and on that basis the current model of SSR is assumed to be rational and implementable. But reality does not work like that. In contrast, the kind of holism proposed in this chapter is 'fractious' – marked by a high degree of tolerance for contestation, instability and difference of identity.[136]

Firstly, I propose using the tool of intersectionality as a way to address the tricky issue of hypermasculinity in the military. As discussed earlier (see endnote 14), intersectionality as a critical feminist concept refers to the 'forms of inequality that are routed through one another and which cannot be untangled to reveal a single cause'.[137] Given the high prevalence of identity wars in Africa and their fallout, it is only logical to widen the post-conflict scope to look at not just gender, but rather to focus on the complex and overlapping intersection of gender, ethnicity, race, class and sexuality. It is important to bear in mind that in conflict areas women are not always most concerned about gender equality. Often family and community take priority in their daily lives.

The liberal approach to SSR privileges Western rationalist versions of masculinity and femininity and silences other forms of femininity and masculinity.[138] The 'target' in addressing militarist culture in SSR should therefore shift from patriarchy as an essentialised and monolithic concept to challenging hegemonic masculinity while highlighting the multiplicity and variation in masculinities.[139] Hegemonic masculinity refers to 'a male-centered order that gives men, instead of women, primary access to power and privilege' which can differ across local contexts and on the basis of the historical legacy of external factors (e.g. colonialism, presence of foreign peacekeepers, donors, etc.).[140] Essentialised versions of men as protectors and aggressors imbued with values of physical ability, courage, endurance, self-control, professionalism and heterosexuality are therefore not the only versions of masculinity. Within this system there are allied masculinities which benefit and others which do not, together with some femininities.[141] Some women in the military may also benefit from the unequal gender arrangements as long as they do not challenge entrenched male power. The risk is, however, that hegemonic masculinities could be replaced by caring versions of masculinity as the 'new' hegemony. Would women in the security sector necessarily benefit from caring (read: paternalist) masculinities?[142] In this regard I argue that the answer lies in developing a 'non-hegemonic' or 'anti-hegemonic' strategy, as it is sceptical of any attempt to build unity, consensus or coordinated political advocacy.[143] Instead, it celebrates a politics of plurality and multiplicity of locations and subjectivities. 'Democratising' a highly authoritarian institution through an emphasis on pluralism and tolerance could be effective in the long run if attention is strategically diverted away from an exclusive masculine versus feminine debate. Allowing space for multiple experiences also helps to shape alternative explanations of why SGBV occurs, how its security and developmental dimensions could be addressed, and finally how this would impact on contextualised approaches to SSR.

As already discussed, the liberal approach to gender mainstreaming overemphasises individual differences and underplays differences at the institutional or structural level and – on its own – is unsuitable to address women, peace and security issues in an African context. Intersectionality as an approach allows space for both individual identity constructions and structural and cultural analyses.[144] The aim is to look at women's issues of development and security in terms of the feminisation of poverty and violence (women cannot work/work the land). The point is that there needs to be greater understanding of the fact that when structural factors intersect with identity, some people become more vulnerable than others.[145] Structural analysis, especially in view of Africa's colonial legacy and marginalised position within the global political economy, must therefore inform policy-making related to Africa. An intersectional lens is useful to study the ambivalent power relations between women's organisations and donors, possibly revealing hidden forms of agency. Intersectionality also has specific utility value for DDR/SSR policy-making. Intersectional theory has the potential to expose the limitations of policies designed to benefit specific target groups. Sometimes both gender- and race-targeted policies fail to meet the needs of groups, because planners were unaware of or overlooked the complexities emanating from an interface of multiple identities. Intersectionality, proposed here as a tool for SSR, is thus different from a conventional multi-causal approach. It is not a linear or layered process where the most important identity issue is isolated in a given context, or where the forms of oppression are ranked. Issues need to be dealt with all at once. This makes the intersectionality puzzle different, as authenticity of experience can only be found in the intersection of many categories. Clearly this approach goes against the traditional notion of targets and deadlines, and would require a change of mindset for many SSR practitioners, but the benefits in terms of meeting the needs of specific target groups outweigh the complexity of a circular and holistic methodology. It has the potential to move solutions beyond a simple focus on different needs, and contextualises needs by linking difference with disadvantage. Target groups would therefore have to be reconceptualised. Intersectional analyses of existing SSR cases as well as deeper insight into what drives local cooptation could be included in a future research agenda to illustrate its utility value and address the charges from policy-makers that such tools are difficult to operationalise.

Gender advocacy cannot only be policy advocacy. It has to include moral advocacy as well.[146] Together they combine as a political expression of solidarity through care. For that reason a second normative tool is proposed, namely an emphasis on a feminist ethic of care. It provides an alternative lens through which to view the role of SGBV in moral debates about human security, humanitarian intervention and development. This ethic of care does not view security as something individuals earn or are being granted. Instead, it is seen in interdependent terms, founded on the relations of care that exist at community level and through the actions of women and others as caregivers.[147] It thus challenges the autonomous thinking and doing of neoliberal patterns of global governance. The notion of 'care' is not something that is extended in a clean and simple relationship between loved ones, and only by women and girls. In reality caring reflects a range of power relations, dependencies and vulnerabilities.[148] In the everyday social reality, autonomous/individualised and relational intersections cut across race, gender and ethnicity. Robinson draws our attention to the fact that care is an act of democratic citizenship, a public issue (at the core of good governance) and therefore theoretically open to all.[149] Addressing SGBV is consequently not the preserve of female peacekeepers or women's organisations. Care is not just a women's issue, but a human issue central to the notion of public service.

Conclusion

This chapter has sought to expose the value-laden nature of labels such as security, development and violence by looking at them through the lens of gender – which in and of itself is also not a neutral construct. Securitisation is a fundamentally value-laden exercise with very real political consequences. As such the gap between policy and practice is largely the result of an adherence to the rhetoric and assumptions of liberal security governance. Expectations may have been created that SSR has the potential to effect fundamental social and cultural change. However, this is too much to expect from a system that is guided by a flawed link between security and development. Its deficits are most clearly illustrated when the tensions between SGBV and VAW come into full view. In spite of the efforts of practitioners to make people the centre of security, overarching frameworks still privilege state security. In spite of efforts to secure all human beings, some groups remain vulnerable. A critical feminist lens could take security thinking beyond a deepening of levels and a broadening of referent objects. It offers an opening to make room for other explanations and solutions. As Wibben reminds us, 'An opening of the agenda, thus, needs to begin by understanding how security has traditionally worked ... and how meanings of security are fixed in certain narratives that make up security studies.'[150] My proposed alternative of an ethic of care is therefore not just a community-level solution, but extends to the international and state levels. How? The kind of global ethic that I envision would allow alternative ethical frameworks such as *ubuntu* into the mainstream, to coexist with 'good governance' principles. *Ubuntu* as a concept and practice defines the individual in relation to others. This will require a difficult but necessary rethink of global governance as a whole. In terms of security governance a differently applied gender-sensitive SSR model would be one that is less concerned about parity in numbers (equality), and more committed to equity

– improving the quality of relations, from the local to the national to the global.

A fixation on adding women at the expense of challenging root causes of inequality in SSR has kept policy-makers very busy, but as long as SSR continues to privilege 'doing things right' rather than 'doing the right things', it will remain an industry lacking legitimacy. It is time for the engineers of SSR to rethink their bridge-building project. A road construction which takes account of intersectional and multiple identities as well as an ethic of care may use these as two markers along this new and unexplored route.

Notes

1. Henri Myrttinen, 'Poster Boys No More: Gender and Security Sector Reform in Timor-Leste', DCAF Policy Paper no. 31 (Geneva: DCAF, 2009): 12.
2. I want to stress that SSR practitioners as a rule are not intent on abusing the system. The so-called 'abuse' sets in because of the lack of consideration given to other models of governance in post-conflict situations, e.g. neopatrimonialism as a hybrid of formal bureaucratic and informal patron-client relations. Blanket assumptions about the inherent virtue of liberal democracy, simply because it worked in developed world contexts, are thus to blame.
3. Chiyuki Aoi, Cedric De Coning and Ramesh Thakur, 'Unintended Consequences, Complex Peace Operations and Peacebuilding Systems', in *Unintended Consequences of Peacekeeping Operations*, eds Chiyuki Aoi, Cedric De Coning and Ramesh Thakur (New York: United Nations University Press, 2007): 8–9.
4. Shahrbanou Tadjbakhsh and Anuradha M. Chenoy, *Human Security: Concepts and Implications* (London and New York: Routledge, 2007).
5. Mary Caprioli, 'Primed for Violence: The Role of Gender Inequality in Predicting Internal Conflict', *International Studies Quarterly* 49 (2005): 161–178.
6. Chris Dolan, 'War is Not Yet Over', in *Community Perceptions of Sexual Violence and Its Underpinnings in Eastern DRC* (London: International Alert, 2010): 20.
7. Essentialism is a theoretical approach stating that certain properties (e.g. human traits) of a group have universal validity, rather than being social, ideological or intellectual constructs which are context-dependent. For instance, this approach assumes that women share the same basic experiences or interests, regardless of differences in class, race, etc. See Janie L. Leatherman, *Sexual Violence and Armed Conflict* (Cambridge: Polity Press, 2011): 13.
8. Yaliwe Clarke, 'Security Sector Reform in Africa: A Lost Opportunity to Deconstruct Militarised Masculinities?', in *Feminist Africa 10: Militarism, Conflict and Women's Activism*, eds Amina Mama and Margo Okazawa-Rey (Cape Town: Africa Gender Institute, 2008): 53.
9. Ecoma Alaga and Emma Birikorang, 'Security Sector Reform and Gender: Beyond the Paradigm of Mainstreaming', paper presented at Annual International Studies Association Convention, Montreal, 2011.
10. Margarete Jacob, 'Engendering Security Sector Reform: Sierra Leone and Liberia Compared', in *Engendering Security Sector Reform: A Workshop Report*, eds Margarete Jacob, Daniel Bendix and Ruth Stanley (Berlin: Freie Universität, 2009): 64.
11. Myrttinen, note 1 above: 12.
12. Ibid.: 27, 31.
13. Dolan, note 6 above: 22–23, 34–38.
14. The term 'intersectionality' was coined by Crenshaw, who argued that justice for black women can only be achieved by studying their experiences at the intersection of racism and sexism. Hancock distinguishes between three approaches to the study of race, gender, class and other categories of difference. Firstly, the unitary approach works with one category of identity at a time, based on an implicit hierarchy. Secondly, the multiple approach acknowledges the need to reflect on more than one category at a time, but the categories are viewed as static, matter equally in a predetermined relationship to each other and remain conceptually independent. It means identifying multiple sites of oppression in the name of inclusivity, with the assumption that there is a correlation

between the number of oppressions and the extent of marginalisation of the individual in a compound sense. Finally, intersectionality is described by Kantola and Nousiainen as viewing the relationship between the categories as 'an open empirical question and the categories themselves are conceptualized as resulting from dynamic interaction between the individual and institutional factors'. Kimberlé Crenshaw, 'Mapping the Margins: Intersectionality, Identity Politics and Violence against Women of Color', *Stanford Law Review* 43, no. 6 (1991): 1241–1299; Ange-Marie Hancock, 'When Multiplication Doesn't Equal Quick Addition: Examining Intersectionality as a Research Paradigm', *Perspectives on Politics* 5, no. 1 (2007): 63–79; Johanna Kantola and Kevät Nousiainen, 'Institutionalizing Intersectionality in Europe', *International Feminist Journal of Politics* 11, no. 4 (2009): 469.

[15] Rita Abrahamsen and Michael C. Williams, 'Security Sector Reform: Bringing the Private in', *Conflict, Security and Development* 6, no. 1 (2006): 2.

[16] Michael Brzoska, 'Development Donors and the Concept of Security Sector Reform', DCAF Occasional Paper no. 4 (Geneva: DCAF, 2003): 46.

[17] David Chandler, *Hollow Hegemony: Rethinking Global Politics, Power and Resistance* (New York: Pluto Press, 2009).

[18] Frances Stewart, 'Development and Security', Centre for Research on Inequality, Human Security and Ethnicity Working Paper no. 3 (Oxford: University of Oxford, 2004): 1–27; Chandler, ibid.: 33; Neclâ Tschirgi, Michael S. Lund and Francesco Mancini, 'The Security-Development Nexus', in *Security and Development: Searching for Critical Connections*, eds Neclâ Tschirgi, Michael S. Lund and Francesco Mancini (London: Lynne Rienner, 2009): 3; Neclâ Tschirgi, 'Security and Development Policies: Untangling the Relationship', paper prepared for European Association of Development Research and Training Institutes Conference, Bonn (21–24 September 2005): 10; Michael Brzoska and Peter Croll, 'Investing in Development: An Investment in Security', *Disarmament Forum* (2005), available at www.unidir.org/pdf/articles/pdf-art2391.pdf.

[19] Cited in Tschirgi et al., ibid.: 2.

[20] Chandler, note 17 above: 33.

[21] Paul Collier, Lani Elliott, Håvard Hegre, Anke Hoeffler, Marta Reynal-Querol and Nicholas Sambanis, *Breaking the Conflict Trap: Civil War and Development Policy* (Oxford: Oxford University Press, 2003); DFID, *Fighting Poverty to Build a Safer World: A Strategy for Security and Development* (London: Department for International Development, 2005).

[22] DFID, ibid.: 20.

[23] Keith Krause and Oliver Jütersonke, 'Peace, Security and Development in Post-conflict Environments', *Security Dialogue* 36, no. 4 (2005): 458.

[24] Caroline Thomas, 'Global Governance and Human Security', in *Global Governance. Critical Perspectives*, eds Rorden Wilkinson and Steve Hughes (London: Routledge, 2002).

[25] OECD/DAC, 'OECD/DAC Handbook on Security Sector Reform. Supporting Security and Justice' (2007), available at www.oecd.org/dac/conflict/if-ssr: 1–2.

[26] Ibid.: 1; DFID, note 21 above: 7.

[27] Colette Harris, 'What Can Applying a Gender Lens Contribute to Conflict Studies? A Review of Selected MICROCON Working Papers', MICROCON Research Working Paper no. 41 (Brighton: MICROCON, 2011): 10.

[28] Tschirgi, note 18 above.

[29] Albrecht Schnabel, 'Ideal Requirements versus Real Environments in Security Sector Reform', in *Security Sector Reform in Challenging Environments*, eds Hans Born and Albrecht Schnabel (Münster: LIT Verlag, 2009): 17–18; Tschirgi et al., note 18 above: 4.
[30] E.g. Mark Duffield, *Global Governance and the New Wars: The Merging of Development and Security* (London: Zed Books, 2001).
[31] Ibid.: 122.
[32] Ibid.: 11; Karin M. Fierke, *Critical Approaches to International Security* (Cambridge: Polity Press, 2007): 154–156; Jason Franks and Oliver P. Richmond, 'Coopting Liberal Peace-building: Untying the Gordian Knot in Kosovo', *Cooperation and Conflict* 43, no. 1 (2008): 83.
[33] Roger Mac Ginty, 'Indigenous Peace-making Versus the Liberal Peace', *Cooperation and Conflict* 43, no. 2 (2008): 145.
[34] Timothy Donais, 'Inclusion or Exclusion? Local Ownership and Security Sector Reform', *Studies in Social Justice* 3, no. 1 (2009): 120.
[35] Schnabel, note 29 above: 26.
[36] Vivienne Jabri, 'Feminist Ethics and Hegemonic Global Politics', *Alternatives* 29, no. 3 (2004): 274; Tschirgi, note 18 above.
[37] Chris Smith, 'Security-sector Reform: Development Breakthrough or Institutional Engineering?', *Conflict, Security & Development* 1, no. 1 (2001): 9.
[38] Roger Mac Ginty, 'No War, No Peace: Why So Many Peace Processes Fail to Deliver Peace', *International Politics* 47, no. 2 (2010): 154.
[39] Chandler, note 17 above: 31.
[40] Ibid.: 26–52.
[41] Ibid.: 45.
[42] Ibid.: 42.
[43] UN Security Council, 'Resolution 1325 on Women, Peace and Security' (2000), available at www.un.org/events/res_1325e.pdf.
[44] Natalie F. Hudson, *Human Security and the United Nations. Security Language as a Political Framework for Women* (New York: Routledge, 2009): 95–118.
[45] Chandler, note 17 above: 47–48.
[46] UN Security Council, 'Resolution 1820 on Women and Peace and Security' (2008), available at http://daccess-ods.un.org/access.nsf/Get?Open&DS=S/RES/1820%20(2008)&Lang=E&Area=UNDOC.
[47] Megan Bastick and Kristin Valasek, eds, *Gender and Security Sector Reform Toolkit* (Geneva: DCAF, OSCE/ODIHR, UN-INSTRAW, 2008).
[48] Margaret Verwijk, *Developing the Security Sector: Security for Whom, by Whom?* (The Hague: Ministry of Foreign Affairs, 2007): 5.
[49] UN Security Council, 'Report of the Secretary-General: Women and Peace and Security' (2009), available at http://daccess-ods.un.org/access.nsf/Get?Open&DS=S/2009/465&Lang=E&Area=UNDOC.
[50] Ibid.
[51] Ibid.
[52] Clarke, note 8 above.
[53] Megan Bastick, 'Integrating Gender in Post-conflict Security Sector Reform', DCAF Policy Paper no. 29 (Geneva: DCAF, 2008): 4.
[54] Myrttinen, note 1 above: 13.

[55] Amy Barrow, 'UN Security Council Resolutions 1325 and 1820: Constructing Gender in Armed Conflict and International Humanitarian Law', *International Review of the Red Cross* 92, no. 877 (2010): 222.

[56] Bastick, note 53 above: 1; Purna Sen, 'Development Practice and Violence against Women', *Gender and Development* 6, no. 3 (1998): 7.

[57] Andrew Morrison, Mary Ellsberg and Sarah Bott, 'Addressing Gender-based Violence: A Critical Review of Interventions', *World Bank Research Observer* 22, no. 1 (2007): 30; Jennifer L. Ulrich, 'Confronting Gender-based Violence with International Instruments: Is a Solution to the Pandemic within Reach?', *Indiana Journal of Global Legal Studies* 7, no. 2 (2000): 634.

[58] Laurie Nathan, *No Ownership, No Commitment: A Guide to Local Ownership of Security Sector Reform* (Birmingham: University of Birmingham, 2007): 11.

[59] Verwijk, note 48 above: 10; Kristin Valasek, 'Security Sector Reform and Gender', in *Gender and Security Sector Reform Toolkit*, eds Megan Bastick and Kristin Valasek (Geneva: DCAF, OSCE/ODIHR, UN-INSTRAW, 2008): 4–5.

[60] UN-INSTRAW, 'UN-INSTRAW Virtual Discussion on Gender Training for Security Sector Personnel: Good and Bad Practices' (2007), available at www.un-instraw.org/es/peace-and-security/gestion-del-conocimiento/gender-training-for-security-sector-personnel.html.

[61] Amalia Sa'ar, 'Postcolonial Feminism, the Politics of Identification, and the Liberal Bargain', *Gender and Society* 19, no. 5 (2005): 689.

[62] Laura J. Shepherd, 'Power and Authority in the Production of United Nations Security Council Resolution 1325', *International Studies Quarterly* 52 (2008): 383–404; Barrow, note 55 above: 231. Gender mainstreaming is a strategy that allows prioritisation of the needs and experiences of both men and women in all planned actions, legislation, policies or programmes and also considers the implications of these interventions for women and men. United Nations, 'Gender Mainstreaming. An Overview' (2002), available at www.un.org/womenwatch/osagi/pdf/e65237.pdf.

[63] Ulrich, note 57 above: 649. In the context of this article 'gender' refers to the socially constructed roles of men and women in the security sector, and the unequal ordering of gender power that disadvantages women practically and systemically as a result of societal understandings and attitudes towards men's and women's roles as security and development actors/victims. These roles are not static.

[64] Tarja Väyrynen, 'Gender and Peacebuilding', in *Palgrave Advances in Peacebuilding: Critical Developments and Approaches*, ed. Oliver P. Richmond (New York: Palgrave Macmillan, 2010): 137.

[65] Ibid.: 140.

[66] Robert Johnson, 'Not a Sufficient Condition: The Limited Relevance of the Gender MDG to Women's Progress', *Gender and Development* 13, no. 1 (2005): 57, emphasis added.

[67] As discussed in R. Charli Carpenter, 'Recognizing Gender-based Violence against Civilian Men and Boys in Conflict Situations', *Security Dialogue* 37, no. 83 (2006): 86.

[68] Ibid.: 87.

[69] DCAF, *Gender and Security Sector Reform: Examples from the Ground* (Geneva: DCAF, 2011): 16–18.

[70] Ulrich, note 57 above: 652.

[71] Meghana Nayak and Jennifer Suchland, 'Gender Violence and Hegemonic Projects', *International Feminist Journal of Politics* 8, no. 4 (2006): 469, emphasis added.

[72] Bastick, note 53 above: 14.

[73] Eirin Mobekk, 'Gender, Women and Security Sector Reform', *International Peacekeeping* 17, no. 2 (2010): 280, 286; Bastick, note 53 above: 13–14.

[74] Heidi Hudson, 'Peacebuilding through a Gender Lens and the Challenges of Implementation in Rwanda and Côte d'Ivoire', *Security Studies* 18, no. 2 (2009): 306.

[75] DCAF, note 69 above: 22.

[76] Valasek, note 59 above: 6–7; Cheryl Hendricks and Mary Chivasa, 'Women and Peacebuilding in Africa: Workshop Report' (Pretoria: Institute for Security Studies, 2008); Verwijk, note 48 above: 26.

[77] Valasek, note 59 above: 8.

[78] OECD/DAC, 'Security System Reform and Governance', DAC Reference Document (Paris: OECD, 2005), available at www.oecd.org/dataoecd/8/39/31785288.pdf: 39.

[79] Dolan, note 6 above: 46, 60.

[80] Bastick, note 53 above: 16.

[81] Jacob, note 10 above: 61.

[82] Myrttinen, note 1 above: 13.

[83] Ibid.: 24–25, 27.

[84] Teresa Rees, 'Reflections on the Uneven Development of Gender Mainstreaming in Europe', *International Feminist Journal of Politics* 7, no. 4 (2005): 555–574.

[85] Valasek, note 59 above: 5.

[86] Jacob, note 10 above: 61.

[87] Kristin Valasek, 'Gender and SSR Toolkit – Origin, Challenges and Ways Forward', in *Engendering Security Sector Reform: A Workshop Report*, eds Margarete Jacob, Daniel Bendix and Ruth Stanley (Berlin: Freie Universität, 2009): 42.

[88] Diane Sainsbury and Christina Bergqvist, 'The Promise and Pitfalls of Gender Mainstreaming: The Swedish Case', *International Feminist Journal of Politics* 11, no. 2 (2009): 216–234.

[89] Bastick, note 53 above: 11.

[90] Jennie E. Burnet, 'Gender Balance and the Meanings of Women in Governance in Post-conflict Rwanda', *African Affairs* 107, no. 427 (2008): 361–386.

[91] Karen Barnes, 'Reform or More of the Same? Gender Mainstreaming and the Changing Nature of UN Peace Operations', YCISS Working Paper no. 41 (Toronto, ON: YCISS, 2006).

[92] Brzoska and Croll, note 18 above: 12.

[93] NGO Working Group on Women, Peace and Security, 'From Local to Global: Making Peace Work for Women, Security Council Resolution 1325 – Five Years On' (2005), available at www.wilpf.int.ch/publications/1325Five_Year_On.pdf.

[94] Fionnuala N. Aoláin and Eilish Rooney, 'Underenforcement and Intersectionality: Gendered Aspects of Transition for Women', *International Journal of Transitional Justice* 1, no. 3 (2007): 338–354.

[95] Eve Ayiera, 'Sexual Violence in Conflict: A Problematic International Discourse', *Feminist Africa* 14 (2010): 10–11.

[96] Ibid.: 14–15.

[97] Margot Wallström, 'Ending Sexual Violence: From Recognition to Action', *New Routes* 16, no. 2 (2011): 49.

[98] Myrttinen, note 1 above: 16–18.

[99] Dolan, note 6 above: 11.

[100] Barrow, note 55 above: 234.

[101] Robert Jenkins and Anne-Marie Goetz, 'Addressing Sexual Violence in Internationally Mediated Peace Negotiations', *International Peacekeeping* 17, no. 2 (2010): 263, 275.

[102] See Valasek, note 59 above: 4, box 2 on the female- and male-specific forms of gender-based violence; Valasek, note 87 above: 36–37.

[103] In Valasek, note 59 above: 1.

[104] UN Inter-Agency Working Group on Disarmament, Demobilization and Reintegration, *Operational Guide to the Integrated Disarmament, Demobilization and Reintegration Standards* (New York: United Nations, 2006): 12, 193.

[105] Ibid.: 195.

[106] Valasek, note 87 above: 40.

[107] Ulrich, note 57 above: 631–632.

[108] DCAF, note 69 above: 90.

[109] Lansana Gberie, 'Liberia: The 2011 Elections and Building Peace in the Fragile State', ISS Situation Report (Pretoria: ISS, 2010).

[110] Aili Mari Tripp, Isabel Casimiro, Joy Kwesiga and Alice Mungw, *African Women's Movements: Changing Political Landscapes* (Cambridge: Cambridge University Press, 2009): 204–205.

[111] Bastick, note 53 above: 9.

[112] Ibid.

[113] Doris Murimi, 'Liberia, Africa's Inspiration on Gender Equality', *ISS Today* (24 June 2009), available at www.iss.co.za/pgcontent.php?UID=14928.

[114] Jenkins and Goetz, note 101 above: 274.

[115] In contrast, the SSR process in South Africa, through the involvement of civil society – women's organisations in particular – and women parliamentarians, managed to bring about relatively gender-sensitive security structures (DCAF, note 69 above: 24). The challenge, however, remains: how to deepen constitutional and structural transformations so that they reflect the everyday security and development concerns of both men and women.

[116] Bastick, note 53 above: 10; Jacob, note 10 above: 54.

[117] Jacob, note 10 above: 65–66.

[118] Veronika Fuest, '"This Is the Time to Get in Front": Changing Roles and Opportunities for Women in Liberia', *African Affairs* 107, no. 427 (2008): 203.

[119] Ibid.

[120] Mobekk, note 73 above: 284–285.

[121] The UNICEF findings are cited by Niels Nagelhus Schia and Benjamin De Carvalho, '"Nobody Gets Justice Here!" Addressing Sexual and Gender-based Violence and the Rule of Law in Liberia', *Security in Practice* 5, NUPI Working Paper no. 761 (Oslo: NUPI, 2009): 8; Mavic Cabrera-Balleza and Tina Johnson, 'Security Council Resolution 1820: A Preliminary Assessment of the Challenges and Opportunities', paper prepared for 1820 Strategy Session, International Women's Tribune Centre (New York, 21-25 September 2009).

[122] Republic of Liberia, 'The Liberia National Action Plan for the Implementation of United Nations Resolution 1325' (2009), available at www.un.org/womenwatch/ianwge/taskforces/wps/nap/LNAP_1325_final.pdf.

[123] DCAF, note 69 above: 87, 89.

[124] Republic of Liberia, note 122 above: 15–16.

[125] Ibid.: 19–21.
[126] Ibid.: 20.
[127] UN-INSTRAW, 'On the National Action Planning Process on Women, Peace and Security in Liberia', UN-INSTRAW interview with Honourable Minister of Gender and Development Vabbah Gayflor (2008), available at http://www.un-instraw.org/data/media/documents/Interview_VG.pdf.
[128] Hudson, note 44 above: 109.
[129] Nagelhus Schia and De Carvalho, note 121 above: 12–14.
[130] Cabrera-Balleza and Johnson, note 121 above: 25.
[131] Nagelhus Schia and De Carvalho, note 121 above: 19.
[132] The case of gender and justice reform in Somalia does, however, show that the international community (UNDP and the Danish Refugee Council in their collaboration with the Somalian justice sector and a local NGO) is aware of the challenges related to contextualisation. In 2009 a conference of traditional leaders was held, aimed at harmonising customary law with international standards and strengthening the interface between formal and informal laws. For more information see DCAF, note 69 above: 37–38.
[133] Hypermasculinity is built on an exaggeration of masculine qualities of toughness.
[134] Morrison et al., note 57 above: 42. This approach places less emphasis on absolute moral rules or the rights and obligations of actors. It is a more relativistic approach which should not be confused with the argument that the ends always justify the means. See in this regard Jack Donnelly, 'An Overview', in *Human Rights and Comparative Foreign Policy*, ed. David P. Forsythe (Tokyo: United Nations University Press, 2000): 327.
[135] Richard Fanthorpe, 'On the Limits of Liberal Peace: Chiefs and Democratic Decentralization in Post-war Sierra Leone', *African Affairs* 105, no. 418 (2005): 43–44.
[136] Anne Sisson Runyan, 'The "State" of Nature: A Garden Unfit for Women and Other Living Things', in *Gendered States. Feminist (Re)visions of International Relations Theory*, ed. Spike V. Peterson (Boulder, CO: Lynne Rienner, 1992), 135–136.
[137] Emily Grabham, Didi Herman, Davina Cooper and Jane Krishnadas, 'Introduction', in *Intersectionality and Beyond: Law, Power and the Politics of Location*, eds Emily Grabham, Davina Cooper, Jane Krishnadas and Didi Herman (Abingdon: Routledge-Cavendish, 2009): 1.
[138] Hegemonic forms of femininity include Western feminists imposing their ideas of global sisterhood on non-Western feminists or controlling research agendas through access to donor funding.
[139] Raewyn Connell, *Masculinities* (Cambridge: Polity Press, 1995).
[140] Leatherman, note 7 above: 17, 18.
[141] Daniel Bendix, 'A Review of Gender in Security Sector Reform: Bringing Post-colonial and Gender Theory into the Debate', in *Engendering Security Sector Reform: A Workshop Report*, eds Margarete Jacob, Daniel Bendix and Ruth Stanley (Berlin: Freie Universität, 2009): 18.
[142] Ibid.
[143] William K. Carroll, 'Hegemony, Counter-hegemony, Anti-hegemony', *Socialist Studies* 2, no. 2 (2006): 30.
[144] Hancock, note 14 above: 74.
[145] Leatherman, note 7 above: 15.
[146] Ibid.: 173.

[147] Fiona Robinson, 'Feminist Ethics and Global Security Governance', in *The Ethics of Global Governance*, ed. Antonio Franceschet (Boulder, CO: Lynne Rienner, 2009): 104.
[148] Ibid.: 106.
[149] Ibid.: 108.
[150] Annick T. R. Wibben, *Feminist Security Studies: A Narrative Approach* (New York: Routledge, 2011).

Chapter 4

Learning from Others' Mistakes: Towards Participatory, Gender-sensitive SSR

Rahel Kunz and Kristin Valasek

Introduction

From its inception, security sector reform (SSR) was intended to promote development and reduce poverty.[1] This primary mission has been reiterated time and again, most notably by the Organisation for Economic Co-operation and Development's Development Assistance Committee (OECD/DAC), which in a 2007 handbook calls for a 'developmental approach to SSR'.[2] This is to be achieved through various routes, such as reducing military spending and increasing economic and social expenditure; disarmament, demobilisation and reintegration (DDR) programmes; small-arms control; and fighting crime and corruption.[3] Yet current practice of SSR has been criticised for not adequately promoting development. It has been shown that SSR processes often lead to an increase in military spending, and neither necessarily reduce crime nor increase security, as is illustrated in the case of Afghanistan. Nor does SSR automatically alleviate poverty, as shown in the cases of Sierra Leone and Liberia, which after many years of SSR rank 158 and 162 respectively on the Human Development Index.[4] This has prompted statements such as by Call: 'Security system reform policy has been castigated as an "idealistic" and "unrealistic" development project, whose hubris far outstrips its achievements.'[5] Other critics have denounced SSR as a mere institution-building exercise with little to no impact on people's security and access to justice.[6] Why is it that SSR processes are seemingly unable to fulfil their original purpose?

A number of reasons have been suggested for SSR's failure to contribute adequately to development, such as a lack of resources, resistance to engaging in 'development' activities, the development community's fear of securitising development and inter-agency struggles for authority and resources between development and SSR communities.[7] In this chapter, we argue that beyond these institutional impediments there are two key elements that render it difficult for SSR, as currently practised, to contribute adequately to development: state-centrism and gender-blindness. As experience from the field of development shows, participatory and gender critiques have transformed development thinking and practice, and today participatory, gender-sensitive approaches are seen as indispensable to development and poverty reduction.[8] We suggest that SSR would do well to learn from this experience, namely through transforming discourse and practice in response to the critiques of feminist development literature and the lessons learned from the participatory turn in development. It seems that until now the majority of traditional SSR initiatives have only paid lip-service to concepts of 'local ownership' and 'civil society participation', and if gender issues are addressed at all, it is mainly in the shape of 'adding women'. Thus we argue that unless the fundamental challenges posed by gender inequality and lack of participation are taken seriously, SSR will continue its path of inadequate contribution to development.

But what is development? In current SSR discourse, the meaning of development is rarely discussed. Development is taken as something pre-defined that will flourish once security is provided. As stated in the OECD/DAC handbook, SSR aims at 'promoting an environment in which individuals and communities feel safe and secure, within which the rule of law is respected and in which sustainable development can flourish'.[9] Yet the concept of development has a long history and has been used to refer to various things, including a material condition, an immanent element of societies or individuals, a process, a prospective state to be attained, an intervention or a political project.[10] Thus it is not a fixed reality, but socially constructed and inherently contested. Similarly, security has been termed an 'essentially contested concept'. Just like development, it has a long history and can take on various meanings. The specific definitions of these terms depend on the context and the referent object, and are situated in power structures that determine who gets to define meaning.[11] Generally speaking, dominant conceptualisations of development and security have followed similar paths from state-focused macro definitions towards human- or community-based understandings with the emergence of 'human development' and 'human security'.[12] Similar to the complexity surrounding the definitions of security and development, the security-development nexus is also a contested concept. Although often assumed to be common sense, this nexus has been imbued with various meanings and used as a political strategy justifying particular interventions and interests.[13]

Acknowledging the essentially contested meanings of development, security and their nexus, instead of working with pre-established definitions, we insist that these terms only make sense in context and it should be left to individuals and collectivities to fill them with meaning. However, for the purpose of this chapter, we adhere to the notions of development and security as processes of empowerment, wherein 'beneficiaries' should define development and security needs and voice ways in which to address them.[14] As such, development and security as empowerment is more of a meta-definition: it refers to the overall goals of development and security, while leaving the concrete content to local meaning-making. Similarly, regarding the security-development nexus, we emphasise the importance of empirically establishing the interlinkages between security and development in context-specific ways. As we shall see below, the understanding of development and security we adopt here, as any other, is fraught with challenges and contradictions. Yet this definition allows us to move beyond a pre-defined, donor-driven approach.

With regard to definitions of gender and security sector reform, for the purpose of this chapter we define gender as 'socially constructed roles, behaviours, activities and attributes that a given society considers appropriate for women and men [as well as boys and girls]'.[15] Currently, the most widely endorsed definition of SSR is that of the United Nations: 'Security sector reform describes a process of assessment, review and implementation as well as monitoring and evaluation led by national authorities that has as its goal the enhancement of effective and accountable security of the State and its peoples without discrimination and with full respect for human rights and the rule of law.'[16] This is clearly a definition based upon the political landscape of the United Nations; however, an in-depth discussion of definitions is not within the scope of this chapter (see Chapter 2).

The next section of the chapter reviews the participation and gender critiques of development, and extracts key lessons learned. The third section outlines how SSR processes have generally made similar mistakes and therefore have been critiqued for state-centrism and gender-blindness. The final section explores how SSR practice can take into account the lessons learned from the field of development in the realms of needs assessment, actor identification, activity selection and monitoring and evaluation. We include brief examples from the field of SSR to illustrate what a more development-oriented practice of SSR could look like. We suggest that adopting a participatory, gender-sensitive approach would allow SSR practitioners to understand development and security in context, and address them more adequately. Thus the aim of this contribution is to examine a number of lessons that SSR could learn in order better to listen to and incorporate context-specific ways of defining 'development' and 'security', and using this as a basis for designing and implementing future SSR initiatives.

Lessons from the field of development

Since its inception, the field of development has faced a number of conceptual and practical challenges. Arguably two of the most significant of these came from the feminist and participatory movements, which have fundamentally transformed the field. This section reviews the critiques, and the transformations that development practice has undergone as a result, as well as the key lessons learned. We also touch upon a few of the contradictions and tensions of participatory and gender-sensitive approaches to development, while continuing to highlight how important these approaches are to meaningful development.

The participatory turn

A widespread frustration with the dominant state-focused, top-down and donor-driven approaches to development research and practice gave birth to a participatory turn in development. The general tenet was that development projects failed because people were left out. In a spirit of 'handing over the stick',[17] the aim was to increase the involvement of socially and economically marginalised people in decision-making over their own lives.[18] The argument was that participation would lead to more equitable and sustainable development, increase effectiveness and promote human-centred development and empowerment.[19] This was based on the assumption that 'participatory approaches empower local people with the skills and confidence to analyse their situation, reach consensus, make decisions and take action, so as to improve their circumstances'.[20]

Participatory approaches to development have taken various forms throughout their long history.[21] Early initiatives that stressed empowerment and collective local action include, for example, the New Deal in India in the 1930s and community development programmes in Latin America in the 1950s.[22] Yet it was only in the 1970s and 1980s that governments and non-governmental organisations (NGOs) were finally convinced to become interested in participatory development, resulting in a 'participation boom'.[23]

With the institutionalisation of participation, it lost much of its initial radical empowerment agenda and became a formula for making people central to development by encouraging 'beneficiaries' to participate in interventions that affect them and over which they previously had limited control or influence, i.e. to increase ownership of 'the community' over its own development.[24] In what has been called the 'participation imperative' era of the 1990s, participatory development methods became the synonym for good and sustainable development, and a key condition for funding.[25] Thus, for example, the Human Development Report of 1993, entitled *People's Participation*, states 'people's participation is becoming the central issue of our time';[26] the Brundtland Commission's report concluded that one of the main prerequisites of sustainable development is securing effective citizen participation;[27] and participation has become central to the World Bank repertoire.[28] International development agencies, donor governments and NGOs alike have come to promote participatory approaches to development.

Yet the first generation of participatory approaches institutionalised in the 1980s has been critiqued in a number of ways. Some criticise the inadequate implementation of participatory methods, including the lack of self-reflection and cultural sensitivity of development practitioners.[29] More fundamental critiques have protested against the so-called 'tyranny' of participation.[30] They argue that participatory approaches are based on simplistic notions of 'community', 'power' and 'participation'.[31] There is a long-standing debate around the concept of community.[32] Community is a 'warmly persuasive word', as Williams famously put it, with many positive connotations but no clear definition.[33] Yet community is also a highly problematic concept. As Booth notes, 'to discuss community is to enter tricky conceptual and political waters'.[34] Community is often dealt with as something natural or given, a 'harmonious and internally equitable collective', based on a 'mythical notion of community cohesion'.[35] This obscures community diversity – including age, economic, religious, caste, ethnic and gender differences – and hides a 'bias that favours the opinions and priorities of those with more power and the ability to voice themselves publicly'.[36] Critics have pointed out that 'the notion of community can be adequately and usefully apprehended only in particular historical and geographical contexts'.[37] Thus they highlight the fact that development interventions should engage with communities in context-specific ways, taking into account internal differences and discriminations.[38] This involves paying attention to the various social dynamics that structure communities.

Although some of the initial participatory efforts were based on challenging dominant power structures, institutionalised participatory development approaches have been criticised for addressing political issues through technical management solutions.[39] This happened as a result of the standardisation of approaches during a 'manual and method-oriented mania', which contradicted the original aim to move away from the limitations of blueprint planning and implementation towards more flexible and context-specific approaches.[40] This points to a more basic failure of mainstream participatory approaches to realise the complexities of power relations, which resulted in harmful outcomes of development initiatives, including manipulation, political cooptation and the reinforcement of marginalisation and the interests of the already powerful.[41] Thereby, supposedly participatory projects masked power structures within communities and between donors and 'beneficiaries', and failed to recognise the importance of power structures in determining 'whose reality counts'.[42] These projects have also been critiqued for lacking an awareness that 'local knowledge' is often shaped by local people's perceptions of what they think the agency in question is expecting to hear and deliver; hence, 'local knowledge' is never unmediated.[43] As a result of these critiques, subsequent participatory initiatives began to acknowledge the need to address power relations. If development was to be empowering, as Hickey and Mohan suggest, 'understanding the ways in which participation relates to existing power structures and political systems provides the basis for moving towards a more transformative approach to development'.[44]

Finally, practitioners and scholars highlighted the minimal consideration of gender issues and the inadequate involvement of women in first-generation participatory approaches to development (see the next section).[45] This was linked to the simplistic, gender-blind notion of 'community' on which such approaches were based, as Maguire notes: 'Gender was hidden in seemingly inclusive terms: "the people", "the oppressed", "the campesinos", or simply "the community". It was only when comparing … projects that it became clear that "the community" was all too often the male community.'[46] Thus participatory approaches often obscured women's worlds, needs and contributions to development. Instead, it was argued that greater involvement of women and attention to gender-differentiated needs and agency hold the promise of more effective and equitable processes of participatory development.[47]

These critiques of first-generation participatory approaches to development have been, at least partially, addressed through increased awareness of power dynamics and the role of the development practitioner, and developing more complex understandings of 'communities' and the social dynamics that structure them. Subsequent participatory approaches have also sought to address the gender-blindness of earlier initiatives by enabling marginal voices to be raised and heard, and 'taking account of the power effects of difference'.[48] A variety of forms of participation can be seen in today's development activities, categorised through the creation of numerous typologies of participation.[49] For instance, Cornwall distinguishes between four modes of participation: functional, instrumental, consultative and transformative (Table 1).[50] This typology illustrates the variety of participatory approaches and their implications in terms of development and empowerment. The first three modes – functional, instrumental and consultative – are merely geared towards getting people involved in already

Table 1: Modes of participation

Mode of participation	Associated with...	Why invite/involve?	Participants viewed as...
Functional	Beneficiary participation	To enlist people in projects or processes, so as to secure compliance, minimise dissent, lend legitimacy	Objects
Instrumental	Community participation	To make projects or interventions run more efficiently by enlisting contributions, delegating responsibilities	Instruments
Consultative	Stakeholder participation	To get in tune with public views and values, garner good ideas, defuse opposition, enhance responsiveness	Actors
Transformative	Citizen participation	To build political capabilities, critical consciousness and confidence; to enable people to demand rights; to enhance accountability	Agents

existing projects, delegating work or responsibility and collecting ideas, in order to secure compliance, minimise dissent and make projects more efficient and effective. In contrast, transformative participation seeks to create the conditions for meaningful participation and open possibilities for people to realise their rights and exercise voice. In this mode, participants are viewed as agents and not merely as passive 'beneficiaries'. In order to achieve development as empowerment, the aim is to move towards transformative participation. Nobody claims this is an easy task. Yet, as has been shown in the field of development, context-specific strategies to overcome the difficulties and contradictions of such an endeavour have been successfully implemented.[51]

Realising gender

Since the early 1970s feminists and women's rights activists have posed critical questions to dominant development theory and practice.[52] Initially focused on the issue of inclusion, the discourse shifted from 'women in development' (WID) to a 'gender and development' (GAD) approach in the 1990s. Early critiques have been assimilated and are now standard fare for development actors, from the UN Development Programme (UNDP) and the World Bank to USAID (the US Agency for International Development) and NGOs. However, whether or not gender and development lessons learned have really been put into practice remains the fodder of lively debates.

The first resounding gender critique of development was that women were not being consulted and included within development initiatives. As famously demonstrated by Ester Boserup, whose research served as a basis for the WID approach, early development initiatives excluded women and often had harmful impacts on their lives, including through increasing their workload.[53] The WID approach was problematically adopted by the international development community and transformed into the rationale that 'women are an untapped resource who can provide an economic contribution to development'.[54] This approach has been critiqued for focusing on women in isolation and promoting measures to have women incorporated into the market economy – thus instrumentalising them and effectively creating a double or triple burden of work.[55] The underlying assumption that 'women' are a homogeneous category of development actors has also been challenged.[56] Women's experiences of, and roles in, development differ based not only on sex, but on a multitude of factors such as age, ethnicity, geography, class, caste, sexuality and religion. As such, the focus has shifted from an essentialist inclusion of women to a broadened discussion on the participation of contextually specific marginalised groups of women, men, boys and girls in development.

The GAD approach critiqued WID's 'add women and stir' doctrine and highlighted the need to shift from a focus on women towards a comprehensive gender critique of development that takes into account the interlinked and contextual gender roles of men and women, girls and boys, and women's and girl's consequent subordination.[57] As a practical example, the GAD approach called for creating space at community level to discuss contentious gender issues which often do not make it on to the agenda of participatory meetings, such as family planning or domestic violence.[58] The GAD approach recognises that not only does development need to be participatory, but it should also contribute to the goal of gender equality rather than entrenching discriminatory gender roles. As such, development programming should address gendered power dynamics and engage both men and women to transform oppressive gender roles. However, this approach has been critiqued as too proscriptive in its quest to develop appropriate gender tools and frameworks for development, turning gender mainstreaming into a technical issue and thus failing to address overarching issues such as cultural gender stereotypes or institutionalised discrimination. Additionally, these tools and frameworks can once again be seen as externally imposed and top-down rather than enabling people to articulate and analyse their own situations.[59]

Finally, postmodernist and post-colonial feminists have critiqued the discourse and practice of development for its Eurocentrism and universal pretentions of modernity.[60] They highlight how early development initiatives employed neocolonial stereotypes of 'third world' women as tradition-bound, oppressed, exotic and backward. This trend has continued throughout the WID and GAD eras, where women in developing contexts continue to be framed as the inherently vulnerable 'other', the helpless victim in need of Western salvation.[61] Fundamentally, postmodern feminists question the neocolonialist discourse of modernity – the imperative of helping the poor, vulnerable Southern woman become 'modern like us'. Instead, they call on the field of development to acknowledge existing power relations, both in defining what development is and in relations between 'developers' and 'developed'. They call for a move towards non-orientalising discourse and practice that seek to empower women, men, boys and girls through focusing on the contextualised voices of marginalised groups and creating equitable dialogue between them and development practitioners, so they can articulate their own needs and agendas.[62]

These feminist critiques have transformed development thinking and practice. While the first reaction was to 'add women and stir', subsequent practice moved towards gender mainstreaming, such as gender budgeting initiatives. Increasingly, women have become the explicit target of particular development initiatives, including microcredit programmes. Though these initiatives have also been critically assessed for their potential reproduction of gender stereotypes and subordination, today there is a general realisation in the field (even though not adopted by all development actors) that gender is indeed relevant. In addition, it is recognised that gender-sensitive development is about more than adding women, that gender is only one (although often the dominant) among many axes of discrimination and that development should be a process of empowerment.

In summary, these two key challenges to development, participation and gender, have resulted in a fundamental transformation of development from a state-based, top-down and donor-led undertaking towards more participatory, gender-sensitive approaches. There has been a rethinking of the meaning of development, with a general shift from macro definitions focusing on economic growth towards an emphasis on human development and empowerment. Even though there are various critiques of this move, and numerous development institutions only partially adhere to this shift, there is acknowledgement that participation and gender sensitivity are conditions for meaningful development. In the following section we argue that the challenge for SSR is to learn lessons from the participatory and gender-sensitive approaches already tested in the field of development, without falling into the same pitfalls as early approaches.

Making the same mistakes?

Turning to the field of SSR, we realise that it has faced critiques similar to those levelled against development, namely for being state-centric and gender-blind. It seems that SSR has in many ways failed to learn from the past 40 years of development work. We argue that by taking into account the recent breakthroughs in the field of participatory and gender-sensitive development, SSR practice may be able to make a more meaningful impact upon development. This section analyses the SSR critiques of state-centrism and gender-blindness and begins making the link to lessons learned from development practice.

SSR as state-centric

SSR has been criticised as being state-centric in two main ways: regarding its definition of (in)security and the referent object of security, and regarding its definition of security providers. SSR practice has tended to focus on security as national security, the state being its key referent object.[63] This understanding of security tends to be expert-led and 'one size fits all', and its conception of security needs is top-down, with a focus on state- and institution-building.[64] SSR theory and practice have also been accused of being largely donor-driven.[65]

SSR has long been characterised by an almost exclusive focus on reforming or building formal state institutions. Indeed, non-state actors,[66] such as customary tribal authorities or local councils of elders, were not initially part of SSR thinking and practice, but have recently been included as a conceptual afterthought.[67] The initial failure to recognise the variety of actors neglects the extent to which people in post-conflict and developing contexts rely on non-state security and justice providers.[68] Evidence shows that non-state actors provide the majority of justice and security in 'fragile' and post-conflict environments. For example, according to the OECD/DAC, in sub-Saharan Africa non-state providers are estimated to deliver at least 80 per cent of justice services.[69] In many contexts, local and non-state (or customary) security and justice providers have been shown to be more trusted, accessible and efficient, and their enforcement capacity superior to that of state providers.[70] While customary actors are often framed as oppositional to the state, in many contexts, including Colombia, Liberia and Sierra Leone, customary actors are state-sanctioned, for instance through recognition in the constitution.[71] In such 'hybrid societies', interlinked state and customary security and justice actors, such as national police and tribal police, exist in parallel. It has to be noted, however, that this observation is not valid for all customary actors. Moreover, these actors have also been heavily criticised, including for perpetrating violence and/or discriminating against specific groups, in particular women.

Due to state-centrism, most SSR funding from international donors has gone to 'reforming' formal state security institutions, such as the police and armed forces, rather than focusing on oversight bodies such as parliament or civil society, or non-state security and justice providers.[72] Though the rules of the international assistance game make it difficult for donors to work with non-state actors that have a conflictual relationship with national government, donors can and do fund initiatives to work with armed opposition groups, and donors, in particular development agencies, have the ability to provide support to community-level customary and civil society actors.[73] Critiques of SSR state-centrism shed light on how reform of state institutions does not necessarily translate into improved access to justice and security at the community level. SSR initiatives often fund institutional reforms that may have a limited impact in the capital city, but barely touch the lives of the majority of the rural population. This approach also sidelines civil society actors, which, as demonstrated by Anderlini and Conaway, make important contributions to both security sector oversight and security and justice provision.[74] Hence, emphasising the different layers of authority in post-colonial states, Scheye encourages SSR donors to support the strengthening of local and non-state security and justice networks.[75]

These critiques have resulted in changing SSR discourse. The role of non-state actors in security provision is increasingly acknowledged. This can be illustrated through a brief comparison of the OECD/DAC SSR guidelines from 2005, where there are only six references to non-state actors, with the 2007 OECD/DAC handbook, where references to non-state actors abound.[76] The OECD/DAC handbook calls for donors to 'take a balanced approach to supporting state and non-state security and justice service provision', and warns that 'programmes that are locked into either state or non-state institutions, one to the exclusion of the other, are unlikely to be effective'.[77] In addition, the notions of 'local ownership' and 'inclusiveness' have entered SSR rhetoric. For instance, according to the OECD/DAC guidelines, SSR should be 'people-centred, locally-owned and based on democratic norms and internationally accepted human rights principles and on the rule of law'.[78] An emphasis on local ownership has also increased the mention of 'participation', which can be seen in the multiplication of OECD/DAC references from the 2005 guidelines to the 2007 handbook.

Although current SSR discourse recognises the role of non-state actors and local ownership, SSR practice is a different picture. Despite appropriating the development rhetoric of local ownership and participation, in practice SSR initiatives often fall back on to state-centric approaches. Non-state actors are frequently excluded and the norm of 'local ownership' is applied selectively, as we discuss in more detail below.[79] In addition, participation is often limited to specific civil society groups, such as capital-based NGOs that have existing contacts with the government and/or international actors. Furthermore, participation seems limited to the incorporation of particular actors in pre-defined SSR projects or in security provision, i.e. a functional mode of participation. Community or even district-level consultations regarding the definition of (in)security needs and priorities are largely absent. The individuals and communities who experience everyday insecurity tend to be excluded from the picture. Furthermore, in cases where local or non-state actors are acknowledged and included, they are often portrayed as mere service providers, which risks instrumentalising and alienating them, i.e. employing an instrumental mode of participation. This can be partly attributed to a gap between SSR discourse and practice. However, it is also a question of not having applied lessons from development practice regarding community, power and participation. It seems that in SSR practice, when participatory approaches are adopted, it is predominantly the functional, instrumental or consultative modes of participation (see Table 1). Thus in SSR participation is mainly justified on the grounds of efficiency and effectiveness, to secure compliance and sustainability,[80] rather than being part of an equitable process of empowerment. SSR processes have a few basic lessons to learn from development practice in order to move towards more transformative forms of participation.

SSR as gender-blind

In the last five years a small body of literature has emerged which specifically challenges SSR for its lack of inclusion of women and gender issues. Though very little has been written on the nexus between SSR, development and gender, clear parallels can be seen between gender critiques of SSR and those levelled against development. SSR practitioners thus have a valuable opportunity to learn from the past 40 years of gender and development experience rather than repeating many of the same mistakes. Key gender critiques of SSR include the lack of equitable participation in security needs assessment, decision-making and provision; insufficient focus on meeting the different security and justice needs of women, men, girls and boys; and the failure to transform institutional culture, including cultures of violent masculinities.

Gender critiques of SSR have their roots in decades of feminist activism, including the work of women's rights and peace activists as well as academic critiques of security and development. From deconstructing neorealist security discourse to innovative street protests against invasion, soaring food prices or violence against women, dominant understandings of security and development are being questioned and challenged (see Chapter 3). Writers on gender and SSR generally tend to draw from the schools of liberal or postmodern/post-structural feminist theory, in many ways reflecting the WID/GAD division. The former focuses on women's equal rights as a platform to call for women's equal participation in security decision-making and their equal right to security. This school often highlights the need for women to join security sector institutions (SSIs), including combat positions within the military, in order to become full citizens and security providers rather than the stereotypical 'victim needing protection'.[81] In contrast, postmodern/post-structuralist feminists argue against uncritically advocating for women's participation in security discourses and practice without an analysis of the gendered power dynamics of these discourses and structures.[82] A common critique of SSIs is their perpetuation of a culture of violent, militarised masculinities. As in the field of development, postmodern feminist critiques of SSR also challenge the simplistic categories of 'women' and 'men', and note that in addition to gender, other factors influence security access and agency, such as ethnicity, class, sexual orientation, location, religion and ability.

Under the banner of 'equal participation and full involvement',[83] SSR has been critiqued for perpetuating women's underrepresentation in security decision-making and SSIs. The lack of female participation is framed as manifest in three different realms. The most obvious is underrepresentation in SSIs, including security sector oversight bodies.[84] Second is the often even lower rate of women in positions with security decision-making power, such as inspector general of the police, minister of defence or chair of the parliamentary committee on defence and security. Third is the minimal level of women's external involvement in SSR through civil society oversight and activism. According to Sanam Anderlini, 'women are in fact highly-relevant local stakeholders seeking to influence and drive SSR processes to meet local needs … Yet international SSR practitioners, as well as local political and military leaders, have tended to sideline such groups, or ignore their relevance, as if women or civil society more generally were not central to discussions of security.'[85] In response to the critique of women's underrepresentation, security sector actors have often implemented isolated initiatives focused on increasing the number of female security sector personnel. These initiatives have been criticised for equating gender sensitivity with numerical representation and ignoring the fact that simply including more female personnel in institutions imbued with sexism is likely to endanger or coopt women rather than transform the institution.[86] In addition, arguments of 'operational effectiveness' used to promote female participation have been critiqued for bordering on essentialism and instrumentalisation. Finally, the approach of 'adding women' to SSIs fails to address women's underrepresentation in the other two realms of decision-making and civil society oversight.

A parallel gender critique, largely emerging from literature, practice and activism on the prevention of and response to gender-based violence (GBV), is that there needs to be a fundamental shift in SSR. Rather than starting with the objective of building effective and accountable SSIs, the point of departure should be the diverse security and justice needs of people.[87] Men, women, boys and girls have different security and justice needs based upon a wide range of intersecting social, cultural, political and economic factors. In order to stay faithful to SSIs' mandate of guaranteeing the security of the people and their state, SSR initiatives should be developed in response to these needs. The locus lies in the person or group of persons who have suffered violence or had their rights violated, and reforms should be constructed with the aim of improving their access to justice and security. Gender critics come to the same conclusion as critics of state-centrism: current SSR initiatives focus on institution-building rather than prioritising identifying and responding to diverse justice and security needs. In particular, feminist theorists and women's rights activists lambast SSR initiatives for not taking into account the urgent need to prevent and respond to pervasive GBV. This approach has in some cases been questioned for its discourse of victimisation.[88] By focusing (in particular) on women as potential or actual victims of violence, women's agency and resistance are marginalised – not to mention their perpetration and perpetuation of violence.[89] In addition, this line of critique is often silent on the specific security and justice needs faced by men and boys, including those who are victims of GBV (see Chapter 3).

The third central critique is that SSR processes do not go far enough with their efforts to transform the institutional culture of SSIs. Rampant corruption, impunity, human rights violations and misogyny are some of the challenges facing many institutions undergoing SSR processes – especially the armed forces, police and border guards. Forms of militarised, violent masculinities are institutionally cultivated in the name of military conduct and group loyalty, resulting in discriminatory institutional policies, structures and practices, including high rates of sexual harassment and exploitation.[90] Despite the grave need for SSR initiatives to address misogynistic and xenophobic institutional cultures, 'Even in post-conflict situations, security sector reform processes do not necessarily lead to any questioning of militarism, or of the cultures of masculinities sustained within military institutions.'[91] However, very little practical research exists on how SSR processes should go about this seemingly Herculean task.

Reviewing the gender critiques levelled against SSR, it is disappointing to see similar mistakes being repeated from the field of development. However, these similarities create an opportunity to learn from development discourse and practice. For instance, in order to steer clear of instrumentalising or essentialising women's representation, there needs to be a shift towards SSR processes that enable the participation of marginalised groups, rather than simply focusing on women's participation. This broader focus on participation recognises the multiplicity of factors influencing security and justice needs, priorities and actions. For example, depending upon the specific context, it may be that young, poor, indigenous men are most excluded from security decision-making. Active, equitable participation where 'beneficiaries' of SSR and development can set the agenda to meet their needs is an antidote to the ongoing trend of imposing development and SSR as a top-down, Eurocentric project. However, in the path towards a context-specific understanding of participation, the objectives of empowerment and gender equality should not be lost. Perhaps due to the differences in development versus security culture, gender critiques of SSR have been more tentative about outright declaring gender equality as a central objective of SSR. SSR could greatly benefit from such a conceptual shift, which at the practical level means enlisting both men and women in the effort to transform oppressive gender roles and discriminatory institutional culture.

What lessons can SSR learn from development practice?

This final section provides a few glimpses of what participatory, gender-sensitive SSR could look like. Through applying lessons learned from the field of development, practical steps can be taken to address the critiques of state-centrism and gender-blindness. We argue that this 'redeemed' form of SSR will be better able to fulfil its original purpose, namely reducing poverty and contributing to development. Though seemingly steeped in jargon, re-imagining SSR as participatory and gender-sensitive entails concrete changes in current practice, including in the realm of needs assessment, the identification of key SSR actors, SSR activities and monitoring and evaluation. To illustrate these points, we provide brief examples from the field.

Participatory and gender-sensitive security and justice needs assessment

Rather than building an SSR programme based solely on a political and institutional-level assessment, SSR processes should be solidly grounded in a participatory and gender-sensitive assessment of national and community-level security and justice needs. Though currently lip-service is paid to seeking 'the direct views of local people who are the consumers of justice and security services and who should be the ultimate beneficiaries of SSR programmes',[92] in practice, SSR assessments rarely take the time for local-level consultations. It is seen as a time-consuming, complex and costly activity that is beyond the scope of SSR assessment. As such, little emphasis is given to its importance, and SSR practitioners are offered a convenient 'way out' by the OECD/DAC handbook on SSR: 'Where community level consultations are not possible in an initial assessment, perception surveys should be included in the design of assistance programmes to provide a means of tracking progress.'[93] Paradoxically, the handbook is using the language of participatory development programming to obscure a fundamental rejection of a participatory approach to SSR in practice. By turning its back on the lessons learnt regarding participatory development practices, SSR is doomed to be a top-down, externally imposed process that fails to acknowledge the agency and authority of people who experience (in)security. Rather than aiming for transformative participation, SSR initiatives are settling for functional participation in order to gain 'buy-in', minimise dissent (i.e. 'winning hearts and minds') and apply a veneer of local ownership.

In other words, SSR donors are missing the point. If the objective is to reduce insecurity and increase access to justice, and thereby contribute to development, the starting point must be to establish a comprehensive picture of the contextual security and justice needs of different groups of people, making sure to take into account those who are most marginalised. People identifying their own security and justice needs, as well as actions to take in response to these needs, can then be the basis for developing SSR programming objectives. Participatory community-level needs assessments are not an impossibility – they are currently being undertaken by a variety of peace-building and development actors, including the UNDP and international NGOs such as International Alert and Saferworld, as well as various local civil society organisations.

When the community problem-solving group (CPSG)[94] in Tirana, Albania, got together to discuss human security and safety issues in the local community, they identified their key problems as loud music, anarchic parking, domestic violence and building safety.[95] This illustrates how locally identified concerns might differ from pre-defined donor understandings of key security threats. It also highlights that when the point of departure is a participatory needs assessment, often the theoretical distinction between requirements for development and SSR-related activities is blurred. As a result, 'traditional' development activities such as youth employment programmes or awareness-raising campaigns on domestic violence in schools may be identified as priority activities to improve community-level security. This was the case in Albania, where the UNDP-supported CPSGs identified a broad range of security/safety issues that require a mixed response of both 'security' and 'development' activities. However, community-level security and justice needs assessments must avoid the common pitfalls that have been identified in the field of participatory development, and recognise the diversity and power relations within communities and between communities and SSR practitioners as well as the often-occurring contradictions between donor agendas and community-level priorities. They also need to ask the crucial question of 'whose reality counts?'

As can be seen with the following example from the post-apartheid South Africa defence reform process, participatory, gender-sensitive community security and justice assessments can also serve as a basis for national-level SSR-related programming. The 1996 white paper on National Defence for the Republic of South Africa, exemplary in its inclusion of gender content and language, called for a defence review to take place.[96] The initial review was set to focus on operations, including doctrine, force design, logistics, armament and human resources. However, at the insistence of female parliamentarians on the Joint Standing Committee on Defence, the review process was expanded to include national consultation on defence priorities. Rather than simply holding a parliamentary hearing on the topic, the decision was made to have district-level meetings and workshops to ensure a broad range of public participation. Military planes and buses were used to transport religious and community leaders, NGO activists and representatives from women's organisations to these consultative meetings. As a result of consultation with grassroots women's organisations, previously ignored security threats were brought to the fore, namely the plight of dispossessed communities whose land had been seized for military usage, the environmental impact of military activities and sexual harassment perpetrated by military personnel. This participatory process revealed the link between security and development issues at the local level, as well as demonstrating the gendered character of security needs. In response, two new subcommittees were formed within the Defence Secretariat to address these issues and concrete reforms to the policies and practices of the armed forces were made, including efforts to reduce sexism and sexual harassment. As an outcome of the two-year participatory defence review process, the human security focus voiced at the community level was institutionalised in the defence reform. In addition, national consensus and public legitimacy had been generated for the new defence policies and structures.[97]

Participatory assessment methodologies, as demonstrated in these two examples, can serve as a basis for transformative SSR-related programming that is far more likely to meet the specific security, justice and development needs of communities through a process of empowerment.

Recognising diverse security sector actors

Based upon the results of participatory security and justice needs assessments, a broader range of actors can be identified to meet these needs. Instead of being restricted to focusing on national-level state actors, such as the police service and the formal justice system, a participatory SSR approach takes into account a range of key actors, including local and non-state actors – which is something participatory development activities have been doing for decades. Yet there are many myths about the supposed challenges of working with customary justice and security providers such as tribal authorities or community leaders.[98] They are often depicted as corrupt, politicised, lacking accountability and expertise, violating human rights and involved in battles over resources.[99] Support for non-state actors is also seen as potentially strengthening local elites,[100] as well as shifting the obligation of service provision away from the state. Even otherwise gender-blind writings on SSR are quick to mention that customary justice systems are often discriminatory against women. However, precisely the same charges can be levelled at state security and justice institutions – and this does not stop them from being seen as valid targets of SSR initiatives and funding. As customary non-state actors are currently the largest providers of security and justice in most developing and post-conflict contexts, working with them at community, regional and national levels in order to improve access to security and justice for women, men, boys and girls should be a key component of participatory SSR processes.

Though this argument is commonly voiced on paper, it is rare to see SSR initiatives engaging with customary security and justice providers. However, a three-year International Alert project in West Africa did just that. Its Human Security Project in Guinea, Liberia and Sierra Leone focused on enhancing community-level security through preventing GBV.[101] According to the project evaluation: 'Women and girls have the least access to protection and recourse to justice because of the perceived weak justice and security services in the three countries. Customary law remains the most accessible recourse for seeking justice in rural areas.'[102] Using community mapping to identify key local security and justice providers and 'beneficiaries', the project brought together community-level activists, police, judiciary, customary leaders, survivors of GBV, women's groups, youth groups, civil society organisations and community radio journalists. Project activities involving customary leaders included dialogue and focus group discussions on women's human security issues and GBV with male community leaders, chiefs (especially those who chair the local courts which are part of the customary justice system), elders, teachers, pastors and imams. In addition, three judicial training workshops brought together (for the first time) customary and statutory justice personnel in order to 'exchange ideas on how to work together to increase access to justice for women and girls at the community level'.[103]

In Colombia there is also a long-standing tradition of customary justice and security actors, many of which are recognised under national law. According to a 2011 Clingendael publication researched in cooperation with the local Centro de Recursospara el Analisis de Conflictos, there are a number of different community-level non-state justice and security providers.[104] These include *juntas de acción comunal* (local development councils), neighbourhood watch groups, *jueces de paz* (justices of the peace), *conciliadores en equidad* (mediators) and indigenous peoples' administrations. With the exception of the neighbourhood watch groups, all these actors are recognised in either the Colombian constitution or national legislation.[105] The research demonstrates that these community security and justice actors – even though they face a number of problems, such as threats from armed groups – are efficient, legitimate and trusted, whereas there is little trust in the national police and armed forces. The assessment also includes a gender dimension, exploring the gendered provision of justice and security of these non-state security actors. Finally, the assessment highlights the challenges and potential entry points for SSR initiatives as well as for external donors to fund these community-level actors. This could be a starting point for participatory, gender-sensitive SSR activities.

Activities to build trust and strengthen collaboration between community, customary and state actors

After assessing needs, identifying the key actors and establishing programmatic SSR objectives based on the needs assessment, the next step is to identify activities to meet these needs and objectives. Traditional SSR activities, such as training security sector personnel or restructuring ministries of interior, should be reoriented towards meeting the security and justice needs of the people, as well as being linked to activities with customary security providers. In addition, a participatory, gender-sensitive approach opens up the field of SSR to non-traditional community-level activities, which are typically categorised as development or peace-building activities, such as training on security issues for radio journalists or micro-grants to women's organisations. In particular, building formal and informal mechanisms for dialogue and interaction on security issues between state, customary and community groups becomes an essential activity for participatory SSR.

A Conciliation Resources pilot project in Kenema, Kailahun and Freetown districts of Sierra Leone, entitled Strengthening Citizens' Security, recognised the importance of strengthening mechanisms for interaction between SSIs and community actors.[106] From 2007 to 2008 this project involved a diverse group of local actors, including women's organisations, in defining, developing and delivering a wide range of activities aimed at making the Sierra Leonean security sector more accessible and accountable to 'ordinary' people. Activities included weekly radio episodes on security issues, roundtable discussions, a student debate series and training for women's groups and radio journalists on engaging effectively with state and customary security and justice providers. In one district the project also supported civilian visits to military barracks and football matches between civilians and armed forces personnel. In addition to these informal trust-building and information-sharing activities, the project sought to strengthen the local policing partnership boards through exchange visits in order to study practices, successes and lessons learned. These boards are a formal mechanism for interaction between civil society and the police. They consist of non-partisan, inter-religious groups that monitor police performance and act 'as a general forum for discussion and consultation on matters affecting policing and enhance public-police cooperation on crime prevention'.[107] Strengthening both formal and informal mechanisms for trust-building and collaboration between SSIs and community actors can sustain a participatory approach to the ongoing process of defining and addressing community-level security and justice needs.

These collaboration mechanisms come in many different forms, from the CPSGs in Albania or the Holywood Neighbourhood Policing initiative in Northern Ireland to the provincial and district-level security committees in Sierra Leone. They can provide a forum for state and customary security and justice providers, as well as community-based organisations and community members, to identify and discuss local security and justice needs on an ongoing basis, coordinate their activities and cooperate on specific initiatives. Yet lessons learnt from the field of participatory development show that in order to avoid marginalisation, particular care needs to be taken to ensure that women and representatives from women's organisations and other marginalised groups are able to participate fully in these mechanisms. Also, due to the power imbalance between the different participants, measures should be taken to create a safe and productive environment for discussion, to avoid community actors becoming instrumentalised, alienated or disempowered. Nevertheless, there is a need to allow conflicting ideas and dissenting voices. Thus realising a participatory and gender-sensitive approach to SSR can include activities such as trust-building exercises, creating forums for dialogue between different actors and formal mechanisms for interaction between community-level representatives of security sector institutions, local government, customary authorities and civil society organisations.

SSR monitoring and evaluation (M&E) activities can also benefit from a participatory, gender-sensitive approach. SSR M&E is often implemented in a top-down way by senior managers or external experts, based on externally pre-defined indicators of success. Participatory approaches to M&E would instead be based on flexible indicators of success, locally defined by the 'beneficiaries' and assessed by methods such as (small-scale) surveys, interviews and oral histories.[108] UNIDIR (the UN Institute for Disarmament Research) has adapted a participatory M&E methodology to evaluate its weapons collection and weapons for development programmes.[109] Findings from a UNIDIR project that tested this methodology in Mali, Cambodia and Albania indicate that it 'represents a compelling tool to ensure improved accountability and transparency in DDR and arms reduction activities'.[110] The results from Albania show that 'the use of inclusive participatory approaches can increase communities' confidence and thereby lead to better results in retrieving illegally held weapons from post-conflict societies'.[111] The caveat, as noted above, is that particular attention should be paid to the mode of participation, i.e.

instrumental and functional participation should be avoided, to encourage transformative participation and gender sensitivity.

Conclusion

This chapter began with the assertion that SSR, as currently practised, does not adequately contribute to development despite the fact that it was originally intended to promote development and reduce poverty. We argue that in order to transform the practice of SSR so it has a positive impact upon development, it is essential that SSR learns from the past 40 years of development practice, namely by adopting a gender-sensitive and participatory approach. Sadly, when comparing critiques of development with critiques of SSR, it becomes clear that SSR processes have committed many of the same mistakes made in the field of development. By charting how development theory and practice have been transformed in the face of these critiques, we identify key practices that can serve as lessons learnt for SSR.

Following in the footsteps of current development practice, we hold that SSR needs to shift away fundamentally from lip-service to local ownership and 'adding women' to a transformative approach based on self-identified security and justice needs, priorities and actions. Rather than basing SSR programming on political and institutional needs assessments, a transformative participative approach would begin with community security and justice assessments – creating a space for the 'beneficiaries' of SSR to set the agenda. A reflective assessment process can take into account existing power dynamics and ensure that the voices of marginalised groups are heard. Moreover, participation should not just be advocated in the name of increased effectiveness and efficiency, but as a path for empowerment and transformation. Participatory, gender-sensitive assessment reveals that SSR processes should move away from state-centrism to work with the full range of local-level security and justice providers, including women's organisations and customary authorities. It also emphasises the importance of support to formal and informal mechanisms for local-level collaboration between representatives of SSIs, customary authorities, community leaders and civil society groups. Yet we are not arguing for completely substituting SSR initiatives at the national and regional levels with community-level activities. Instead, we propose to take individuals and communities as a starting point for assessing the various context-specific security and development needs, and to improve linkages between different levels and sites of SSR.

Finally, a participatory, gender-sensitive approach enables us to see the links at community level between 'security' and 'development'. As illustrated in the examples from the field, the meanings of development and security are highly context-specific. Listening to the contextual understandings of development and security needs highlights the ways in which they are interlinked. This provides an opening for traditionally separated 'security' and 'development' activities to collaborate or merge, with the mutual goal of empowerment. In sum, a participatory, gender-sensitive approach would allow SSR to understand development and security and their linkages in context, and contribute to addressing these needs in ways that support the empowerment of men, women, girls and boys. In order to make this happen, SSR practitioners need to do their development homework.

Notes

[1] Clare Short, 'Security, Development and Conflict Prevention', speech at Royal College of Defence Studies, London (13 May 1998).

[2] OECD/DAC, *Handbook on Security System Reform, Supporting Security and Justice* (Paris: OECD, 2007): 13.

[3] Michael Brzoska, 'Development Donors and the Concept of Security Sector Reform', DCAF Occasional Paper no. 4 (Geneva: DCAF, 2003); Julius Ward, 'Can Security Sector Reform Alleviate Poverty?', *Polis* 3 (2010): 4–6.

[4] Ward, ibid.

[5] Charles Call, 'Conclusion', in *Creating Justice and Security after War*, ed. Charles Call (Washington, DC: USIP Press, 2006): 326.

[6] See, for example, Lauren Hutton, 'A Bridge Too Far? Considering Security Sector Reform in Africa', Institute for Security Studies Paper no. 186 (Pretoria: ISS, 2009): 2–3; Eric Scheye, *Pragmatic Realism in Justice and Security Development: Supporting Improvement in the Performance of Non-state/Local Justice and Security Networks* (The Hague: Clingendael, Netherlands Institute of International Relations, July 2009): 2.

[7] Brzoska, note 3 above: 44.

[8] Robert Chambers, 'Foreword', in *The Myth of Community: Gender Issues in Participatory Development*, eds Irene Guijt and Meera Kaul Shah (London: Intermediate Technology Publications, 1998): xvii–xx.

[9] OECD/DAC, note 2 above: 15.

[10] A detailed overview of the meaning and history of development goes beyond the scope of this chapter.

[11] Gunhild Hoogensen and Kirsti Stuvoy, 'Gender, Resistance and Human Security', *Security Dialogue* 37, no. 2 (2006): 207–228.

[12] Mark Duffield, *Human Security: Linking Development and Security in an Age of Terror* (Bonn: German Development Institute, 2006); Maria Stern and Joakim Öjendal, 'Mapping the Security-Development Nexus: Conflict, Complexity, Cacophony, Convergence?', *Security Dialogue* 41, no. 5 (2010): 5–29. Even though this shift has been judged as progressive by many, there are also voices that criticise the universalising, exclusionary, interventionist and disciplinary tendency of human security and human development discourse and practice.

[13] Stern and Öjendal, ibid.

[14] Bill Cooke and Uma Kothari, eds, *Participation: The New Tyranny?* (London and New York: Zed Books, 2001): 14.

[15] Council of Europe, Convention on Preventing and Combating Violence against Women and Domestic Violence (Istanbul, 11 May 2011), available at http://conventions.coe.int/Treaty/EN/Treaties/Html/210.htm.

[16] United Nations, 'Securing Peace and Development: The Role of the United Nations in Supporting Security Sector Reform', Report of the Secretary-General, UN Document A/62/659–S/2008/39 (New York: UN General Assembly, Security Council, 23 January 2008): 6.

[17] Cooke and Kothari, note 14 above: 2.

[18] Irene Guijt and Meera Kaul Shah, 'Waking Up to Power, Conflict and Process', in *The Myth of Community: Gender Issues in Participatory Development*, eds Irene Guijt and Meera Kaul Shah (London: Intermediate Technology Publications, 1998): 1.

[19] Ibid.; Majid Rahnema, 'Participation', in *The Development Dictionary: A Guide to Knowledge as Power*, ed. Wolfgang Sachs (London: Zed Books, 1992): 116–131.
[20] Gujit and Kaul Shah, note 18 above: 1.
[21] For an overview see Samuel Hickey and Giles Mohan, 'Towards Participation as Transformation: Critical Themes and Challenges', in *Participation: From Tyranny to Transformation?*, eds Samuel Hickey and Giles Mohan (London and New York: Zed Books, 2004): 5.
[22] Gujit and Kaul Shah, note 18 above: 3.
[23] Ibid.: 4.
[24] Ibid.: 3; Cooke and Kothari, note 14 above: 5.
[25] Gujit and Kaul Shah, note 18 above: 4.
[26] UN Development Programme, *People's Participation*, Human Development Report (New York: UNDP, 1993).
[27] Brundtland Commission, *Our Common Future* (Oxford: Oxford University Press, 1987).
[28] Paul Francis, 'Participatory Development at the World Bank: The Primacy of Process', in *Participation: The New Tyranny?*, eds Bill Cooke and Uma Kothari (London and New York: Zed Books, 2001): 72–87.
[29] Cooke and Kothari, note 14 above.
[30] Ibid.
[31] Gujit and Kaul Shah, note 18 above: 2.
[32] See for example Gerald Creed, 'Reconsidering Community', in *The Seductions of Community: Emancipations, Oppressions, Quandaries*, ed. Gerald Creed (Santa Fe, NM, and Oxford: School of American Research Press, 2006).
[33] Raymond Williams, *Keywords: A Vocabulary of Culture and Society* (London: Croom Helm, 1976): 76.
[34] Ken Booth, *Theory of World Security* (Cambridge: Cambridge University Press, 2007): 134.
[35] Gujit and Kaul Shah, note 18 above: 1.
[36] Ibid.
[37] Creed, note 32 above: 24.
[38] Bina Agarwal, 'Participatory Exclusions, Community Forestry, and Gender: An Analysis for South Asia and a Conceptual Framework', *World Development* 29, no. 10 (2001): 1623–1648; Andrea Cornwall, 'Whose Voices? Whose Choices? Reflections on Gender and Participatory Development', *World Development* 31, no. 8 (2003): 1325-1342.
[39] Gujit and Kaul Shah, note 18 above: 3.
[40] Ibid.: 5.
[41] Cooke and Kothari, note 14 above: 6.
[42] Robert Chambers, *Whose Reality Counts? Putting the First Last* (London: Intermediate Technology Publications, 1997): 101.
[43] Cooke and Kothari, note 14 above: 8.
[44] Hickey and Mohan, note 21 above: 5.
[45] Bina Agarwal, 'Re-sounding the Alert: Gender, Resources and Community Action', *World Development* 25, no.9 (1997): 1373–1380.
[46] Patricia Maguire, 'Proposing a More Feminist Participatory Research: Knowing and Being Embraced Openly', in *Participatory Research in Health: Issues and Experiences*, eds Korrie de Koning and Marion Martin (London: Zed Books, 1996): 29–30.
[47] Gujit and Kaul Shah, note 18 above: 3.

[48] Cornwall, note 38 above: 1338.
[49] See for example Agarwal, note 38 above: 1624; Cornwall, note 38 above: 1327.
[50] The table has been taken from Cornwall, ibid.
[51] Ibid.
[52] For a detailed discussion of feminist theories of development see for example Naila Kabeer, *Reversed Realities: Gender Hierarchies in Development Thought* (London: Verso, 1994); Nalini Visvanathan, Lynn Duggan, Nan Wiegersma and Laurie Nisonoff, eds, *The Women, Gender and Development Reader*, 2nd edn (New York: Zed Books, 2011).
[53] Ester Boserup, *Women's Role in Economic Development* (Sydney, NSW: Allen & Unwin, 1970).
[54] Caroline O. N. Moser, *Gender Planning and Development: Theory, Practice and Training* (New York: Routledge, 1993): 2.
[55] Supriya Akerkar, 'Gender and Participation: Overview Report' (Brighton: BRIDGE, Institute of Development Studies, 2001): 2.
[56] Chandra Talpade Mohanty, 'Under Western Eyes: Feminist Scholarship and Colonial Discourses', *Feminist Review* no. 30 (Autumn 1988): 61–88.
[57] Moser, note 54 above: 3–4.
[58] Andrea Cornwall, 'Making a Difference? Gender and Participatory Development', IDS Discussion Paper 378 (Brighton: IDS, 2000): 19.
[59] Akerkar, note 55 above: 3.
[60] Jane L. Parpart and Marianne H. Marchand, 'Exploding the Canon: An Introduction/Conclusion', in *Feminism/Postmodernism/Development*, eds Marianne H. Marchand and Jane L. Parpart (Abingdon: Routledge, 1995): 11–12.
[61] Mitu Hirshman, 'Women and Development: A Critique', in *Feminism/Postmodernism/Development*, eds Jane L. Parpart and Marianne H. Marchand (Abingdon: Routledge, 1995): 42–55.
[62] Parpart and Marchand, note 60 above: 11–20.
[63] Bruce Baker and Eric Scheye, 'Multi-layered Justice and Security Delivery in Post-conflict and Fragile States', *Conflict, Security and Development* 7, no. 4 (2007): 505.
[64] See for example OECD/DAC, note 2 above.
[65] Robin Luckham, 'Introduction: Transforming Security and Development in an Unequal World', *IDS Bulletin* 40, no. 2 (2009): 1–10.
[66] The term non-state actor, as well as the term civil society, which is sometimes used as a synonym, is problematic in many respects. As Scheye points out, the distinction between state and non-state actors is often purely analytic and does not represent reality on the ground, where many security and justice providers are hybrid forms of actors. Instead, the situation on the ground is often one of 'legal pluralism'. Eric Scheye, *Local Justice and Security Programming in Selected Neighbourhoods in Colombia* (The Hague: Clingendael, Netherlands Institute of International Relations, 2011): 1.
[67] Sanam Naraghi Anderlini and Camille Pampell Conaway, 'Security Sector Reform', in *Inclusive Security, Sustainable Peace: Toolkit for Advocacy and Action*, available at http://international-alert.org/sites/default/files/library/TKSecuritySectorReform.pdf; Peter Albrecht and Lars Buur, 'An Uneasy Marriage: Non-state Actors and Police Reform', *Policing & Society* 19, no. 4 (2009): 392; Scheye, ibid.
[68] Rita Abrahamsen and Michael C. Williams, 'Security Sector Reform: Bringing the Private in', *Conflict, Security and Development* 6, no. 1 (2006): 3.

[69] OECD/DAC, note 2 above: 11.
[70] Scheye, note 6 above: iii.
[71] Ibid.
[72] This was identified as one of the main shortcomings of SSR programmes analysed in DCAF's 2009 yearly book project. See Hans Born and Albrecht Schnabel, eds, *Security Sector Reform in Challenging Environments* (Münster: LIT Verlag, 2009).
[73] See for example the work of Geneva Call, available at www.genevacall.org.
[74] Anderlini and Conaway, note 67 above.
[75] Scheye, note 6 above: ii.
[76] OECD/DAC, 'Security System Reform and Governance', DAC Guidelines and Reference Series (Paris: OECD, 2005); OECD/DAC, note 2 above.
[77] OECD/DAC, note 2 above: 11.
[78] OECD, note 76 above: 21.
[79] Brzoska, note 3 above: 39.
[80] OECD/DAC, note 2 above.
[81] April Carter, 'Should Women Be Soldiers or Pacifists?', in *The Women and War Reader*, eds Lois Ann Lorentzen and Jennifer Turpin (New York: New York University Press, 1998): 33–34.
[82] See authors such as Cynthia Cockburn, Carol Cohen and Cynthia Enloe. See also Vanessa Farr, Henri Myrtinnen and Albrecht Schnabel, eds, *Sexed Pistols: The Gendered Impacts of Small Arms and Light Weapons* (Tokyo: United Nations University Press, 2009).
[83] UN Security Council, 'Security Council Resolution 1325 on Women, Peace and Security', UN Doc. S/RES/1325 (31 October 2000).
[84] See Kristin Valasek, 'Security Sector Reform and Gender', in *Gender and Security Sector Reform Toolkit*, eds Megan Bastick and Kristin Valasek (Geneva: DCAF, OSCE/ODIHR, UN-INSTRAW, 2008): 7–8.
[85] Sanam Naraghi Anderlini, 'Gender Perspectives and Women as Stakeholders: Broadening Local Ownership of SSR', in *Local Ownership and Security Sector Reform*, ed. Timothy Donais (Geneva: DCAF, 2008): 105–106.
[86] See Yaliwe Clarke, 'Security Sector Reform in Africa: A Lost Opportunity to Deconstruct Militarised Masculinities?', *Feminist Africa* 10 (2008): 49–66.
[87] See Annalise Moser, *Case Studies of Gender Sensitive Police Reform in Rwanda and Timor Leste* (New York: UNIFEM, 2009).
[88] Hoogensen and Stuvoy, note 11 above.
[89] See Albrecht Schnabel and Anara Tabyshalieva, eds, *Defying Victimhood: Women and Post-conflict Peacebuilding* (Tokyo: United Nations University Press, forthcoming in 2012), particularly Kristin Valasek's chapter on 'Combating Stereotypes: Female Security Personnel in Post-conflict Security Sector Reform'.
[90] See Sandra Whitworth, 'Militarized Masculinities and the Politics of Peacekeeping', in *Critical Security Studies and World Politics*, ed. Ken Booth (Boulder, CO: Lynne Rienner, 2005).
[91] Clarke, note 86 above: 63.
[92] OECD/DAC, note 2 above: 49.
[93] Ibid.
[94] Community problem-solving groups were established in ten communities across Albania in 2003 by the UNDP Support to Security Sector Reform Programme. Group members are supposed to be representative of the local community and work with local government

and the police in order to enhance human security and safety issues in the local neighbourhood. Sean E. DeBlieck, *Representation, Relevance and Interest: An Assessment of the SSSR Programme's Community Problem Solving Group* (Tirana: UNDP Albania, 7 May 2007): 1–2.

[95] Ibid.: 9.

[96] Department of Defence, *White Paper on National Defence for the Republic of South Africa: Defence in a Democracy* (Cape Town: SAN Publications, May 1996), available at www.info.gov.za/whitepapers/1996/defencwp.htm: para. 24. On gender content and language see paras 5, 11.5, 14, 18, 36 and 37.

[97] Sanam Naraghi Anderlini and Camille Pampell Conaway, *Negotiating the Transition to Democracy and Reforming the Security Sector: The Vital Contributions of South African Women* (Washington, DC: Initiative for Inclusive Security, 2004): 23–25.

[98] Scheye, note 6 above: iii.

[99] Albrecht and Buur, note 67 above: 397.

[100] Ibid.: 395.

[101] Sandra Ayoo and Alemu Mammo, 'Final Evaluation of the Human Security Inter-country Project: Liberia, Guinea and Sierra Leone' (London: International Alert, 2010).

[102] Ibid.: 3.

[103] Ibid.: 7.

[104] Scheye, note 66 above.

[105] Ibid.: 13.

[106] Conciliation Resources, 'Strengthening Citizens' Security', Peace, Security and Development Update no. 5 (Freetown: Conciliation Resources, 2008).

[107] Rosalind Hanson-Alp, 'Civil Society's Role in Sierra Leone's Security Sector Reform Process: Experience from Conciliation Resources West Africa Programme', in *Security Sector Reform in Sierra Leone, 1997–2007: Views from the Front Line*, eds Peter Albrecht and Paul Jackson (Geneva: DCAF, 2010): 201.

[108] International Institute for Sustainable Development, 'Rapid Rural Appraisal' (2000), available at www.iisd.org/casl/caslguide/rapidruralappraisal.htm.

[109] Geofrey Mugumya, *From Exchanging Weapons for Development to Security Sector Reform in Albania: Gaps and Grey Areas in Weapon Collection Programmes Assessed by Local People* (Geneva: UNIDIR, 2005): 1.

[110] Robert Muggah, *Listening for a Change! Participatory Evaluations of DDR and Arms Reduction Schemes* (Geneva: UNIDIR, 2005): xv.

[111] Mugumya, note 109 above: 1.

PART III

SSR AND DEVELOPMENT – REGIONAL PERSPECTIVES

Chapter 5

Security Sector Reform, Crime and Regional Development in West Africa

Tim Goudsmid, Andrea Mancini and Andrés Vanegas Canosa[1]

Introduction

Since the end of the Cold War, globalisation has brought many benefits. However, it is becoming obvious that it also has dark sides, such as the spread of cross-border crime.[2] Crime networks benefit from global economic integration processes and have grown and diversified in response to opportunities in various parts of the world. While the threats posed by serious crime are diverse and adversely affect states, communities and people in various ways, it is possible to make some generalisations regarding the effects. The term 'serious crime' is used here to describe criminal offences, such as violent crime, property crimes and organised crime, of sufficient gravity to mandate prolonged incarceration. According to the UN Office on Drugs and Crime (UNODC), serious crime brings violence, distorts local economies, corrupts institutions and fuels conflict.[3]

While serious crime is a global phenomenon, the threats it poses are manifested most prominently in fragile states and post-conflict situations, where security and governance institutions are limited in their capacity to respond. Some prominent examples of specific crime threats and affected countries and regions cited by the UN Security Council include opium production and trafficking in Afghanistan,[4] resource theft in the Democratic Republic of the Congo,[5] piracy off the coast of Somalia[6] and illicit drug trafficking and criminal activities in West Africa.[7] The strong links between conflict, crime and slow development are elaborated upon by the World Bank in the 2011 *World Development Report*.[8]

In December 2009 UN Secretary-General Ban Ki-moon expressed his concern regarding the threat to international peace and security posed by serious crime, after noting the association of drug organisations with brutal insurgencies, violent criminal groups and corruption, and said that 'Cooperation between Governments is lagging behind cooperation between organized crime networks.'[9] He further concluded that increased international cooperation is needed to tackle the transnational challenges posed by serious crime, since no state can face these challenges alone.

Serious crime is a complex social phenomenon with multiple causes, often linked to the development-security nexus. It therefore requires a comprehensive approach which takes into account governance, development and security aspects. In this chapter we argue that regional (and international) cooperation on security sector reform (SSR) can play a key role in combating serious crime. Because serious crime is a complex, context-specific phenomenon we perform an in-depth analysis of a single case, namely the threat posed by serious crime in West Africa and the response by the Economic Community of West African States (ECOWAS) West Africa Coast Initiative (WACI), a regional SSR initiative. By closely examining this case, we seek to draw more general conclusions on how the core principles of SSR can be practically implemented in the struggle against serious crime, in order to enhance both security and development.

Defining serious crime

To study the effects of serious crime in West Africa, it is important to define the concept first. While the term 'crime' is widely used, the concept itself remains context-specific, and there is no internationally accepted definition of what constitutes a crime. The 2004 UN Convention against Transnational Organized Crime (also known as the Palermo Convention) attempts to sidestep the issue and defines 'serious crime' as 'conduct constituting an offence punishable by a maximum deprivation of liberty of at least four years or a more serious penalty', which makes the term very context-specific.[10]

In common parlance the terms organised crime, serious crime and violent crime are used interchangeably. However, there are some key differences between them. Organised crime implies a degree of coordination and professional hierarchy, which is often lacking in the context of West Africa, where crime networks tend to be small and flexible.[11] Violent crime is also an unsuitable description, since many of the most significant criminal activities, such as drug trafficking, are not inherently violent acts, although they are generally associated with violence. We therefore choose to use the broader term 'serious crime', in line with the Palermo Convention. In the context of West Africa, crimes which can be labelled as 'serious' include drug trafficking, large-scale resource theft, human trafficking, various types of violent crime, property crime and organised crime.

The case of West Africa

The relation between serious crime, security and (regional) development consists of a complex interplay between the various factors. The use of a single case study allows for a more in-depth and nuanced approach to the issue. West Africa is an ideal case study for a number of reasons.

Firstly, the region is facing a variety of significant crime threats, including cocaine trafficking, resource theft and human trafficking, but also localised violent crime and threats posed by illicit pharmaceuticals. The scope and magnitude of some of these threats are immense in comparison with the licit economy of West Africa. The profits derived from cocaine trafficking alone are larger than the gross domestic product (GDP) of some West African states.[12] The states individually lack the capacity to deal with these threats: there is a need to tackle them at the regional level in addition to the state level.

Secondly, in recent decades West African states have been plagued by conflict, instability and underdevelopment. The security sector agencies responsible for delivering security and combating crime and violence often act as spoilers of security and development rather than as facilitators of a conducive development environment. As such, there is a significant need for improvement in security sector governance in order to support development.

Thirdly, it has been noted that SSR needs to flow from the bottom up to be effective. WACI, initiated by ECOWAS, is a unique initiative which aims to put SSR methods into practice and foster regional cooperation on SSR to tackle serious crime threats at the regional level. The goal is to improve the security and development potential in the region. Because WACI started in 2008, evidence regarding concrete outcomes and results is still too limited to give a comprehensive overview of the achievements and failures of the initiative. However, mid-term evaluations and interim reports allow us to highlight some of the key successes and disappointments thus far, and give an overview of the potential of the initiative. While caution should be exercised when drawing conclusions from a single case, approaches which have proven to be successful in the challenging security and development environment of West Africa might serve as an example to be replicated in less challenging environments.

A major challenge in writing this chapter is that reliable quantitative data on crime in West Africa are limited, especially with regard to community-level crimes. In addition, data collection standards are non-existent, meaning that all available data are of limited reliability. Information on the practical functioning and concrete results of WACI was garnered through a series of interviews with relevant stakeholders and others associated with the initiative.

The first step in our case study is to examine the ways in which serious crime affects security and development in West Africa at the human, state and regional levels. We subsequently examine the history, purpose and structure of WACI and conduct an evaluation of the impact it has had thus far, its weaknesses and its future potential. This will enable us to draw more conclusions on SSR, regional development and serious crime.

The serious crime threat to security and development in West Africa

West Africa is one of the poorest and least stable regions on earth. All but four of the 16 countries in this region are on the UN list of least developed countries.[13] Despite remarkable GDP growth figures posted by countries such as Ghana and Liberia in recent years, structural problems persist and limit prospects for sustainable and equitable growth and development. Brenton et al. note that much of the growth has been driven by a rise in global demand for primary commodities.[14] The historical record has shown that reliance on primary commodities in general and natural resources in particular can prove to be a curse rather than a blessing in areas with weak governance structures. Collier and Hoeffler find that countries dependent on oil and other minerals that can be illegally trafficked are prone to conflict and civil war,[15] a view echoed by the 2011 *World Development Report*[16] and exemplified in West Africa by Nigeria, where conflict is financed by oil 'bunkering', and the conflict in Liberia, which was largely paid for with conflict diamonds and wood.[17] All these factors also attract unscrupulous economic operators, facilitate the establishment and development of local and transnational criminal networks and foster a cultural model in which money can buy everything, including impunity, political power, social status and respectability.[18]

In addition to its economic frailty, West Africa has a history of political instability. Since independence it has experienced close to 50 successful coups and a comparable number of failed coup attempts.[19] Some of these are very recent, such as the 18 February 2010 coup in Niger. At present, of the 15 ECOWAS nations, about half are experiencing some form of instability.[20] Long-standing insurgencies are found in Côte d'Ivoire,

Senegal, Mali, Niger and, arguably, Nigeria. Both Sierra Leone and Liberia are recovering from brutal civil wars. Mauritania and Guinea recently experienced *coups d'état*. According to the UNODC, virtually every political conflict in Guinea-Bissau has criminal undertones, and it is not alone in this respect.[21] Conflicts in the region are funded with proceeds from criminal activity and crime thrives in circumstances of conflict, leading to a security-development milieu characterised by a vicious cycle of insecurity and underdevelopment.

In this context of underdevelopment and insecurity, any viable change in the governance of the security sector needs to address issues of poverty alleviation and human security, which includes protection from crime. However, the concept of human security, while very appealing at a conceptual and rhetorical level, confronts real difficulties at the level of operationalisation and implementation. The chasm between efforts to reform the security sector and the final goal of good governance within the sector can only be bridged by democratic governance processes and institutions as well as licit development opportunities.[22]

Identifying the most important crime threats

Economic weakness, political instability and insecurity make West Africa particularly vulnerable to serious crime threats. According to a UNODC threat assessment, the trafficking of cocaine has become the most important serious crime threat to security in West Africa. Over the last decade West Africa has become a stopover point for Latin American drug cartels looking to exploit the lucrative European cocaine market while avoiding European law enforcement agencies. Based on seizure data, it is estimated that 35 tonnes of cocaine are trafficked to West Africa annually,[23] with an estimated wholesale value of around US$2 billion.[24] Approximately two-thirds of this amount is subsequently transported to Europe and the rest is distributed over Africa and, in some cases, trafficked back across the Atlantic Ocean.

Some experts argue that drugs were initially traded into West Africa in small boats, after which propeller planes were used to land on deserted islands and abandoned runways before the cargo was carried by land north towards the European consumption market. Then the trade evolved with the use of cargo planes,[25] leading West Africa to become a hub not only for cocaine trafficking from Latin America to Europe but also for other trafficking flows. While some interpret recent declines in drug seizures as evidence of a reduction in the drug flow, it is also possible to attribute the decline to increased professionalisation of drug traffickers.

The second-largest threat is the theft of oil and other natural resources. In Nigeria alone, 55 million barrels of oil a year (a tenth of the total production) are lost through 'bunkering',[26] particularly in the Niger Delta. This theft represents a source of pollution, corruption and revenue for insurgents and criminal groups. West Africa also faces challenges related to human trafficking of illegal migrant workers and trafficking related to prostitution. To give an illustration, it is estimated that every year between 3,800 and 5,700 West African women are trafficked to Europe for sexual exploitation.[27]

Additional serious crime challenges are small-arms trafficking, toxic waste dumping and counterfeit cigarettes and medication, as well as various forms of violent crime. As much as 80 per cent of the cigarette market in some West and North African countries is illicit, meaning that cigarette sales in those countries chiefly profit criminals; 50–60 per cent of all medications used in West Africa may be substandard or counterfeit, contributing to health risks in a region where there is high demand for anti-infective and anti-malarial drugs; and West Africa also faces challenges related to the dumping of electronic waste (including old computers and mobile phones) which contains heavy metals and other toxins.[28]

To appreciate the magnitude of the problem, it is important to note that, in some cases, the value of those illicit flows through the region surpasses the GDP of West African states. For example, the illicit income generated from illegal cocaine trafficking or oil bunkering (approximately US$1 billion each) rivals the GDP of Cape Verde or Sierra Leone. The value of 45 million counterfeit anti-malarial tablets (US$438 million) is greater than the GDP of Guinea-Bissau; the revenue from cigarette smuggling (around US$775 million) is greater than the GDP of the Gambia. These criminal activities affect West African development potential in various ways and at human, state and regional levels.

The effects of serious crime on the human and community levels

One of the chief features of West Africa is that the criminal organisations are mirroring the licit existing power frameworks, and exploit advantageous positions close to official power to gain benefits and opportunities for illicit activities. In Latin America, by contrast, crime reflects more of a divide in the wealth distribution system, and is characterised by the emergence of a new entrepreneur class which better represents the impetus of the market and economic liberalism.

There is a strong link between violence and serious crime in West Africa, which threatens the basic physical security of individuals.[29] In a survey conducted in various African countries, 20 per cent of respondents from Liberia reported that they and/or their relatives had been the victim of one or more violent attacks over the previous year.[30] The corresponding figure for Burkina Faso was 13 per cent. In contrast, the percentage of assault victims in the United States, by no means the least violent developed country, is around 1.2 per cent.[31] The life-threatening impact of violent acts is compounded by the limited access to medical facilities. Because of the lack of social security in the region, the incapacitation or death of a working relative can be economically devastating for a family.[32] It has been noted that women carry a large share of this burden, which is difficult to bear in the patriarchal culture societies in West Africa.[33]

The UNODC report on 'Crime and Development in Africa' gives an analysis of how crime erodes social and human capital. The fear of crime and violence impedes the development of local communities, since asset accumulation is discouraged and movement is limited to reduce exposure to crime threats. As a consequence it is more difficult for those living in crime-riddled environments to find employment and educational opportunities.[34] Those who are able to find opportunities abroad will be more inclined to leave, resulting in a brain drain which makes the community even more vulnerable.

Economic opportunities are further eroded by the fact that businesses have to face increased costs from crime threats, reducing their competitiveness. Firms in sub-Saharan Africa lose a higher percentage of sales to crime and spend a higher percentage of sales on security than any other region.[35] As businesses close and local economic activity stagnates, individuals who are left in desolate crime-ridden communities with no significant opportunities and resources are easy prey for human traffickers.

The situation is worsened by the fact that many West African states are fragile post-conflict environments where poverty and social upheaval are common problems. Crime is a profitable post-conflict livelihood for ex-combatants and young people trained for violence, especially when no alternatives are available and law enforcement capacity is limited. Moreover, as discussed in more detail in Chapter 8, the criminal linkages and profitable smuggling routes created during periods of conflict tend to remain intact in peacetime.[36]

It could be argued that cross-border serious crime, such as cocaine trafficking in West Africa, fosters economic growth by bringing large amounts of capital to the less well off and into underdeveloped communities.

However, in addition to the direct damage associated with criminal activities, serious crime is likely to have a negative impact on long-term economic growth through a so-called 'resource curse' effect. As Reuter et al. argue, although criminal activities can generate profits in the short run, serious crime leads to the weakening of institutions and an increase in corruption, which in turn leads to a reduction in growth on the macroeconomic level.[37]

How serious crime threatens West African states

In West Africa the threats to security and development posed by serious crime activities at the state level can roughly be divided in three categories.

First, criminal activities can undermine the ability of states to achieve security and justice and promote development. Theft of natural resource wealth by both criminal groups and officials (or a combination of the two) is problematic in West Africa, where many states are reliant on resource extraction for their revenue stream. A concrete example of the damage done by serious crime is the case of Nigeria, which was reported to have lost approximately US$3 billion due to oil theft by criminal organisations and corrupt officials in the first nine months of 2008.[38]

The capacity of police and security institutions in West Africa is limited by resource constraints, but they are expected to combat criminals engaging in hugely profitable enterprises. As a consequence, they are vulnerable to corruption and graft. This is reflected by public perception of police officers. According to a poll in Nigeria, 71 per cent of those surveyed suspect that most or all police officers engage in corruption.[39] The same survey was also conducted in Liberia, where 52 per cent of those surveyed believed the same with regard to their police force. While accurate figures are hard to come by, it is clear that corruption is a significant problem in West Africa and, as this poll shows, public confidence in the integrity of government officials is extremely low.

Crime undermines the ability of the state to promote development by weakening the social contract between people and institutions.[40] Lack of protection from crime due to corruption destroys the trust relationship between the state and the people. Rent-seeking at all levels of government further undermines governance structures by fostering a culture of unrepresentative government, which jeopardises democratic legitimacy.[41] Such rent-seeking also comes at the expense of government spending on healthcare, education and basic social welfare, as well as security provision.

When people lose confidence in the criminal justice system, they may resort to vigilantism, which further undermines the state. One of the best-known examples in this regard is the Nigerian Bakassi Boys, a vigilante gang originally formed by local shoemakers to defend merchants against criminals. Their violent campaign against local criminals led to a reduction in crime rates and they were subsequently invited to other communities. The Bakassi Boys were eventually recognised and funded by the Nigerian government.[42] While this example might be seen as a success story, it should be noted that such gangs have crude governance structures and can easily be transformed from a pillar of the community into a scourge when they themselves start engaging in illegal activities. The endorsement and funding of vigilante groups allow local politicians to create their own private armies, which are allegedly used to intimidate political opponents and secure their positions as sources of patronage. A good example of this is the Askarawan Kwankwaso (Kwankwaso's Police) security outfit recently created by Governor Rabiu Musa Kwankwaso of Kano state in Nigeria.

Second, groups engaging in criminal activities actively challenge the government, as in the Niger Delta where groups involved in anti-government violence fund their activities with oil bunkering. These organisations exemplify the dangerous convergence of crime and politically oriented violence all over the region, which is having a disastrous direct impact on the community in terms of humanitarian consequences, but also at the state level, where resources are spent on counterinsurgency activities rather than development objectives. In addition, counterinsurgency operations executed by insensitive central government actors might engender indifference or hostility in local communities destroyed by conflict, as is happening in Nigeria.

Third, the state itself can become engaged in criminal activities, as crime infiltrates the state. There are some well-known cases of senior politicians with links to serious crime, such as a member of the Ghanaian parliament who was arrested in New York in 2005 trying to smuggle 67 kilograms of heroin.[43] Collusion of government and security institutions with serious crime activities also occurs on a larger scale. A prominent example of state involvement is Guinea-Bissau, where senior political and military figures are reportedly involved in large-scale drug trafficking. As a consequence, Guinea-Bissau is often labelled as a 'narco-state'.[44]

A graver case is the conflict linked to Liberia's Charles Taylor, who is accused of financing a civil war in Sierra Leone with criminal activities including trafficking diamonds, timber, arms and humans. The Coalition for International Justice estimates that Taylor amassed US$105–450 million through his criminal enterprises.[45] At the height of the conflict in Sierra Leone, illegal exports accounted for more than 90 per cent of its diamond trade – more than US$200 million in 2002.[46]

Regional consequences of serious crime

The conflicts in Liberia and Sierra Leone exemplify the regional implications of threats associated with serious crime and how such crime can threaten stability and security at a regional level. Though both started off as political conflicts, each was facilitated and prolonged by serious crime activities. The criminal links laid between Liberia, Sierra Leone and Côte d'Ivoire during the war persist.

Serious crime also limits development and economic integration at the regional level. The weakening of government structures through corruption and government engagement in crime limits the trust in the competence and integrity of regional partners, which might lead to a limited willingness to transfer national sovereignty to the authority of a supranational entity.

In the case of West Africa, the key regional authority is ECOWAS. Although initially conceived as a body to foster economic cooperation and development, it had to grapple with subregional conflict and insecurity from the start.[47] The effectiveness of ECOWAS in achieving development at the regional level continues to depend on the willingness of national political actors to delegate authority. However, in recent years West African leaders have generally opted for increased cooperation rather than isolation as the best response to the problem.

The West Africa Coast Initiative: SSR at the regional level

WACI is the outcome of the need of regional countries to confront the threats posed by serious crime hindering development and licit economic activities. The initiative was established in October 2008, when the ECOWAS Commission, the UNODC and the UN Office for West Africa in partnership with the European Union convened a ministerial conference in Praia to address the serious threat of drug trafficking to subregional security. These parties found that only coordinated and complementary efforts at the national, regional and international levels would make a significant impact on the activities of the cartels. WACI's initial focus was on accountable law enforcement capacity-building in Sierra Leone, Liberia, Côte d'Ivoire and Guinea-Bissau.

Responding to the need for a comprehensive and multi-stakeholder approach, with complementing mandates, WACI works in synergy to support the implementation of the ECOWAS Regional Action Plan to Address the Growing Problem of Illicit Drug Trafficking, Organised Crime, and Drug Abuse in West Africa. WACI is established as a transnational authority which overlooks the safety of people in the region and is aligned to ECOWAS plans.

Objectives of WACI

WACI is a joint programme[48] that entails a comprehensive set of activities targeting capacity-building at both national and regional levels. WACI foresees joint technical assistance programmes in the fields of law enforcement, forensics, border management, anti-money laundering and the strengthening of criminal justice institutions, contributing to peace-building initiatives and security sector reforms. WACI includes the setting up of specialised transnational crime units (TCUs) with means commensurate to the threat posed to West African states by drugs and crime, as well as the establishment or strengthening of financial intelligence units.

The overall objectives of WACI are closely aligned with the 2006 Strategy for Regional Integration for Growth and Poverty Reduction in West Africa. The Abuja Declaration, which lies at the basis of WACI, explicitly states that drug trafficking and other serious crimes in the region are not only law enforcement challenges, but also 'serious threats to the regional and national security, political, economic and social development of Member States'.[49] In addition, the document mentions that serious crime undermines the rule of law, democratic institutions and transparent governance in ECOWAS member states.

WACI was created in response to these threats. The initiative uses the comparative advantage of each partner organisation to create a comprehensive and multilateral approach, including strengthening judicial procedures to prosecute, convict and sentence criminals and ensure that existing laws against illicit drug trafficking and serious crime are in accordance with relevant international conventions, as well as measures which ensure transparency, accountability and good governance in general.

WACI aspires to enhance the safety and security of individuals by supporting a well-managed security sector that is responsive to the needs of the population. Furthermore, it seeks to promote and protect human rights, and prevent and minimise abuses and violations by the security sector through the development and maintenance of a democratic culture rooted in respect for the rule of law and human rights within security institutions. It aims to promote peace-building by ensuring that tensions in post-conflict situations are not exacerbated by impunity for perpetrators of human rights violations in the security sector. It is also important to note that WACI supports development goals by ensuring that SSR is linked to broader priorities and needs, and that effective and transparent management of the security sector minimises the risk of imposing unnecessary financial and opportunity costs on society.

In February 2010 the participating countries – Côte d'Ivoire, Guinea-Bissau, Liberia and Sierra Leone – committed to take a series of measures,[50] including signature and ratification of the relevant regional and international legal instruments aimed at combating serious crime (UN Transnational Organized Crime Convention and its three additional protocols, the three drugs conventions, ECOWAS conventions, etc.). They also agreed to implement the technical components of WACI and thereby ensure an operational response to the ECOWAS Political Declaration of December 2008 and to the 2008–2011 ECOWAS regional action plan.

Key elements in practice

The innovative element about WACI is that it is the first regional response to a common threat posed by serious crime to security and development in West Africa. Among other things, the initiative has triggered important cooperation, including political and technical engagement, between the four participating countries. There has been general sensitisation in the region, including at leadership level, to the possible impact of widespread criminal activity on security, stability and development. While the initial focus is on reforming the security sector to combat serious crime and improve security, development aims have driven its establishment and represent the longer-term objective, as is shown by the Abuja Declaration.

WACI has prompted cooperation among security institutions in tandem with the promotion of a better environment to enable development. The Freetown Commitment refers to 'the need to encourage and accelerate the economic and social development of our States in order to improve the living standards of our people',[51] which shows that fostering development is a key priority of ECOWAS.

At the technical level, WACI strives to create TCUs to enhance national and international coordination and enable intelligence-based investigations. The initiative combines regional coordination with international mentorship and local ownership to maximise both effectiveness and accountability. TCUs[52] are elite inter-agency units, trained and equipped to fight transnational serious crime and coordinate their activities in an international framework. Through the creation of a TCU, national and international cooperation is centralised in one inter-agency unit, making use of a wide array of law enforcement expertise and benefiting from the synergies of this cooperation. As experiences in the Pacific region, the Caribbean and Central and Southeast Asia have shown, this approach can significantly enhance the efficiency and effectiveness of law enforcement operations aimed at fighting transnational organised crime, which is an important step in creating a development-friendly environment.

Intelligence and border control management are fundamentally interlinked and constitute the basis for a comprehensive response to transnational security threats, especially serious crime and terrorism. Moreover, WACI facilitates South-South cooperation in this regard, and can assist in improving regulatory systems to control the manufacturing and smuggling of firearms and light weapons.

However, the key challenge for SSR is not only to ensure that the various agencies active in combating serious crime have the appropriate skills and tools to deal with serious crime threats, but also to ensure that these agencies are accountable to local communities, judicial bodies and democratic oversight mechanisms. WACI focuses on specialised capabilities while factoring in universal criteria and standards, such as respect for human rights, to promote effective interlinkages with other law enforcement agencies. Another important element in WACI is that some components of the respective national budgets are dedicated to supporting the initiative.

Evaluation of WACI

Latest developments

This section assesses the main achievements so far. Although the case study is still too immature for a conclusive analysis, it shows how the combination of development and security has improved conditions in both sectors. With ECOWAS behind the initiative and representing the framework for sustainable development in one of the poorest and most war-torn regions of the world, it is a good test of the best avenues to induce regional development through security sector reform. Based on the main lessons learned to date, security sector cooperation is necessary for economic integration, which leads to growth.

At the WACI High-Level Policy Committee inaugural session on 20 June 2011 in Dakar, the achievements and challenges of implementation in the pilot countries were discussed. Participants welcomed the commitment of subregional authorities in mobilising West African countries in the fight against transnational organised crime and the close connection established between WACI and the ECOWAS enterprise in this field. The participants emphasised the need for deeper subregional coordination as well as wider international support. They also identified a series of strategic priorities and discussed statutory issues before pledging to implement a set of recommendations, as stated below.[53]

Conscious that transnational organised crime in general and illicit drug trafficking in particular constitute a threat to peace and security, with concomitant impacts on sustainable development, WACI has been working towards dismantling criminal groups which are increasingly weakening the rule of law, democratic institutions and governance, and undermining economic development in West African states. Sierra Leone is the first country to have established a fully operational TCU. WACI has ensured functional and operational coordination between the national authorities and a good level of interconnection between WACI structures and the ECOWAS Commission by maintaining coherence with the ECOWAS action plan.

WACI so far has managed to organise a special technical and needs assessment mission in selected countries and set up the overarching WACI management structure providing coherence and better cooperation. WACI is considering its operational expansion to other West African countries, beginning with Guinea in light of its political transition, and requires wide international financial support for concrete activities in the plans directly supporting the implementation of the ECOWAS action plan. Currently, the opportunity and feasibility of creating a WACI fund are being discussed to render it more effective in mainstreaming its main components into SSR programmes and peace-building strategies in line with national priorities.

Achievements and impact of WACI

The young initiative shows some interesting developments at the political and technical levels, as well as a synergy between the two. However, at this early stage of implementation it is hard to draw conclusions on the impact of the initiative or its effects on regional or national development. Nonetheless, some of the contributions demonstrate that progress is being achieved in combating serious crime, corruption, money laundering and terrorism. Others highlight challenges, threats and the impact of these types of criminality on the development of the region, and more broadly the continent.[54] SSR initiatives implemented within the framework of the WACI programme are strictly sticking to outputs, whereas its impact is part of a theory of change entailing other variables and effects conducive to regional economic and political development.[55]

WACI is accomplishing some institutional change through the process of initiation of TCUs as basic units, meaning that different law enforcement agencies meet around a table and agree to cooperate and exchange information and plans.[56] This is considered a significant achievement and an important development in the long run, because by creating such state institutions they will contribute to better security at national and regional levels, leading to better governance and rule of law, encouraging foreign investments and reducing capital flight. With regard to the effectiveness of the TCUs, of the four pilot countries only Sierra Leone has a fully operational TCU at this stage.

The Sierra Leone TCU has made significant progress over the last six months in terms of inter-agency coordination and development of intelligence-gathering and operational capabilities. This initiative still requires huge efforts by the international community, and in particular the United Nations and donors, which are to a large extent responsible for the initial progress.[57] Nonetheless, the first signals are very promising, especially because some regional development initiatives supporting the security programme have been conceived. For instance, in some countries, like Sierra Leone, WACI has started collecting evidence of positive impacts on the security of the local society. It has reported seizures, and several crime and drug-related cases have been investigated. In addition, the UN Integrated Peacebuilding Office in Sierra Leone, working closely with local NGOs, has launched several community-based drug prevention and educational activities. Awareness and knowledge of illicit drugs among certain groups of youngsters have been increased, fostering a better sense of community and shared concern about threats to their development.

In Guinea-Bissau WACI helped to establish security institutions – which never existed after independence – namely prisons and judiciary police. Higher rates of political engagement, better cooperation, improved economic development indicators, greater income from taxation and better border control and security in selected areas suggest that WACI has an impact beyond SSR and the potential to help break the vicious cycle of insecurity, poverty and a lack of rule of law. However, progress is still fragile and many challenges persist, including poverty, unemployment and political instability.

Regional results of WACI include improvement of the capacity to fight crime on a regional scale, which is having positive effects on security at the local (individual) level. WACI contributes indirectly to disarmament, demobilisation and reintegration by making crime less feasible, removing sources of activity that create fragility and conflict and empowering younger generations within licit economic schemes. Moreover, more efficient anti-crime measures allow for increased liberalisation of cross-border trade and increased potential for development through economic cooperation and integration. Capital investment in Liberia and Sierra Leone is on the rise, with significant interest from Chinese and European companies. While it would be wrong to attribute these increases to the success of WACI alone, it does indicate that the economic environment is improving, thereby giving hope that escape from the vicious cycle of conflict and underdevelopment is now a concrete possibility.

WACI also functions as an important impetus for state-building in West Africa. WACI's approach to SSR on the regional level forces ECOWAS member states to become more accountable to each other, as it instils peer pressure and introduces common goals and standards. With WACI an advanced stage of cooperation is witnessed in areas primarily related to security institution capacity-building. Through this framework, countries of the region are also supported to adopt common procedures to ensure effective and democratic governance of the security sector. Democratic governance of the sector should ensure that security agencies and their staff meet expected standards of performance and behaviour as defined in laws, policies, practices and relevant social and cultural norms, both as agencies and as individual staff members.

Future challenges and potential of WACI

The future challenges for WACI, in terms of its potential contribution to stability, governance and development, are twofold: to advance the political and technical-level efforts of participating countries and neighbours, and to increase the efficacy of the regional organisation, ECOWAS, to capitalise on WACI and further promote trade and development initiatives in the region. In particular, a continuous commitment at all levels by each participating country, the sustainability of the reforms and the maintenance of technical cooperation, support and guidance are essential in the initial phase, in order to overcome the inertia of regional cooperation and pave the way for the creation of a beneficial development environment.

With regard to programme sustainability, there remain some challenges. The national governments are overdependent on official development assistance funds and support from international parties, while local institutions and authorities have not invested sufficiently in this initiative financially. The timid ownership and scarce investments resulting from a dependency culture may lead to the ultimate failure of the initiative after external sponsorship ceases.[58] The success of the initiative will also depend on the goodwill of governments, as well as on the skill of development partners in ensuring that the scheme can be taken over and managed within the region. Although all governments have expressed strong support for WACI, political and social instability may force them into erratic decision-making, which could jeopardise the project.

Therefore, in the context of its progressive maturity, the main challenges for WACI are political commitment, for which regional pressure and advocacy should be exercised; personnel capacity, for which continued multi-year international support is required; the capacity of ECOWAS to be more involved and acquire a leading position in triggering development dynamics in the West African region; and, lastly, the ability of donors and the United Nations to ensure that the aid provided continues to have positive rather than negative effects, and to create a satisfactory and successful exit scheme.

Conclusion and future perspectives

The West Africa Coast Initiative displays the flexibility of SSR by demonstrating the applicability of the concept in a hitherto unexplored context and using a unique modality. While SSR is generally regarded as a primarily national affair, WACI shows that its core principles can be used on a regional level to combat cross-border threats, such as serious crime. At the same time, the approach taken with WACI, its achievements thus far and its future potential can be traced back to the roots of the SSR debate and some of the core SSR principles: to ensure that the security sector acts as a pillar of the community rather than a scourge, and facilitates the establishment of a conducive development environment.

Lessons from WACI

The first key lesson to be drawn from this case study is the continuing relevance of the security-development nexus as an analytical framework, and with regard to challenges posed by serious crime. Analysis of crime threats in West Africa shows strong linkages between security and development on the human, state and regional levels. Moreover, the analysis shows that these threats are mutually reinforcing and can lead to a vicious cycle of weakness, insecurity and underdevelopment that is not confined within national borders.

The ECOWAS response to these threats, WACI, mirrors these challenges and aims to combat crime in a comprehensive manner, focusing not just on efficiency, but also on accountability, delivery of justice and community support. In addition to this security element, development initiatives have been attached to the programme to support local communities and reverse the vicious cycle of crime, insecurity and underdevelopment.

The second lesson which can be drawn from WACI is the importance of recognising the regional dimension of development and security threats. The historical record of West Africa shows that external security stresses can end up amplifying internal tensions and undermining initial settlements, preventing stability. Porous borders can provide rebels and organised criminal gangs with escape routes from security forces. This lesson is extremely valuable for many other regions of the globe, for example South and Southeast Asia, West and Central Africa, the Horn of Africa, the Middle East, Central America and the Andean region, where states face cross-border attacks by anti-government elements and non-state groups, support of neighbouring states for internal rebels, or traffickers and transnational terrorists. Many zones of insecurity and violence are concentrated in border areas. In fact, several internal conflicts are generated or fuelled by cross-border or global dynamics.

In the case of West Africa, some of the crime threats, such as cocaine trafficking, have global significance. Recognition of responsibility at the regional and global levels supported the creation of WACI. WACI's design acknowledges that accountability of the security sector is no longer a national affair, but is also required at the regional level through inter-state peer pressure and the introduction of common goals and standards. Accountability is extended to the human level through ECOWAS, which is moving towards being a community of people rather than a community of states, and consequently approaches security from the human rather than the state level.

The third lesson is that regional approaches can be more cumbersome than national or local initiatives. Coordination and cooperation on the regional level and the creation of regional governance structures acceptable to state parties are difficult and time-consuming processes which do not always have successful outcomes. As a consequence, expectations regarding concrete short-term results should be limited to operational and organisational gains rather than direct development and security outcomes. This can be problematic when funding depends on external partners and the goodwill and continued interest of donor governments.

The fourth lesson drawn from the case study reconfirms the conclusion drawn by Born regarding the discrepancy between 'ideal-type SSR' and SSR in practice.[59] That is to say, efforts aiming to make the security sector more effective are often at the core of SSR, while accountability and transparency remain more elusive in practice. Currently, the various security sector agents in West African are not oriented towards the people, often remain disarticulated from the larger society and maintain anachronistic structures.

Future perspectives

In this chapter we have provided an analysis of the threats posed by serious crime in West Africa, and a limited evaluation of a regional SSR-based approach to serious crime through the case study of WACI. While it is not possible to draw definitive conclusions on WACI at this stage, since the initiative is relatively young, we do think that the case highlights some of the key dynamics and the utility of an SSR-based approach to serious crime threats in developing country environments.

The threats to security and development in West Africa, including those posed by serious crime, are significant and make the region one of the most challenging development environments in the world. In such an environment, the final success of initiatives such as WACI, which aim to achieve results in the medium and long term, remains contingent on various endogenous and exogenous factors. However, the approach taken with WACI does show that there is significant potential, and evidence thus far indicates that WACI is improving security and governance in West Africa, as well as having an impact on the development environment.

The regional SSR-based approach to serious crime taken by ECOWAS through WACI might serve as an example for other states in regions facing transnational serious crime threats to security and development, such as Afghanistan and South and Central America. Looking at crime from a security-development perspective opens up opportunities to tackle crime challenges in a more comprehensive manner.

Notes

1 The views expressed in this chapter are those of the authors and do not necessarily reflect those of the United Nations or the UN Office on Drugs and Crime.

2 H. E. Eduardo Medina-Mora, 'Organized Crime: The Dark Side of Globalization', *LSE IDEAS Today* no. 6 (December 2010): 12.

3 UN Office on Drugs and Crime, 'The Globalization of Crime: A Transnational Organized Crime Threat Assessment' (Vienna: UNODC, 2010): 19.

4 UN Security Council, 'Resolution 1974 Extending Mandate of Afghanistan Mission' (22 March 2011).

5 UN Security Council, 'Resolution 1991 Extending Mandate of MONUSCO' (28 June 2011).

6 UN Security Council, 'Resolution 1976 Regarding the Situation in Somalia' (11 April 2011).

7 UN Security Council Presidential Statement, 'Peace Consolidation in West Africa', S/PRST/2009/20 (10 July 2009).

8 World Bank, *World Development Report 2011* (Washington, DC: World Bank): 51–68.

9 UN Security Council SC/9867, Presidential Statement (24 February 2010), available at www.un.org/News/Press/docs/2010/sc9867.doc.htm.

10 United Nations Office on Drugs and Crime, 'United Nations Convention Against Transnational Organized Crime and the Protocols Thereto' (New York: United Nations, 2004): 5.

11 Mark Shaw, 'Crime as Business, Business as Crime: West African Criminal Networks in Southern Africa', *African Affairs* 101, no. 404 (2002): 291–316.

12 UN Office on Drugs and Crime, 'Drug Trafficking as a Security Threat in West Africa', (Vienna: UNODC, 2008): 40.

13 The LDCs in West Africa are Benin, Burkina Faso, Gambia, Guinea, Guinea-Bissau, Liberia, Mali, Mauretania, Niger, Senegal, Sierra Leone and Togo. Non-LDCs in the region are Cape Verde ('graduated' in 2008), Côte d'Ivoire, Ghana and Nigeria.

14 Paul Brenton, Nora Dihel, Ian Gillson and Mombert Hoppe, 'Regional Trade Agreements in Sub-Saharan Africa: Supporting Export Diversification', *Africa Trade Policy Notes* 15 (March 2011).

15 Paul Collier and Anke Hoeffler, 'On Economic Causes of Civil War', *Oxford Economic Papers* 50, no. 4 (1998): 563–573.

16 World Bank, note 8 above: 54.

17 Emily Harwell, 'Forests in Fragile and Conflict-affected States' (Washington, DC: Program on Forests, 2010).

18 Antonio Mazzitelli, 'Transnational Organized Crime in West Africa: The Additional Challenge', *International Affairs* 83, no. 6 (2007): 1073.

19 Patrick J. McGowan, 'Coups and Conflict in West Africa, 1955–2004', *Armed Forces and Society* 32, no. 1 (2005) lists 44 successful coups and 43 failed attempts. Since then, there have been (attempted) coups in Guinea (2008), Guinea-Bissau (2010), Mauretania (2005, 2008), Togo (2005) Côte d'Ivoire (2006) and Niger (2010).

20 UN Office on Drugs and Crime, note 12 above: 235.

21 Ibid.: 236.

22 Adedeji Ebo, 'Security Sector Reform as an Instrument of Sub-regional Transformation in West Africa', in *Reform and Reconstruction of the Security Sector*, eds Alan Bryden and Heiner Hänggi (Münster: LIT Verlag, 2004): 5.

[23] UN Office on Drugs and Crime, 'World Drug Report' (Vienna: UNODC, 2010).

[24] UN Office on Drugs and Crime, 'Transnational Trafficking and the Rule of Law in West Africa: A Threat Assessment' (Vienna: UNODC, 2009).

[25] This trend was first discovered in November 2009 when a Boeing 727 landed in the Malian desert, miles from the nearest town or commercial airport. When authorities arrived, the aircraft had already been set alight, prompting authorities to speculate that it was being used to carry cocaine.

[26] 'Bunkering' is the term used for oil theft and smuggling.

[27] UN Office on Drugs and Crime, note 24 above: 41.

[28] UN Office on Drugs and Crime, note 23 above.

[29] Abdullahi Shehu, 'Drug Trafficking and Its Impact on West Africa', paper presented at Meeting of Joint Committee on Political Affairs, Peace and Security/NEPAD and Africa Peer Review Mechanism of the ECOWAS Parliament (Nigeria, 28 July – 1 August 2009), available at www.giaba.org/media/speech/09-8-4-EFFECTS_OF_DRUG_TRAFFICKING_FOR__PARLIAMENTARIANS,_09-comrs.pdf.

[30] Subah-Belleh Associates and Michigan State University, 'Afrobarometer Summary of Results: Round 4 Afrobarometer Survey in Liberia' (2008), available at www.afrobarometer.org/index.php?option=com_docman&task=doc_download&gid=751: 6

[31] UN Interregional Crime and Justice Research Institute, 'Correspondence on Data on Crime Victims' (Turin: UNICRI, March 2002). The reported figures are representative of the total population.

[32] Clive Bailey, 'Extending Social Security Coverage in Africa', ILO ESS Paper no. 20 (Geneva: ILO, 2004).

[33] Abigail Gyimah, 'Gender and Transitional Justice in West Africa: The Cases of Ghana and Sierra Leone', African Leadership Centre Research Report no. 4 (Nairobi: ALCR, 2009).

[34] UN Office on Drugs and Crime, 'Crime and Development in Africa' (2005), available at www.unodc.org/pdf/African_report.pdf.

[35] World Bank, 'Enterprise Surveys' (Washington, DC: World Bank, 2010).

[36] UN Office on Drugs and Crime, note 34 above: 37.

[37] Peter Reuter, Justin L. Adams, Susan S. Everingham, Robert Klitgaard, J. T. Quinlivan and K. Jack Riley with Kamil Akramov, Scott Hiromoto and Sergej Mahnovski, 'Mitigating the Effects of Illicit Drugs on Development', RAND Corporation Project Memorandum Series (Santa Monica, CA: RAND, 2004).

[38] Technical Committee on the Niger Delta, 'Technical Committee Report 2008', vol. 1 (November 2008), available at www.mosop.org/Nigeria_Niger_Delta_Technical_Committee_Report_2008.pdf.

[39] Practical Sampling International and Michigan State University, 'Afrobarometer Summary of Results: Round 4 Afrobarometer Survey in Nigeria' (2008), available at www.afrobarometer.org/index.php?option=com_docman&task=doc_download&gid=270: 28.

[40] Mazzitelli, note 18 above: 1086.

[41] Antonio L. Mazzitelli, 'The New Transatlantic Bonanza: Cocaine on Highway 10' (Miami, FL: Western Hemisphere Security Analysis Center, 2011).

[42] Kate Meagher, 'Hijacking Civil Society: The Inside Story of the Bakassi Boys Vigilante Group of Southeastern Nigeria', *Journal of Modern African Studies* 45, no. 1 (2007): 89–115.

[43] Stephen Ellis, 'West Africa's International Drug Trade', *African Affairs* 108, no. 431 (2009): 171–196.

[44] Ibid.

[45] Coalition for International Justice, 'Following Taylor's Money: A Path of War and Destruction' (Washington, DC: Coalition for International Justice, 2005).

[46] World Bank, note 8 above.

[47] Jimam Lar, 'The ECOWAS SSR Agenda in West Africa: Looking Beyond Normative Frameworks', KAIPTC Occasional Paper no. 24 (Accra: KAIPTC, 2009): 3.

[48] The ECOWAS Commission supports member states in the implementation of the regional action plan against drugs and crime. The UN Office for West Africa and DPA raise awareness, mobilise political support, coordinate and provide advisory services. The UNODC provides programme design, management and support, including the delivery of specialised advisory services and technical expertise to respond to the rising challenges of drugs, crime and terrorism. The UN Department of Peacekeeping Operations provides guidance and expertise on police matters based on its experience in the region. Interpol oversees the strengthening of existing national central bureaus, and where appropriate provides specialised training and operational support to law enforcement. See UNOWA factsheet on WACI, available at http://unowa.unmissions.org/Portals/UNOWA/WACI/110818%20Waci_AN_UNOWA%20oklight.pdf.

[49] ECOWAS, 'Political Declaration on the Prevention of Drug Abuse, Illicit Drug Trafficking and Organized Crimes in West Africa', Ordinary Session of the Authority of Heads of State and Government of ECOWAS (Abuja, December 2008).

[50] West Africa Coast Initiative, 'Freetown Commitment on Combating Illicit Trafficking of Drugs and Transnational Organized Crime in West Africa' (Freetown, Sierra Leone, 17 February 2010).

[51] Ibid.

[52] A cornerstone of WACI is the establishment of a transnational crime unit in each country. The national inter-agency unit gathers and analyses information, and develops operational intelligence to support its lead investigative role in the most complex crime cases. TCUs are elite units, manned with staff seconded from national law enforcement agencies, trained and equipped to fight transnational organised crime. The UNODC, UN Department of Peacekeeping Operations Police Division and Interpol provide advisory and mentoring services to selected and vetted national staff.

[53] High-Level WACI Policy Committee, 'Report of the West Africa Coast Initiative 1st Meeting' (Dakar, 20 June 2011).

[54] Annette Hübschle, 'Crime and Development – A Contentious Issue', *African Security Review* 14, no. 4 (2005): 3.

[55] William Shiyin, drug control and crime prevention officer, UN Integrated Peacebuilding Office in Sierra Leone, interview, July 2011.

[56] Aisser Al Hafedh, associate programme officer (team leader for West and Central Africa), UNODC, interview, July 2011.

[57] Darío Rubén García Sanchez, counter-narcotics and organised crime adviser, UN Integrated Peacebuilding Office in Sierra Leone, interview, July 2011.

[58] Shiyin, note 55 above.

[59] Hans Born, 'Security Sector Reform in Challenging Environments: Insights from Comparative Analysis', in *Security Sector Reform in Challenging Environments*, eds Hans Born and Albrecht Schnabel (Münster: LIT Verlag, 2009): 260.

Chapter 6

The Failure of Security Sector Reform to Advance Development Objectives in East Timor and the Solomon Islands

Derek McDougall

Introduction

This chapter focuses on security sector reform (SSR) in East Timor and the Solomon Islands with a view to determining how such reform affects development in these countries. Since the 1990s SSR has become a significant issue in many developing countries, particularly in post-conflict situations. It is important to assess SSR not just in terms of changes occurring in the security sector, but also from the perspective of the impact on the wider society. It is possible to conduct such an assessment at a global or regional level, but also in relation to particular situations where SSR has taken place. Here the focus is on East Timor and the Solomon Islands as two post-conflict situations where SSR has been relevant. The aim is to assess the extent to which SSR has occurred in each situation, showing in turn how such change and the limitations associated with change have affected development.

For the purposes of this assessment, the security sector refers to the military, police and the wider justice system, encompassing not just the organisations themselves but also the relevant aspects of the executive government and the legislative system (all located within the polity and society as a whole). In the Solomon Islands there is no official military force, hence the assessment of the security sector in that country relates primarily to the police and justice systems. In this chapter, development refers to the way in which human well-being is improved in any given situation. Given the broad nature of the concept, more specific areas of analysis have emerged, such as economic development, political development, social development and cultural development; another area of focus is international development, encompassing the international aspects of these more specific areas.[1] Of course, these 'specific' areas are still quite broad. However, for each area, and for development as a whole, there have been many debates and much research, all giving expression to the quest for improvements in human well-being in some form. At an official level we also have expressions of this quest in such measures as the Human Development Index of the UN Development Programme and objectives such as the UN Millennium Development Goals. This chapter adopts a comprehensive approach to development in the context of the security-development nexus: any aspect of development as outlined here that is affected by the situation in the security sector is relevant.

East Timor and the Solomon Islands are both suitable as case studies for assessing the impact of SSR on development because they are polities where issues relating to the security sector have been central in the breakdown of order at particular times. Of course many polities have had such experiences, and I do not claim that East Timor and the Solomon Islands are unique. However, they are situations that I have studied, and both conflicts have received significant attention in Australia where I reside. Australia has played the leading role in external involvement in both East Timor and the Solomon Islands, and has been a significant influence on SSR in both situations. The impact of SSR on development and hence on long-term security is very relevant to Australian policy towards these countries. If Australian governments wish to achieve security in its neighbouring region, some measure of development in the various Southwest Pacific island states and East Timor is imperative; one component of development in this context is the way in which it is affected by SSR.

In both East Timor and the Solomon Islands changes relating to the security sector have been difficult, but particularly so in the former case. Given the centrality of the security sector in the two countries, these difficulties in turn have a negative impact on the wider societies, undermining prospects for achieving development. While it is important to understand the particularities of the two case studies, East Timor and the Solomon Islands can also be taken as good examples of the way in which security relates to development, i.e. the security-development nexus. The situation of the security sector is not the only factor affecting development, but it is a very important one. Approaching SSR from the perspective of how it affects the wider society can result in a positive contribution to development in each context. It is thus important to assess clearly both the advances and the obstacles relating to SSR in any given situation, with particular reference to its impact on development. An important point to emerge in this study is that the attempts to reform security institutions in

East Timor have frequently focused on the institutions alone, without sufficient reference to the security sector as a whole (particularly through security governance) and with little reference to the implications for development. In the Solomon Islands there has been relatively greater emphasis on overall coordination, but the security institutions are also less important than in East Timor. In both situations the security-development link has not received a high level of explicit attention. This situation can be attributed to the failure of the key political actors in the two situations to appreciate sufficiently the importance of the link between SSR and development. At one level this has been a failure on the part of the respective governments, attributable to internal divisions and more pressing political preoccupations. At another level external actors, particularly the United Nations in East Timor and Australia in both East Timor and the Solomon Islands, have given insufficient attention to the SSR-development linkage; and even where these actors have given some attention to the linkage and have some influence, their ability to effect change is circumscribed by local factors.

After briefly reviewing SSR and the security-development nexus as general issues, this chapter proceeds to an analysis of SSR in East Timor and the Solomon Islands. This is followed by an assessment of the impact of SSR on development in the two situations. To understand this impact it is important in the first instance to understand what has been attempted in the way of SSR; this includes whether or not a specifically SSR approach has been adopted in the course of reforming security institutions. The conclusion highlights the current 'state of play' in relation to the SSR-development nexus in the two sites.

Preliminary comments

Before embarking on an assessment of SSR in relation to the security-development nexus in East Timor and the Solomon Islands it is necessary to make some preliminary comments on SSR and the security-development nexus more generally. As far as SSR is concerned, a more systematic approach to reforming security sectors emerged during the 1990s; a major impetus was the problems associated with such sectors in post-communist Eastern Europe and developing countries that had recently experienced conflict. A definition used in the report of the UN Secretary-General on UN involvement in SSR presented in January 2008 is that SSR

> describes a process of assessment, review and implementation as well as monitoring and evaluation led by national authorities that has as its goal the enhancement of effective and accountable security for the State and its peoples without discrimination and with full respect for human rights and the rule of law.[2]

This definition in turn relates to the characterisation of the security sector as encompassing

> the structures, institutions and personnel responsible for the management, provision and oversight of security in a country … [It] includes defence, law enforcement, corrections, intelligence services and institutions responsible for border management, customs and civil emergencies … [and] Elements of the judicial sector … Furthermore, the security sector includes actors that play a role in managing and overseeing the design and implementation of security, such as ministries, legislative bodies and civil society groups. Other non-State actors that could be considered part of the security sector include customary or informal authorities and private security services.[3]

The concern with the security-development nexus overlapped with the emergence of SSR on the international agenda, although the focus was somewhat broader.[4] Situations of conflict where external intervention under UN or 'coalition of the willing' auspices had occurred raised questions about how long-term security could be achieved. A narrow definition of security might deal with the immediate causes of conflict, but without the deeper causes being addressed. Longer-term approaches generally necessitated a focus on underlying economic, social and political problems. Such a focus could become part of a longer-term development strategy. The longer-term approach was also consistent with the increasing emphasis on human security, not necessarily at the expense of traditional politico-military security but certainly extending the previously dominant approach. The interests of external actors, whether as contributors to peacekeeping forces or as aid donors, were involved in the security-development nexus. While there could be a moral element to this interest (a humanitarian focus on states and societies in need), external actors were also motivated by a concern that failed or failing states could lead to increased numbers of refugees and the creation of havens for organised crime and terrorist groups.

Whether from a local or an external focus, an understanding of the security-development nexus would mean increased attention to the underlying causes of insecurity. Many different areas would need to be considered in formulating and implementing a development strategy designed to overcome or at least manage these causes. SSR could be one of these areas, but even a broad conception of SSR would be insufficient in these circumstances. A 'whole-of-society' approach based on 'whole-of-government' involvement would be necessary. Successful SSR could assist, but other dimensions would also require attention. From the perspective of the security-development nexus SSR that focused simply on reforming the security institutions would not be enough. The impact on society more generally needed to be an important focus. By focusing on development in this context one could think more explicitly about ways of ensuring that SSR improves human well-being more generally in a given situation.

In the next two sections I consider what SSR in East Timor and the Solomon Islands has amounted to, before asking how it has affected development more broadly in the two situations.

Security sector reform in East Timor

Security institutions have played a key role in East Timor's history during both the transition to independence under UN auspices (1999–2002) and subsequently as an independent state. The army, known as Falintil-Forças de Defesa de Timor-Leste (F-FNTL), had its origins in Falintil,[5] the guerrilla force that fought for independence. During the period of UN rule (UN Transitional Administration in East Timor or UNTAET) Falintil was reorganised so that it could become East Timor's official military force on the achievement of independence. In addition to the army, the United Nations established a police force in 2000, known as the Polícia Nacional de Timor-Leste (PNTL).[6] Divisions within the military and between the military and police played a crucial role in the breakdown of order in East Timor in early 2006. In response to the divisions between the military and police, the East Timorese government requested international assistance, and an international force involving Australia (the major contributor), New Zealand, Portugal and Malaysia was despatched. The United Nations also assumed a prominent role in responding to the situation through the replacement of the UN Office in Timor-Leste with the UN Integrated Mission in Timor-Leste (UNMIT) under UN Security Council Resolution 1704 of 25 August 2006. The United Nations was particularly important as an umbrella for international police contributions and in assisting with the reorganisation of policing in East Timor.

Because the key security institutions were so prominent in the breakdown of order in East Timor in 2006, post-conflict peace-building efforts needed to give particular attention to SSR. In assessing these events central questions involve how the security sector contributed to the crisis of 2006, what has been attempted in SSR subsequently and what the outcome has been. In examining the security sector and attempts to achieve SSR I focus in particular on developments relating to the military, the police and the justice system.[7] It should be noted at the outset that one of the key issues in SSR in East Timor is the blurring of the lines of responsibility between the military and the police. SSR is thus not simply a matter of organising each force in terms of effectiveness, but also entails the establishment of clearer lines of responsibility and better coordination between the two. Achieving this goal is difficult because of the army's origins in the pre-1999 struggle against Indonesia and the subsequent establishment of the police as a separate institution during the period of UN tutelage (1999–2002).

What was the involvement of security institutions in the breakdown of order in 2006?

In the period before 2006 the East Timor army and police developed as separate institutions but with overlapping responsibilities. Given the weakness of the executive and legislative bodies, the army and the police (along with the Catholic Church) were among the most significant institutions in the new state. Within a fragile political order there was space for strong-minded leaders to use both the army and the police for their own political ends. Constraints arising from the supposed authority of the legislative and executive institutions over the security sector were limited. At the same time discipline within both security institutions was weak. Such a situation could not only impede them in fulfilling their specific security responsibilities, but could also be a major factor undermining stability in East Timorese society more broadly.

In the case of the F-FDTL, at one level internal tensions related to its origins as a reconstituted version of Falintil, the anti-Indonesian guerrilla force active prior to 1999. Falintil had been dominated by men from the eastern regions of East Timor (known as *Lorosae*). Although recruitment for the F-FDTL encompassed people from all regions of the country, there was a feeling that people from the western regions (*Loromonu*) were being discriminated against.[8] This situation culminated in a group of 159 soldiers petitioning President Xanana Gusmão in January 2006 for redress of their grievances. Failing to achieve an adequate response, the 'petitioners' left their barracks within a few weeks and were joined subsequently by other soldiers. Given the level of discontent within the F-FDTL, its commander,

Brigadier-General Taur Matan Ruak, sacked 594 soldiers (almost half the army) on 16 March.[9]

Despite this action the petitioners remained a significant political group in East Timor. Demonstrations to highlight the grievances of the disaffected military group took place in Dili in late April 2006. Police attempts to control the protest and the accompanying violence involving discontented young men proved inadequate. Prime Minister Mari Alkatiri called on the army for assistance, but it lacked experience in controlling civil disorder; three civilian deaths resulted from the army's deployment on 29 April.[10] The situation in the army was further complicated by the desertion of Major Alfredo Alves Reinado, head of the military police, and 17 of his men on 3 May in protest at the shooting of civilians (this group was known as the 'mutineers').[11] Their disaffection was related primarily to politics within the military rather than the broader development context.

While there had been tensions between the army and the police before 2006, these tensions were exacerbated by the situation that developed in the early months of that year. Given the rivalry between the two, calling in the army to deal with civil disorder amounted to a vote of no confidence in the police. Overall political authority appeared to be lacking. Some political leaders appeared to be manipulating the security institutions and the security situation more broadly for their own ends. There is some suggestion that tensions between President Gusmão and the Fretilin government contributed to the failure to resolve the situation created by the petitioners' grievances in January 2006.[12] Alkatiri as an easterner had an affinity with the eastern-dominated F-FDTL, and this might explain his calling upon the army to restore order. In the case of the police the relevant minister was Interior Minister Rogério Lobato, who appeared to regard the police as a tool to facilitate his own political ambitions. It should also be noted that many of the police came from East Timor's western regions and had previously been police during the period of Indonesian rule;[13] the different regional orientations of the army and the police exacerbated their institutional rivalry.

Tensions and occasional violence between the military and the police culminated in a police attack on army headquarters outside Dili and on Ruak's residence on 24 May; on 25 May the army responded by attacking police headquarters. Police members whose safe passage from their headquarters had been negotiated under UN auspices were attacked by army members, with eight police being killed.[14] With violence spreading in this way, and the security institutions about to unravel, the East Timor government initiated a request for international assistance on 24 May ('reluctantly signed' by Prime Minister Alkatiri[15]). One month later, Alkatiri resigned 'under intense domestic and international pressure' on 26 June and was subsequently replaced by José Ramos Horta.[16] While international intervention would result in the restoration of order in the short term, the question remained as to what reforms would be instituted in the security sector to prevent a recurrence of the breakdown of April–May 2006. From a development perspective the main point emerging from the 2006 crisis concerns the weakness of the political institutions in dealing with the challenges coming from the security organisations. The authority of the government was not respected by those who were attempting to use the security sector for their own ends. While international intervention was a short-term solution to the crisis, a longer-term approach to ensure that SSR could enhance development was needed.

What has been attempted in SSR since 2006?

Gordon Peake summed up the situation after the crisis of 2006 with his statement that 'Events showed the PNTL and F-FDTL were more providers of insecurity than stability.'[17] Even though there had been some discussion before 1999 of an independent East Timor without a military (the 'Costa Rican model'), the political strength of the F-FDTL meant this option was not seriously considered either after 1999 or after 2006.[18] In the post-2006 situation the emphasis was on making both the military and the police more 'effective' as institutions, but without necessarily focusing on their impact on the broader society. The relevance of SSR for long-term development was not explicitly considered. The existence of the military and the police was taken as a given; the institutions could be reformed to enhance their security role, but the benefits for wider society were assumed rather than argued. The assumption was that a well-organised security sector, with each component having a clear role and good overall coordination, would enhance domestic security in East Timor, thus enabling long-term socio-economic and political development to occur.

In the immediate aftermath of the 2006 crisis the UN Secretary-General's report of 8 August 2006 emphasised the importance of SSR in East Timor as a means of dealing with the crisis: 'A holistic approach to the security sector will be required that coordinates reform efforts in the areas of policing and defence.'[19] Also relevant in emphasising the need for SSR was the report of the UN Independent Special Commission of Inquiry established by the UN High Commissioner for Human Rights after a request from the East Timorese foreign minister on 8 June 2006, and published on 2 October of the same year.[20] The focus in these reports was on SSR alone, without explicit reference to development issues. They were prepared by external actors, but in consultation with local actors (both official and non-governmental). Despite the emphasis on a holistic approach to SSR, in practice the focus was on the reform of the institutions themselves, downplaying institutional linkages and the impact on development.

In the context of the calls for SSR a planning document, entitled 'Force 2020', was published by the government of East Timor in May 2007.[21] This focused on the F-FDTL rather than the security sector as a whole. It should also be noted that work on this document began in 2004 at the instigation of army commander Taur Matan Ruak, and it was not a response to the 2006 crisis as such; one observer notes that 'References to the crisis are strikingly absent.'[22] Although one aim of the exercise was to 'Timorise' policy development,[23] a significant source of support for the work came from Australia.[24] There is reference in 'Force 2020' to the F-FDTL as contributing to East Timor's 'identity formation', but the precise security role of the military is not clarified adequately. The question of how the reform of the F-FDTL might relate to East Timor's development objectives more broadly is not discussed. On the assumption that the F-FDTL will continue to be an important part of East Timor's security sector, the report focuses on how the force might develop through to 2020 and beyond. There is a recommendation that F-FDTL be expanded to 3,000 personnel by 2020[25] (compared with 1,435 in January 2006 and 715 after the crisis[26]). The report did not propose conscription as such, but recommended that the law on conscription (approved in January 2007) be assessed in terms of whether it was in East Timor's interest[27] (universal conscription would result in a much bigger force than the 3,000 proposed). In relation to force structure, 'Force 2020' proposed a land force (45 per cent of the total force), a light naval fleet (35 per cent), a support and service unit (15 per cent) and a command unit (15 per cent).[28]

In the case of the police as another important component of the security sector, there has been a more explicit focus on the wider context, including SSR, with the reform process involving cooperation between the United Nations and the East Timor government. The crisis of May–June 2006 led to a withdrawal of the PNTL from Dili. As a result, international security forces had responsibility for security in the capital. An agreement between the United Nations and the East Timor government on 1 December 2006 established two main dimensions for police reform. PNTL members were to be screened and monitored as a basis for 'reconstituting' the force, while a reform, restructuring and rebuilding (RRR) plan provided for 'institutional development and strengthening'.[29] PNTL members who passed the screening would work alongside UN police (UNPOL) for six months before obtaining final certification. PNTL control in a given district would be restored once 80 per cent of PNTL officers were certified (but with UNPOL continuing to perform a support role). With many former police not making it through the screening process, this situation could also become a security problem.[30] It should be noted that in the context of its involvement in SSR, UNMIT established a security sector support unit (SSSU). Given the UN's focus on policing issues, the SSSU has concentrated mainly on that aspect of the security sector and appears to have had limited effectiveness.[31] The more politically ambitious RRR plan, which could be significant for SSR more broadly, has not proceeded very far.[32] The reintegration of former police into East Timorese society was not part of the process.

Another aspect of SSR in East Timor is the situation of the justice system. The violence of 1999 and then the further violence of 2006 have highlighted the limitations of the political order (different though it was at those two times) in upholding security. The inadequacies of the justice system are part of this general failure. The situation of the justice system within the wider security system and the link to development have not been emphasised in the reform packages. Two important reports relating to violence in the 1974–1999 period have been completed, but no action against perpetrators has occurred. The Commission for Reception, Truth and Reconciliation (in Portuguese, Commisão de Acolhimento, Verdade e Reconciliação de Timor-Leste or CAVR) was set up under UN auspices to cover the whole period and reported in December 2005. The Truth and Friendship Commission was a joint initiative of Indonesia and East Timor, established in August 2005 and reporting in July 2008. It focused mainly on 1999 and by its nature was far weaker than the CAVR. In relation to both reports, governments in East Timor have judged that the pursuit of offenders would lead to a deterioration in relations with Indonesia and thus jeopardise East Timor's security as a state. Nevertheless, the failure to pursue offenders has encouraged the development of a widespread culture of impunity in East Timor: acts of violence can be committed with perpetrators assuming that the chances of being apprehended are limited. Apprehension requires not just an effective police force but a legal system that functions properly in deterring crimes and administering justice. The failure to deal with perpetrators from the past encouraged the perception that perpetrators in the present will go unpunished. An independent report in 2009 highlighted the many challenges facing East Timor's justice system, drawing attention to such issues as the incomplete legal framework, the need to strengthen the judiciary and courts (still based on a Court of Appeal and four district courts as instituted by UNTAET) and the requirements for effective law enforcement.[33]

It is clear that SSR in East Timor has been limited in scope. Reforms relating to the army and the police have concentrated on the institutions themselves. While there has been some progress relating to the institutions, the wider issues have received insufficient attention: the SSR-development nexus is generally not dealt with explicitly. Internal rather than external security is the main threat to East Timor, with 'crime, political violence and internal instability' being most significant.[34] The argument for the Costa Rican model in this situation is on national security rather than development grounds, although there would be development implications. In the event of a threat emerging from East Timor's neighbours (meaning Indonesia or Australia), an East Timor military force would be in a weak position to offer resistance. Without a military force East Timor would have additional resources to devote to its domestic development. Tensions between the army and the police would no longer be part of political life; indeed, there would be no scope for military intervention in politics.

The argument for East Timor following the Costa Rican model will, however, remain theoretical given the position of the F-FDTL in East Timor's domestic politics. Therefore, a major challenge is to ensure an effective division of labour between the F-FDTL and the PNTL. The PNTL should have the main responsibility for internal security, with the F-FDTL in a support role; however, defining the precise terms of that support role is important. For instance, the F-FDTL has been assigned the role of protecting East Timor's maritime security. The International Crisis Group has suggested that the army, with appropriate skills, could 'help respond to natural disasters and humanitarian crises and participate in engineering and development work to benefit the population',[35] but this suggestion has not been taken up as a major focus. The government's proposal to transfer border security from the police to the army is judged risky by the International Crisis Group on the grounds that there is a large Indonesian military presence on the western side of the border; an East Timorese military presence could exacerbate tensions.[36] An effective 'division of labour' would be facilitated by stronger oversight of the security institutions by the executive and legislative arms of government. This is also important in ensuring the development and consolidation of the justice system.

The implications of the failure to achieve effective SSR for the security-development nexus in East Timor will be discussed subsequently, using the Solomon Islands as a comparison. Before getting to that point, however, I shall assess SSR in the Solomon Islands, giving particular attention to the development perspective.

Security sector reform in the Solomon Islands

In the case of the Solomon Islands the breakdown of order in the period from about 1998 to 2003 was less related to the role of security institutions as such than was the case in East Timor in 2006.[37] Nevertheless, security institutions did play a role, and hence SSR has been a focus in the period of peace-building since 2003. This section reviews what has been attempted in SSR in the Solomon Islands and what has been achieved, keeping in mind the lesser significance of security institutions as compared with East Timor, but also paying attention to the broader way in which the security sector is conceived on the whole. This allows more scope for linking security issues to questions of development, although mostly the focus is implicit. It should be noted (as indicated previously) that the Solomon Islands does not have an army.

The absence of an army points to a major difference with East Timor. As discussed, the F-FDTL occupies a significant place in independent East Timor because of the prominent role played by Falintil in the struggle against Indonesian rule in the 1975–1999 period; the colonial power before 1975 and formally sovereign until 1999 was Portugal. In the Solomon Islands independence from Britain was achieved peacefully in 1978. Keeping this very significant difference in mind, East Timor and the Solomon Islands are both small island countries (East Timor sharing a border with Indonesia), with similar levels of development and populations that are relatively small in international terms. East Timor and the Solomon Islands were ranked 122nd and 125th respectively in the Human Development Index for 2010 (both medium human development).[38] The population estimates for July 2011 were 1,177,834 for East Timor and 571,891 for the Solomon Islands, predominantly young in both cases.[39] Both countries have societies that are fragmented in terms of social organisation, language and regional identification, and have weak institutions that struggle to overcome this fragmentation.

In relation to fragmentation, the breakdown of order in the Solomon Islands in the period 1998–2003 occurred mainly on the island of Guadalcanal where the capital, Honiara, is located. Over a period of decades people from the neighbouring island of Malaita had been attracted to Guadalcanal because of the greater economic opportunities there. This led to tensions between Guadalcanalese and Malaitans that could be exacerbated in periods of economic downturn, such as the aftermath of the Asian economic crisis of 1997. Clashes occurred on Guadalcanal between the Isatabu Freedom Movement (the political movement representing Guadalcanalese) and the Malaita Eagle Force (MEF, representing Malaitans), both of which had access to arms; over 200 deaths resulted from this conflict.[40] The police force, the Royal Solomon Islands Police Force (RSIPF), was subject to politicisation and did not act to quell the disturbances. In June 2000 rogue elements of the RSIPF combined with the MEF to force the resignation of Prime Minister Bartholomew Ulufa'alu (himself a Malaitan but regarded by the militants as an obstacle), and subsequently a change of government. The Townsville Peace Agreement of October 2000, brokered by the Australian government, was designed to facilitate reconciliation and involved some low-level international monitoring. By 2003 it was clear that the parties in the Solomons were so deeply divided that they were unable to effect long-term reconciliation on their own. Despite initial hesitation by the Australian government, by mid-2003 the decision had been made for Australia to lead an international operation to restore order and ensure more effective long-term governance in the Solomon Islands. While security was the first consideration, in the longer term issues of development strategy would also be important. Operation Helpem Fren, organised under the auspices of the Regional Assistance Mission to Solomon Islands (RAMSI), obtained prior approval from both the Solomons' government and parliament. International legitimacy came from the approval given by the Pacific Islands Forum (PIF) in a meeting of its foreign ministers on 30 June 2003. Although RAMSI initially had a large military component (about 70 per cent of the original commitment of 2,225), in the long term the most important element was the police; most personnel came from Australia, but New Zealand also contributed significantly along with representation from a range of PIF countries. Regional countries were concerned that the situation in the Solomon Islands had regional security implications. In the longer term development issues needed to be addressed to prevent any further breakdown in security, although this was not the immediate focus.

Shahar Hameiri has argued that RAMSI represented not just an attempt to reconstruct the Solomon Islands as a state, but entailed a new form of transnational governance whereby Australia would have a continuing role in ensuring that a certain kind of model was followed in the Solomons.[41] This model involved neoliberal economic assumptions and a centralised 'whole-of-government' approach as practised by the national government in Australia. While it might be argued that this model is neocolonial, it can be contended that the arrangement is subject to the consent of the various parties involved. The approach adopted by RAMSI provides the context for understanding what has been attempted in relation to SSR in the Solomon Islands. In formal terms RAMSI reports to the PIF, but in practice the Australian government is the key actor. RAMSI is organised on the basis of three pillars, covering law and justice, economic governance and the machinery of government (essentially the organisation of government departments); the general aim in each pillar is to build indigenous capacity. Each pillar is headed by an Australian public servant, and there is also a special coordinator from the Australian Department of Foreign Affairs and Trade. The coordinator not only facilitates an integrated approach among the three pillars but also liaises with the Solomons' government, the PIF and its members, and the Australian government's RAMSI interdepartmental committee.[42] Despite the goal of building indigenous capacity, it is clear that through RAMSI Australia remains the driving force in upholding peace and stability in the Solomon Islands. According to a Wikileaks report in August 2011, an assessment by the US embassy in Port Moresby stated that following any withdrawal of RAMSI 'it would take about a week for trouble to break out since none of the underlying issues [which caused widespread ethnic violence] have been addressed'.[43]

As far as SSR is concerned, it does not as such feature in the programme of reforms instituted by RAMSI. However, many issues coming under RAMSI's law and justice pillar are relevant to SSR as normally defined. Achieving change within the military forces is not relevant since the Solomon Islands does not have an organised military. The emphasis rather has been on reforming the police, corrections service and justice system, and (at a cross-pillars level) combating corruption. Generally, the emphasis has been on individual institutions, although there is some attention to overall goals for the Solomon Islands. These goals relate essentially to the achievement of peace and stability in the country, assuming that long-term development (including institutional strengthening) is a necessary part of the strategy required for moving in this direction. While the Solomon Islanders can be assumed to share these goals (whatever their differences over questions of strategy and implementation), it is also clear that Australia through RAMSI regards these goals as very important for its own security and as part of its international 'responsibility'; other Pacific island countries and New Zealand, expressing themselves through the Pacific Islands Forum in particular, similarly view the situation in the Solomon Islands as having regional implications. Although there can be differences with Australia over matters of emphasis and implementation, the goals of peace, stability and development for the Solomon Islands have wide support across the region at a general level.

In the short to medium term RAMSI has been relatively successful in restoring law and order in the Solomons. Initially there was a major focus on disarmament, demobilisation and reintegration programmes. This involved collecting illegal weapons and apprehending militant leaders and criminals; special constables in the RSIPF were demobilised and given opportunities for counselling and training to assist with reintegration into the local economy.[44] Where the local economy was experiencing difficulties, however, such reintegration would clearly be difficult. A longer-term focus for RAMSI was to rebuild the RSIPF, with the removal of corrupt and incompetent officers and the implementation of a mentoring and 'partnership' approach between the RSIPF and the participating police force from PIF countries. RSIPF personnel amounted to 1,050, with 250–300 in the PIF force (predominantly from Australia's International Police Deployment Group).[45] In the case of the corrections service, this system had virtually collapsed in the Solomons. RAMSI has been able to institute a new approach to corrections with rehabilitation as the main aim; a new corrections centre has been built on Malaita in response to the needs of that island, and other facilities refurbished.[46] In relation to the justice system the aim has been to ensure that all Solomon Islanders have access to justice through properly functioning courts from the local level up to the Superior Court of Record and Court of Appeal. Specific issues have included the completion of trials from 'the tensions' (the local term for the 1998–2003 period), and facilitating reforms through the Solomon Islands Law Reform Commission. Anti-corruption activities have underpinned RAMSI's approach in relation to all three pillars, applying as much to economic governance and machinery of government as to law and justice as such. In so far as corruption is antithetical to development, this approach can be expected to have positive implications for development.

The partnership agreement between the Solomons' government and RAMSI as approved by the Solomons' parliament in November 2009 provides a good indication of the stage reached in relation to SSR as well as for other aspects of RAMSI involvement.[47] Assuming that RAMSI's 'law and justice' pillar is the main focus for SSR, there was a continued emphasis in this agreement on both the RSIPF and the correctional service with a view to their functioning 'effectively and independently of RAMSI'. The justice system should be 'capable and independent', with attention to capacity-building and law reform; strengthening 'traditional justice mechanisms in rural areas' should also be a goal (thus contributing to the strengthening of traditional society as a development aim). Anti-corruption would continue to be a goal across all areas. Clearly, over a period of six years there was a judgement by RAMSI and the Solomons' government that progress had been made in relation to the goals originally set for RAMSI, but a continuing role for RAMSI was clearly envisaged. Strengthening local institutions as specified in the document would enable a reduction in this role. However, consistent with Hameiri's analysis, a reading of the agreement suggests that RAMSI's role in the Solomons is likely to be a long-term one. This has implications for SSR in relation to development as well as to the various other activities covered by RAMSI. Overall, perhaps influenced by RAMSI, there has been some attention to long-term strategic goals in relation to institutions within the security sector, as well as other aspects of government. Nevertheless, the emphasis has been on reforming the institutions themselves rather than having a more strongly coordinated 'security sector' approach. The focus on goals has allowed some scope for development issues to be considered, but attention to the security-development nexus as such has been limited.

The implications of this situation for the security-development nexus are examined in the next section in the context of a comparison with East Timor.

Security sector reform in East Timor and the Solomon Islands in relation to the security-development nexus

At this point we need to put the situation relating to SSR in East Timor and the Solomon Islands into the context of the security-development nexus.[48] Whatever happens with the security sector in these two countries will have implications for development irrespective of whether or not the security institutions are explicitly coordinated as a security sector. If discussions of the security-development nexus have a particular concern for post-conflict situations and strategies for peace-building, it is important to ask how SSR contributes to the improvement of people's lives. In other words, it is insufficient to focus on reforms within the security sector as ends in themselves, enhancing the effectiveness of the sector as such: one must ask whether SSR enhances human well-being in any given situation.

In both East Timor and the Solomon Islands there has been a widespread recognition that what occurs in relation to security institutions has implications for development. This can be seen in the recognition that has occurred, albeit at a very general level, of the link between security and development in various development plans and also in the way in which external actors (such as donor countries or security partners) have attempted to highlight this issue. When one examines the way in which these general intentions are implemented the situation becomes more problematic. One needs to assess what has actually occurred in relation to SSR in both East Timor and the Solomon Islands to see whether the impact has been positive or negative in relation to development. In both situations I shall outline the general orientation concerning the SSR-development link, and then attempt a more precise assessment as to how the link has worked out in practice.

In the case of East Timor the circumstances of the 2006 crisis have led to some attention being given to the reform of security institutions in official long-term development plans. The government that came to office in 2007 gave some priority to dealing with a number of the specific issues relating to the security sector that had been prominent in the breakdown of the previous year. At the same time the programme guidelines relegated issues that might be construed as SSR to the latter part of the document outlining the programme.[49] There was a statement that 'Internal stability and the security of the people and properties … [were] crucial elements … for the social peace and serenity of citizens … [and] also a sine qua non condition for the development of any country.'[50] Consequently the government aimed to 'confirm the authority of the State', promote 'national cohesion' and give 'special attention to fighting violence generated by organized groups'.[51] The strategic development plan for 2011–2030, a summary of which was published in April 2010, focused mainly on economic and social issues.[52] Included is a statement that 'Fragile institutions of the state will have to be strengthened',[53] but there is no explicit attention to SSR as such. Among the external actors, through its Development Assistance Framework for 2009–2013 for East Timor the United Nations complements the approach taken by the government of the country. Among the three major outcomes set under this framework, Outcome 1 focuses on 'democratization and social cohesion' and is most relevant to SSR. Outcome 1.3, for instance, specifies that 'Timorese society is better able to internalize democratic principles and use non-violent mitigating mechanisms.'

While there is recognition of the importance of SSR for development in official documents issued by East Timor and the United Nations, in practice the focus has been on effecting changes relating to the military and the police on the assumption that 'putting the house in order' in both cases will be beneficial for development. However, the emphasis has been on the institutions themselves rather than on the link with development. Whether or nor there is explicit attention to the SSR-development linkage, there will be an impact. Assessments of the security situation in East Timor indicate a marked improvement, yet this outcome is not necessarily due to SSR. There appear to be some improvements in socio-economic circumstances, mainly due to the way in which funds flowing from oil and gas development have been used to finance new infrastructure, social programmes and employment opportunities. A 2009 AusAID assessment of progress towards achievement of the Millennium Development Goals in the small states near Australia suggested a bleak picture in relation to East Timor, with the country described as 'off track' on six out of seven goals and 'of concern' in the goal relating to gender equality and the empowerment of women.[54] By 2011 AusAID was reporting that East Timor had improved 42 places in the UN's Human Development Index in 2010 as compared with 2009, and that there were improvements in maternal and child health and school enrolments.[55]

Despite these improvements it remained the case that the reform of security institutions was not contributing effectively to development goals in East Timor. Although the 2009 and 2010 Timor-Leste and development partners' meetings had as their theme 'Goodbye conflict, welcome development', such optimism was premature. At the 2009 meeting UNMIT in its report stated that 'the UN System cannot underscore enough the importance of putting in place a legislative and policy framework that defines the responsibilities of each of the institutions of the sector. A clear demarcation of roles between the PNTL and the F-FDTL is particularly critical.'[56] However, as the previous assessment of SSR in East Timor indicates, there has been limited progress toward achieving this goal. While some socio-economic progress appears to have had beneficial consequences for security, the overall situation remains fragile. The F-FDTL and the police have both been undergoing changes, as indicated, but the F-FDTL remains a strong institution politically and the issue of an appropriate relationship between the military and the police remains unresolved. The fact that political institutions are weak in East Timor makes it difficult to ensure that the security sector conforms to the rule of law; this weakness also extends to the judicial institutions. In these circumstances one cannot be sure that in more difficult socio-economic circumstances powerful individuals within the security institutions might not act to uphold their own positions irrespective of whether this runs counter to the constitutional framework.

Greater security, deriving from SSR but not confined to SSR (since security extends beyond the security institutions themselves), would enhance people's sense of well-being and thus contribute to development. The problem in East Timor is that the F-FDTL is such a strong institution that it is difficult to effect more than modest reforms. While many small states function without military forces, such a goal is not practicable in East Timor. Having one security force alone (perhaps a police force, but with some enhanced capacity to deal with paramilitary situations) would overcome the problem of insecurity arising from police-military tensions. Yet such a goal does not resemble 'practical politics' in East Timor. This situation will therefore detract from development. Maximising cooperation between the two institutions, with clear delineation of areas of responsibility and each institution functioning clearly within that framework and subject to the legislative and executive institutions, is the desirable goal. Such an outcome from SSR would be beneficial for development in East Timor, but the record so far is mixed.

If well-coordinated SSR had paid due attention to the implications for development, one would have above all expected that the prospects for political stability in the country would be greatly enhanced. Given that both the military and the police will continue as major institutions within East Timor, there is a strong need for them to be clearly subordinate to the overall security governance that derives from the political structures of the state. Political stability provides a foundation on which it is possible for East Timor to focus more clearly on establishing strategies for moving towards supporting the various aspects of development that have been prioritised within the national political process, enunciated in such documents as the programme of the 2007–2012 government and the strategic development plan for 2011–2030.[57]

In the Solomon Islands SSR has been approached differently than in East Timor. In East Timor problems in the security sector as such were central to the breakdown of order in 2006. In the Solomon Islands the security sector was relevant to the way in which the situation evolved during the 'tensions' from 1998 to 2003, although it was not a major source of conflict. Under RAMSI, since 2003 there has been an emphasis on restoring security while also giving attention to long-term development designed to overcome and prevent the problems that resulted in the 'tensions'. SSR, or more specifically the reform of security institutions, has been one element of an overall national strategy that pays some attention to development issues. The aim has been to ensure that the security sector (meaning primarily the police and the justice system in this context) can deal effectively with any challenges to security in the Solomon Islands. Achieving this goal would contribute to human well-being and thus development. This approach to SSR has been evident in key documents for development strategy, beginning with the National Economic Recovery, Reform and Development Plan, 2003–

2006, issued by the Solomon Islands government in October 2003.[58] This document identified 'Normalising law and order and security situation' as one of five key strategic areas. One of the objectives listed for this strategic area was the re-establishment of police services in all provinces by 2004. There was an assumption that SSR would contribute to development without this being spelt out in detail.

Achieving SSR in the Solomons proved more complicated than the initial document for development strategy had assumed. In its 2009 annual performance report RAMSI indicated that the goal of the RSIPF of being 'capable of independently carrying out its mandated functions under law and order and targeting corrupt conduct' had 'not yet been reached', although there was 'evidence of considerable progress in that direction'.[59] Matthew Allen argues more directly that 'The RSIPF, deeply fractured and compromised during the 1998–2003 conflict, remains unarmed, and is years away from being able to carry out its mandated functions.'[60] Essentially the Solomon Islands continued to rely on RAMSI for the provision of security services. Regarding the correctional services, the 2009 performance report argued that capacity development had been strong.[61] However, progress in the law and justice sector was judged 'the weakest in RAMSI': 'The law cannot currently be administered without further assistance and it appears unlikely to be achievable by 2013.'[62]

From 2009 a partnership framework between the Solomon Islands government and RAMSI provided a means for the two parties to assess progress towards key goals with a view to determining whether further responsibilities could be transferred from RAMSI to the government.[63] Assessments of the situation in the Solomon Islands suggested the likelihood of an economic downturn over the next few years, linked particularly to the likely decline of the logging industry.[64] A worsening economic situation would likely revive the kinds of tensions that led to the period of conflict from 1998 to 2003. It might also be noted that the Solomon Islands, much like East Timor, has made limited progress towards achieving the Millennium Development Goals. AusAID's 2009 report on progress had the Solomon Islands 'on track' for two goals (child mortality; maternal health), 'of concern' for three goals (eradication of extreme poverty and hunger; achievement of universal primary education; combating HIV/AIDS and other diseases) and 'off track' for two (promotion of gender equality and empowerment of women; ensuring environmental sustainability).[65] Even if SSR in the Solomon Islands had achieved its goals, the positive impact of this situation on development would not compensate for the fact that, as indicated by the 2009 AusAID report, so many aspects of development in Solomon Islands are at risk.

While there has been some attention to the reform of security institutions in the Solomons in relation to national goals, it is also the case that there could have been more effective overall coordination based on an SSR approach. This would have meant primarily more emphasis on establishing stronger security governance by the country's political institutions, although under current circumstances in conjunction with RAMSI. As has been argued in relation to East Timor, a positive outcome would be the enhancement of political stability, thus providing a better framework for formulating and implementing agreed development goals.

In terms of SSR and the security-development nexus in East Timor and Solomon Islands we thus face two different yet related situations. In both cases there has been little explicit focus on the SSR-development linkage as such. The emphasis has been on the reform of security institutions, with greater overall coordination and more attention to development in the Solomon Islands than in East Timor – yet also with significant shortcomings in both cases. In the case of East Timor the problems in achieving effective SSR make it difficult to attain long-term stability; without the assurance of order, progress in relation to the socio-economic and political goals of development becomes more difficult. The strong role of the security institutions in East Timor exacerbates this situation. In the Solomon Islands the security institutions are important but less central (particularly without an army), and relatively more attention is given to the overall justice system. The situation involving transnational governance based on RAMSI could conceivably continue on a long-term basis. SSR is a relatively minor aspect of the various factors affecting development in the Solomons.

Conclusion

This chapter has demonstrated that, while issues relating to SSR have been important in both East Timor and the Solomon Islands, the link between SSR and development has received less attention than it warrants. The main contribution of a strong SSR-development link would be the strengthening of political stability, thus providing a framework within which the goals of socio-economic and political development might be more effectively pursued. In both countries SSR has been approached mainly in terms of reforming specific security institutions rather than on the basis of an integrated approach involving overall security governance. The impact of the reform of security institutions on the wider society has generally been left implicit, although this criticism is less valid in relation to the Solomon Islands than it is with East Timor. More attention to the value of a well-functioning security sector for political stability would facilitate progress towards political and socio-economic development.

While in practice focusing mainly on institutional reform, SSR still has an impact on development irrespective of whether governments and other relevant actors consciously try to determine the nature of the SSR-development linkage. If political stability is the main development issue affected by SSR, then failure to make the linkage a positive one could have deleterious consequences. The focus on the reform of specific security institutions in East Timor – yet with relatively weak overall coordination of the security sector – could result in a well-functioning military force that could also be more effective as a political actor and conceivably undermine the civil authorities under some circumstances. A reformed military would face a reformed police force that could also be an important political actor. At the same time the institutional development of the legal system and the legislative and executive branches of government might remain relatively weak. The overall outcome would potentially be greater instability. Still, one would have to take into account a whole range of factors that would be relevant in the circumstances, such as the socio-economic situation, the involvement of civil society and the policies of external actors. To avoid this situation a more systematic SSR approach is needed that gives due attention to overall security governance. Focusing merely on the reform of specific security institutions is insufficient and could be detrimental to development.

In the Solomon Islands the potential for instability derives mainly from the fragmented nature of the society. There are no potentially powerful security actors waiting in the wings. There is perhaps greater awareness of the significance of the security sector for development goals than is the case in East Timor. The focus has been on building indigenous capacity, on the assumption that this will contribute to political stability. Progress towards strengthening indigenous capacity in the security sector has been mixed, and RAMSI or a similar arrangement appears likely to have a long-term future in the Solomon Islands. Political stability is achieved, but derives to a great extent from the role played by the external actor.

As far as the future is concerned, there is a strong case for developing a more integrated approach to SSR (including a clear role for overall security governance) in both East Timor and the Solomon Islands, while also making the SSR-development linkage much more explicit. The consequences for development of SSR-related policies should be spelt out. If some policies are likely to have negative consequences for development, then clearly they should be avoided. Policies that are likely to promote development should be emphasised. The recommendation for a more explicit approach to the SSR-development linkage is primarily a matter for the governments in the two situations, but it should also be a focus for the major external players.[66] The obstacle to moving in this direction has been the political factors affecting the way in which security sector issues are dealt with in the two countries. Factors such as the political position of the military in East Timor and the weak political institutions in both countries make it difficult to develop strong policies. If the arguments in favour of a stronger SSR-development linkage are appreciated by relevant actors in both situations and acted upon, this can have positive consequences in the long term. In the Solomon Islands RAMSI is strategically well placed to move in this direction. The same can be said for East Timor, although there is no single actor with the same authority and potential influence on national decision-making, and the overall political situation is considerably more complex.

Notes

1 On the difficulties involved in defining development, together with an attempt to ground the concept in the discourse of political economy, see Anthony Payne and Nicola Phillips, *Development* (Cambridge: Polity Press, 2010).

2 UN Secretary-General, 'Securing Peace and Development: The Role of the United Nations in Supporting Security Sector Reform', UN Doc. A/62/659-S/2008/39 (23 January 2008): para. 17.

3 Ibid.: para. 14. For other official statements relating to SSR see OECD, 'Security System Reform and Governance: Policy and Good Practice', Policy Brief (Paris: OECD, May 2004), available at www.oecd.org/dataoecd/20/47/31642508.pdf; OECD/DAC, 'Security System Reform and Governance', DAC Guidelines and Reference Series (Paris: OECD, 2005), available at www.oecd.org/dataoecd/8/39/31785288.pdf; UK Department for International Development, 'Understanding and Supporting Security Sector Reform' (London: DFID, 2002), available at www.securitycouncilreport.org/atf/cf/%7B65 BFCF9B-6D27-4E9C-8CD3-CF6E4FF96FF9%7D/supportingsecurity%5B1%5D.pdf.

4 On the security-development nexus a good starting point is Mark Duffield, *Global Governance and the New Wars: The Merging of Development and Security* (London: Zed Books, 2001); Mark Duffield, *Development, Security and Unending War: Governing the World of Peoples* (Cambridge: Polity Press, 2007).

5 Forças Armadas de Libertação Nacional de Timor-Leste (Armed Forces for the National Liberation of East Timor).

6 For a full-length study of the PNTL see Bu Vicki Elizabeth Wilson, 'Smoke and Mirrors: The Development of the East Timorese Police 1999–2009', unpublished PhD thesis, Australian National University (2010), available at www.regnet.anu.edu.au/program/ people/profile/thesis/BuWilson_Thesis.pdf.

7 Two important and detailed assessments of SSR in East Timor are International Crisis Group, 'Timor-Leste: Security Sector Reform', Asia Report no. 143 (Brussels: ICG, 2008); Gordon Peake, 'A Lot of Talk But Not a Lot of Action: The Difficulty of Implementing SSR in Timor-Leste', in *Security Sector Reform in Challenging Environments*, eds Hans Born and Albrecht Schnabel (Münster: LIT Verlag, 2009): 213–240.

8 The distinction between westerners and easterners should not be exaggerated; it has been described as one of 'mild prejudice' in Sven Gunnar Simonsen, 'The Authoritarian Temptation in East Timor: Nationbuilding and the Need for Inclusive Governance', *Asian Survey* 46, no. 4 (2006): 590. While there appeared to be discrimination against westerners in the F-FDTL, there is no evidence of significant maldistribution in terms of socio-economic development. It should also be noted that this 'mild prejudice' could nevertheless be a strong influence on the way in which people behaved.

9 International Crisis Group, note 7 above: 2.

10 Ibid.

11 Ibid.

12 See Matthew B. Arnold, 'Challenges Too Strong for the Nascent State of Timor-Leste: Petitioners and Mutineers', *Asian Survey* 49, no. 3 (2009): 429–449.

13 Damien Kingsbury, 'Timor-Leste: The Harsh Reality after Independence', *Southeast Asian Affairs* (2007): 367.

14 Svar Gunnar Simonsen, 'The Role of East Timor's Security Institutions in National Integration – And Disintegration', *Pacific Review* 22, no. 5 (2009): 581.

[15] International Crisis Group, 'Resolving Timor-Leste's Crisis', Asia Report no. 120 (Brussels: ICG, 2006): 12.

[16] Kingsbury, note 13 above: 368.

[17] Peake, note 7 above: 219.

[18] See Simonsen, note 14 above: 586. Note also the point made in Chapter 8 by Henri Myrttinen regarding the way in which SSR processes favoured people who were part of the armed resistance in East Timor (and Aceh).

[19] UN Security Council, 'Report of the Secretary-General on Timor-Leste pursuant to Security Council Resolution 1690', UN Doc. S/2006/628 (2006): para. 62.

[20] 'Report of the United Nations Independent Special Commission of Inquiry for Timor-Leste' (Geneva, 2 October 2006), available at www.ohchr.org/Documents/Countries/COITimorLeste.pdf.

[21] Government of East Timor, 'Force 2020' (2007), available at www.etan.org/etanpdf/2007/Forca%202020%20-%202007.pdf.

[22] Simonsen, note 8 above: 589.

[23] International Crisis Group, note 7 above: 8.

[24] Cynthia Burton, 'Security Sector Reform: Current Issues and Future Challenges', in *East Timor: Beyond Independence*, eds Damien Kingsbury and Michael Leach (Clayton, Vic.: Monash University Press, 2007): 101.

[25] Government of East Timor, note 21 above: 134.

[26] International Crisis Group, note 7 above: 5. In 2010 there were 1,250 personnel in the army and 82 in the naval element – see International Institute for Strategic Studies, *The Military Balance 2011* (London: Routledge, 2011): 279.

[27] Government of East Timor, note 21 above: 52.

[28] Ibid.: 116.

[29] International Crisis Group, note 7 above: 6–7.

[30] Ibid.: 7–8.

[31] There is a detailed assessment in Peake, note 7 above.

[32] Recent assessments of the policing situation in East Timor, with particular reference to progress relating to the handover from the United Nations to the PNTL, include International Crisis Group, 'Handing Back Responsibility to Timor Leste's Police', Asia Report no. 180 (Brussels: ICG, 2009); International Crisis Group, 'Timor-Leste: Time for the UN to Step Back', Asia Briefing no. 116 (Brussels: ICG, 2010).

[33] Independent Comprehensive Needs Assessment Team, 'The Justice System of Timor-Leste: An Independent Comprehensive Needs Assessment', Dili (13 October 2009), available at www.laohamutuk.org/reports/UN/UNMIT/JusticeNeedsAssessment Oct09.pdf. This report was commissioned through UNMIT.

[34] International Crisis Group, note 7 above: 6.

[35] Ibid.: 15.

[36] Ibid.

[37] For overviews of the development of the situation in the Solomon Islands during this period see Jon Fraenkel, *The Manipulation of Custom: From Uprising to Intervention in the Solomon Islands* (Wellington: Victoria University Press, 2004); Clive Moore, *Happy Isles in Crisis: The Historical Causes for a Failing State in Solomon Islands, 1998–2004* (Canberra: Asia Pacific Press, 2004).

[38] UN Development Programme, 'Human Development Index (HDI) – 2010 Rankings', available at http://hdr.undp.org/en/statistics/.

[39] Central Intelligence Agency, 'The World Factbook: East & Southeast Asia – Timor-Leste' (last updated 7 September 2011), available at www.cia.gov/library/publications/the-world-factbook/geos/tt.html; Central Intelligence Agency, 'The World Factbook: Australia-Oceania – Solomon Islands' (last updated 23 August 2011), available at www.cia.gov/library/publications/the-world-factbook/geos/bp.html.

[40] Australian Department of Foreign Affairs and Trade, Economic Analytical Unit, *Solomon Islands: Rebuilding an Island Economy* (Canberra: Department of Foreign Affairs and Trade, 2004): 4.

[41] Shahar Hameiri, 'State Building or Crisis Management? A Critical Analysis of the Social and Political Implications of the Regional Assistance Mission to Solomon Islands', *Third World Quarterly* 30, no. 1 (2009): 35–52.

[42] Details of RAMSI's organisation are based on Hameiri, ibid.: 39. See also RAMSI's website at www.ramsi.org/.

[43] Phillip Dorling, 'Solomons' Legacy: "$1bn Wasted"', *Age* (30 August 2011).

[44] Security Sector Reform Resource Centre, 'Solomon Islands' (2010), available at www.ssrresourcecentre.org/wp-content/uploads/2010/04/Country-Profile-Solomon-Islands-April-28.pdf: 4.

[45] Eden Cole, Thomas Shanahan and Philipp Fluri, *Enhancing Security Sector Governance in the Pacific Region: A Strategic Framework* (Suva: Pacific Islands Forum Secretariat/UN Development Programme Pacific Centre, 2010): 118. On policing issues, with particular reference to Australian involvement, see Sinclair Dinnen, Abby McLeod and Gordon Peake, 'Police-building in Weak States: Australian Approaches in Papua New Guinea and Solomon Islands', *Civil Wars* 8, no. 2 (2006): 95–101; Andrew Goldsmith and Sinclair Dinnen, 'Transnational Police Building: Critical Lessons from Timor-Leste and Solomon Islands', *Third World Quarterly* 28, no. 6 (2007): 1101–1105.

[46] RAMSI, 'Corrections', available at www.ramsi.org/our-work/law-and-justice/corrections.html.

[47] 'Partnership Agreement between Solomon Islands Government and Regional Assistance Mission to Solomon Islands' (April 2009), available at www.ramsi.org/Media/docs/SIG-RAMSI_PartnershipFramework-f69fa231-cc6a-47bf-99a3-3e88eba95414-0.pdf.

[48] Derek McDougall, 'The Security-Development Nexus: Comparing External Interventions and Development Strategies in East Timor and Solomon Islands', *Asian Security* 6, no. 2 (2010): 170–190.

[49] Democratic Republic of Timor-Leste, 'Programme of the IV Constitutional Government (2007–2012)', (Dili, 2007), available at http://timor-leste.gov.tl/?p=16&lang=en.

[50] Ibid.: Chapter VI, para. 10.

[51] Ibid.

[52] Democratic Republic of Timor-Leste, Office of the Prime Minister, 'On Road to Peace and Prosperity: Timor-Leste's Strategic Development Plan, 2011–2030: Summary' (7 April 2010), available at www.laohamutuk.org/econ/10TLDPM/RDTLStratDevPlanSumm7Apr2010En.pdf.

[53] Ibid.: 5.

[54] AusAID, 'Tracking Development and Governance in the Pacific', Annex 2 (August 2009), available at www.ausaid.gov.au/publications/pdf/track_devgov09.pdf: 69.

[55] AusAID, 'East Timor' (2011), available at www.ausaid.gov.au/country/country.cfm?CountryId=911. Given Australia's role as a major donor in both the South Pacific and East Timor, the Australian Agency for International Development (AusAID) is well

placed to report on progress or lack thereof towards development goals in this region, albeit from an official Australian perspective.

[56] UN Integrated Mission in Timor-Leste, 'Statement to Timor-Leste Development Partners Meeting' (3 April 2009), available at www.laohamutuk.org/econ/09TLDPM/UN/UNMITTLDPM.htm.

[57] However, note the point made by Ann M. Fitz-Gerald in Chapter 11 that national dialogue in East Timor (and her other case studies of Uganda and Sierra Leone) was stronger in relation to national development and poverty eradication than in relation to national security.

[58] Solomon Islands Government, 'National Economic Recovery, Reform and Development Plan, 2003–2006, Strategic and Action Framework, Final Report' (Honiara: Department of National Reform and Planning, October 2003), available at www.sprep.org/att/IRC/eCOPIES/Countries/Solomon_Islands/42.pdf.

[59] Regional Assistance Mission to the Solomon Islands, 'Annual Performance Report 2009', available at www.ramsi.org/Media/docs/RAMSI-Annual-Performance-Report-2009-c93293d3-a6e8-4de7-81fc-e97c13adda0e-0.pdf: 2.

[60] Matthew Allen, 'Long-term Engagement: The Future of the Regional Assistance Mission to the Solomon Islands', Strategic Insights no. 51 (Canberra: Australian Strategic Policy Institute, 2011): 16.

[61] Regional Assistance Mission to the Solomon Islands, note 59 above: 3.

[62] Ibid.: 5.

[63] See Graeme Wilson (RAMSI special coordinator), 'The Solomon Islands Government-RAMSI Partnership Framework: Towards a Secure and Sustainable Solomon Islands', paper presented in State, Society and Governance in Melanesia Program seminar series, Australian National University (17 December 2009), available at www.ips.cap.anu.edu.au/ssgm/ papers/seminars/09_1217_sp_wilson_RAMSI.pdf.

[64] See for example Allen, note 60 above: 2.

[65] AusAID, note 54 above: 62.

[66] The *World Development Report 2011* argues that a problem for international agencies in contributing to violence prevention and recovery is that they are 'geared to minimizing domestic reputational and fiduciary risk'. See World Bank, 'International Support to Building Confidence and Transforming Institutions', in *World Development Report 2011: Conflict, Security and Development*, available at http://wdr2011.worldbank.org/sites/default/files/WDR2011_Chapter6.pdf.

PART IV

SSR, DDR AND DEVELOPMENT

Chapter 7

Pushing Pieces Around the Chessboard or Changing the Game? DDR, SSR and the Security-Development Nexus

Alan Bryden

Introduction

Disarmament, demobilisation and reintegration (DDR) and security sector reform (SSR) are central pillars of the international community's efforts in support of states emerging from conflict. The DDR-SSR nexus has become the focus of a burgeoning policy literature.[1] Yet if a growing community of academics, policy-makers and practitioners have emerged to promote both sets of activities, the linkages between DDR and SSR remain ill-defined and misunderstood. In practice, activities are often carried out independently, frequently maintain a narrow focus on hard security imperatives and ignore potential synergies that could generate security and development pay-offs.[2]

Critiques of DDR and SSR point to missed opportunities to move beyond 'security first' perspectives and support longer-term development. While the disarmament and demobilisation components of programmes can be extremely resource intensive, the socio-economic reintegration of ex-combatants – the part usually anticipated to deliver the most tangible 'peace dividend' – tends to lag behind, afflicted by inadequate or slowly disbursed funding and poor planning. And despite the roots of the SSR concept in the development donor discourse, interventions are often accused of having a marginal impact on the lives of ordinary people. For both DDR and SSR, gaps between progressive policy frameworks and more limited returns on the ground raise important questions about programme design and implementation in challenging post-conflict environments. This chapter explores how fostering synergies between DDR and SSR can help bridge these policy-practice gaps, and in so doing reinforce the security-development nexus.

A central message of the World Bank's 2011 *World Development Report* (WDR) is that 'strengthening legitimate institutions and governance to provide citizen security, justice, and jobs is crucial to break cycles of violence'.[3] It can be argued that the report is only the latest in a line of admonitions to 'secure development'.[4] Yet the findings of the WDR are significant because they tally with a common shortcoming of internationally supported SSR programmes: disproportionate emphasis is placed on building more effective security providers without commensurate improvements in the accountability or responsiveness of the security sector to people's needs. In other words, SSR practice does not follow through on the relatively uncontested policy prescription that effective security provision needs to be embedded within a framework of democratic security sector governance (SSG).[5] Applying an SSG perspective that reflects development thinking can permit the identification of nuanced, context-specific synergies that contribute to the sustainability and legitimacy of DDR and SSR outcomes. It can also situate discrete activities within broader processes of post-war political transition in which security institutions and cultures should be reoriented towards the needs of the state and its citizens.

The chapter begins by examining the DDR and SSR concepts and outlining key elements of an SSG approach to DDR/SSR as a means to reinforce the security-development nexus. Insights are drawn from the cases of Afghanistan and Burundi to explore different challenges and opportunities associated with ensuring that DDR and SSR support development as well as enhancing security. The penultimate section builds on this analysis by exploring three ways that DDR and SSR can reinforce the security-development nexus: firstly, through addressing critical knowledge gaps; secondly, by applying a human-security-driven approach that anchors activities to individual and community needs; and thirdly, by focusing on SSG concerns in order to develop strong, legitimate security institutions. The chapter concludes by highlighting key messages and points to useful areas for further work that can elaborate the developmental dimensions of the DDR-SSR nexus.

Pushing pieces around the chessboard?

This section interrogates the DDR and SSR concepts to identify critical relationships with development issues and objectives. Some limitations to current approaches are identified. In particular, to address concerns that the developmental impact of activities may be superficial, the relationship between the security-development nexus and an SSG-driven approach to DDR and SSR is examined.

Disarmament, demobilisation and reintegration and security sector reform

According to the UN Integrated Standards for Disarmament, Demobilization and Reintegration (IDDRS),[6] DDR is intended to 'deal with the post-conflict security problem that arises when combatants are left without livelihoods and support networks during the vital period stretching from conflict to peace, recovery and development'.[7] This is done 'by removing weapons from the hands of combatants, taking the combatants out of military structures and helping them to integrate socially and economically into society by finding civilian livelihoods'.[8] The security-development nexus is therefore intrinsic to the definition of DDR accepted across the UN system. Security-related activities to demilitarise ex-combatants (the '2Ds') go hand in hand with the developmental imperative to ensure their social and economic reintegration. Similarly, the security-development nexus provides the rationale for a holistic SSR agenda. The emphasis on linking security provision, management and oversight flows from an essentially developmental insight that a multitude of narrow security promotion measures – from downsizing armed and security forces to providing training and equipment – have in some cases been counterproductive. In fact, improving the performance of the army, police and other security providers in the absence of democratic control has frequently exacerbated insecurities and rolled back development gains.

The security-development nexus is apparent in evolving appreciations of agency within DDR and SSR processes. DDR programmes typically feature a clear primary target caseload including former combatants and their dependants. Importantly, the diversity of participants has become more widely recognised, and they are now understood to include women and children associated with armed groups, disabled ex-combatants and non-combatants associated with fighting forces in previously unrecognised support roles.[9] The need to assist communities affected by DDR initiatives is a major conclusion of the influential Stockholm Initiative on Disarmament, Demobilisation and Reintegration.[10] The findings of this multi-stakeholder initiative stress the need to think about ex-combatants and communities together by developing parallel programmes that mirror measures in favour of ex-combatants with support for the communities that receive them.[11] This emphasis is particularly important because it underlines that DDR measures to address ex-combatants in isolation are inadequate from a developmental perspective. Demobilisation and reintegration packages generally function on the basis of sticks and carrots offered to ex-combatants. Security and development benefits to the community tend to be assumed as a logical by-product of demobilisation. Yet without looking beyond the individual to create closer links between ex-combatants and receiving communities, there will be insufficient trust to enable longer-term recovery and development.

The scope of the SSR agenda is reflected in the wide range of associated actors. Part of the reason for this diversity lies in the emphasis given to joined-up approaches to security provision, management and oversight. It also reflects a more fundamental argument that institutional reforms will be ineffective over the long term if they do not acknowledge and address the cleavages between political elites, the security sector and citizens, and how these exacerbate state fragility. Thus the language of institutional transformation found in the WDR echoes similar calls for a transformative SSR agenda that can support human development through addressing actual and perceived insecurities caused by a weak or illegitimate security sector.[12]

The UNDP 'Practice Note on Disarmament, Demobilization and Reintegration' emphasises that the structure and sequencing of activities depend on the particular circumstances of each country and careful timing is essential.[13] Yet a criticism of certain DDR programmes is that they fail in this, instead adopting top-down approaches that marginalise influential non-state actors and local nodes of resilience.[14] In the process, the needs of vulnerable groups (whether former combatants or civilians) are ignored. DDR may, in other words, ameliorate the immediate security environment by channelling ex-combatants and their weapons along different, non-violent paths, but any gains will be transitory without due attention to the longer-term needs and prospects of concerned groups.[15] The SSR approach folds security-building measures within an SSG framework. At least in principle, this recognises the importance of changing cultures of security in favour of human security needs. However, again, in practice the 'game' remains the same because the preponderance of external support remains tilted towards (re)building more effective security providers.

Increasing innovation has been witnessed from the field in response to perceived shortcomings of 'traditional' DDR interventions. Recent research on interim stabilisation or 'second-generation' DDR[16] has highlighted ways in which programmers and local authorities have jointly developed bottom-up strategies through drawing on local cultural norms and other contextual factors. Second-generation approaches have emerged in particularly fragile and contested societies where there is a need to buy time and space through micro-level measures that can respond quickly to local needs. Examples of activities falling under this label include community violence reduction initiatives, transitional military integration arrangements and the establishment of local dialogue processes. Although not conceptualised as such, these approaches seem to overlap significantly with developments in the policy and practice of SSR. Internationally supported SSR can tap into a growing body of empirical analysis[17] as well as emerging expert networks drawn from within countries undergoing reform. This has led to greater awareness of the need to look beyond formal or government actors in crafting SSR programmes. In particular, it is more and more apparent that SSR must be based on engagement with non-state actors that serve legitimate security and justice functions or fill vacuums left by a non-responsive, dysfunctional state security apparatus.

Emerging good practices notwithstanding, why have the developmental objectives of these activities not been realised on the ground? On one level, this can be attributed to misunderstandings and contestation over the relationship between DDR and SSR. In conceptual terms, some analysts situate DDR as a sub-activity within a broader SSR agenda. However, for institutions (such as the United Nations) engaged in both, this approach has often led to narrow bureaucratic preoccupations over 'turf' at the expense of a genuine engagement with substance. More fundamentally, a lack of traction for SSG concerns inhibits the linking of a more discrete set of activities surrounding ex-combatants with a wider agenda that seeks to address the causes and effects of an ineffective, inefficient and poorly governed security sector. This chapter therefore endorses a pragmatic approach that explores the ways in which SSR and DDR are two separate but related endeavours. This is consistent with the *OECD/DAC Handbook on Security System Reform*, which states that 'the two issues [DDR and SSR] are often best considered together as part of a comprehensive security and justice development programme'.[18] The following subsection outlines an SSG approach to the DDR-SSR nexus as a means to realise common security and development goals.

A security sector governance approach to the DDR/SSR relationship

If bad governance is frequently identified as a major cause of underdevelopment and fragility, governance deficits in the security sector are particularly challenging. In many post-conflict contexts, national security architectures and policies are outdated or founded on inappropriate, externally imposed models. Structural weaknesses are compounded by the close control of security decision-making by political and security elites.[19] The result is that the missions, resources and values underpinning security institutions remain skewed and disjointed, with external and regime security trumping internal and human security concerns. As discussed below, linking DDR to SSG-focused reforms can help move beyond these deficits.

Disarmament, narrowly understood, separates ex-combatants from their weapons. But apparent 'successes' can be illusory: weapons obtained through a disarmament process may be stolen or sold before destruction, disappearing during transit or after they have reached an arms depot. Yet successful disarmament can be both an enabler and an indicator of a secure environment in which development can occur. Through an SSG lens, disarmament represents one part of a wider set of issues surrounding the state's ability to regulate and manage the transfer, trafficking and control of weapons on its national territory. SSR initiatives can therefore play a critical role in building more effective security institutions to support this process.

In simple terms, disarmament reflects the logic that fewer arms in the hands of former fighters equates to less risk of violence. However, this correlation has limited utility from the perspective of human development if it is not linked to civilian weapons holdings and community security concerns. One means to measure the quality of security experienced by communities is therefore to examine the relationship between ex-combatant and civilian disarmament. The latter is highly unlikely to see positive results without progress in the former. Through a combination of prevention and enforcement, SSR efforts to support community security can play an important part in creating an environment conducive to reducing the overall level of arms availability.

Demobilisation is a two-stage process to discharge active combatants. A first stage of demobilisation involves the processing of individual combatants or the massing of concerned troops in cantonment sites or other assembly areas. This is followed by provision of an initial reinsertion support package.[20] While the demobilisation process may seem straightforward and technical, its potential for success will be determined at the political level before the cantonment of a single soldier. Negotiations on agreeing demobilisation terms can be particularly contentious. This issue forms part of the larger question of who wins or loses from the terms of a peace settlement. Concerns over patronage can therefore predominate over a more rational assessment of the kind of security sector the country needs or can afford. This means that early DDR decisions with long-term implications for force size and structure can be far removed from the security needs and priorities of citizens.

Following demobilisation, a filtering process sees ex-combatants either re-enter civilian life through reintegration or become part of reconstituted security forces through integration into reformed security institutions. Development gains may seem most apparent through successful reintegration of ex-combatants into society, which reduces the risk of conflict recurrence while creating social and economic capital within communities. However, this is also the most challenging phase of the process. Jobs are generally scarce and ex-combatants typically have limited skills to transfer to the civilian economy.

Whether or not understood as such, reintegration/integration is an SSR process with choices shaping the operational capabilities and constraints of security institutions. They also influence how security institutions will be regarded by different groups within society. In this sense, the integration of former conflict parties into reformed security forces mirrors a wider process of national reconciliation that needs to take place within and across communities. Integration should make a positive contribution to this process by addressing concerns of parity across lines drawn during the conflict, such as in the attribution of ranks or promotions, in acknowledging the rights of veterans or in the vetting of personnel.

In sum, DDR and SSR can clarify and realign security sector mandates, structures and resource flows. If pursued with sensitivity and openness, this can tangibly increase levels of protection provided to individuals and communities while creating public confidence in security institutions. To do this, processes must be carefully negotiated. This requires flexibility and transparency so that local, regional and national stakeholders help to define, prioritise and undertake reforms. The evident sensitivity of the issues involved means that steps to reshape the security sector governance environment and reform security institutions must be pursued incrementally and with humility. At the same time, as explored in the following section, failure to grasp this nettle can undermine both immediate security and longer-term development objectives.

DDR/SSR challenges in Afghanistan and Burundi

This section analyses opportunities seized (or missed) to link security goals with development pay-offs in implementing DDR and SSR programmes in two contexts facing very different security challenges.[21] The first case study,

Afghanistan, lacks a negotiated political settlement that addresses DDR and SSR. As a consequence, it has been difficult to incentivise armed groups to join the DDR process or assign security roles and responsibilities as part of this process. Conflict-driven insecurity in different parts of Afghanistan is compounded by the effects of the narcotics trade and other forms of crime. In responding to these challenges, DDR and SSR interventions have been primarily developed in order to address immediate security threats. This has narrowed the focus of efforts to training and equipping security forces.

Despite the patchwork nature of the various political and security-related peace agreements, the second case study, Burundi, is in a more stable post-conflict phase. If the political context has not until recently enabled the establishment of a holistic SSR programme, a web of negotiations following the end of the conflict has at least drawn all the major armed groups into a DDR process that has been closely linked to the issue of integrating ex-combatants into a reformed security sector.

Important distinctions notwithstanding, both countries share a common struggle to create the conditions necessary to ensure sustainable security and development for their citizens. The relationship between these objectives is clearly demonstrated in the Afghan government's Millennium Development Goals report for 2010, which adds a ninth goal of addressing insecurity to the eight official Millennium Development Goals (MDGs). As the report states, 'progress towards MDGs is not necessarily a sufficient condition for attaining human development. In Afghanistan, MDG progress is analysed in the context of the political, economic, social, governance and security framework.'[22] In a similar vein, while Burundi's MDGs progress report for the same year restricts itself to the official MDGs, peace and security are identified as 'principal progress factors' for MDGs 1–3.[23] While the obstacles to achieving security and development goals in these contexts are numerous and complex, this discussion of necessity addresses a narrow set of issues. The nexus between DDR and SSR activities is examined through an SSG lens to understand the implications of this approach for the security-development nexus.

Disarmament and citizen security

Whether in the hands of ex-combatants or civilians, reducing the availability of small arms and light weapons can have a positive influence on social and economic development by lowering levels of criminality and improving physical security.[24] Paradoxically, economic and security incentives can also account for disarmament failure. As Sedra points out, in Afghanistan weapons possession is both a means of tapping into and a protection against the illicit economy.[25] By focusing on the community level, the close relationship between the disarmament of ex-combatants, civilian disarmament and development outcomes in Afghanistan and Burundi can be better understood.

In Burundi, tit-for-tat self-arming of Hutu and Tutsi civilians as well as local militias was a feature throughout the civil war. A lack of success in gathering weapons from ex-combatants has a direct bearing on the (paltry) results of civilian disarmament efforts. Community insecurities that underpin high levels of weapons possession are explained by a 2007 public perceptions study, which identifies the rebels, armed forces and police as the three greatest security concerns after general criminality for citizens.[26] Consistent with a key message emerging from the WDR, the ability to move towards recovery and development is constrained in both cases by the absence of collaboration across societal divisions.[27] This shows that disarmament and SSR need to address ethnically driven alliances and enmities that transcend military and civilian spheres. An important part of the relationship between SSR and the development benefits of disarmament therefore lies in their ability to overcome continued mistrust across the ethnic divide.

The Disbandment of Illegal Armed Groups (DIAG) programme in Afghanistan – launched in 2005 as the successor to an earlier DDR programme – was intended to set out and implement requirements for voluntary, negotiated or enforced disarmament of illegal armed groups. The UNDP as the responsible UN agency identifies a direct link between these disarmament activities and 'delivering development projects to enhance socio-economic outcomes in compliant districts'.[28] An important lesson that can be drawn from the DIAG process to date is the cost of ignoring local knowledge and expertise in programme planning, design and implementation. At the community level, a strategy was put in place offering development incentives in the form of cash grants for locally generated projects in order to undermine support for armed groups. However, this approach was based on a false premise of positive relations between communities and armed groups. Instead, the relationship of armed groups to communities has been in many cases predatory and criminally motivated. These groups had no interest in potential development benefits for the community flowing from development-linked weapons collection. Their main concern was retaining a stranglehold on the (much higher) rewards of the illicit economy.[29] Emphasis should have been placed on customised approaches that distinguish 'habitual' membership of armed groups from criminal or anti-government motivations that are less susceptible to development incentives. The reported relapse into insurgency of compliant districts[30] further demonstrates the need to combine community development activities with SSR initiatives to protect communities from the re-emergence of these groups.

Afghanistan's 2010 MDGs progress report includes complementary targets relating to illegal weapons possession and reforms to the Afghan National Police (ANP) that identify the need for greater public confidence in the police and an improved ratio of reported crime to convictions.[31] In practice these activities have been conducted in parallel. Security has been enhanced through police training at regional centres conducted under the Focused District Development (FDD) programme. Earlier individual rather than district-focused police training saw returning trainees forced by commanders to extort money from businesses and travellers.[32] The FDD was thus spurred by the costs to development of improving security provision without reference to the wider institutional environment.

It is certainly the case that the FDD programme has led to improvements in the effectiveness and interoperability of Afghan police units in delivering security within the regions concerned.[33] However, the overall impact of the initiative has not been maximised because it has only been approached through an SSR lens. Visible security benefits from the FDD were not synchronised with renewed disarmament efforts. The FDD could have generated important synergies through linkage with parallel weapons collection initiatives in these regions. Greater public confidence in local law enforcement capacities would undoubtedly have improved the likelihood of disarmament success.[34] On a different level, the Afghan government has criticised police training for its focus on honing paramilitary skills rather than developing civilian police to protect civilians better.[35] If this is unsurprising, given that FDD implementation is supported largely by US military personnel,[36] it nevertheless underlines the need for civilian disarmament to be accompanied by SSR efforts that are directed towards improving community security.

At least in principle, the DIAG programme embodies the DDR-SSR nexus: the ANP is the main instrument mandated to enforce disarmament compliance. Yet in practice the programme has been undermined by the inability of the ANP to perform this task as a result of systemic corruption, mismanagement and capacity deficits. These delegitimised the programme and fuelled non-compliance.[37] The FDD does have a security impact at the point of delivery. But as a stand-alone SSR initiative without a core focus on community security concerns, potential synergies have not been realised.

Reflecting wider SSG deficiencies, this stove-piping effect can also be found in the delinkage of disarmament from efforts to ensure effective control of weapons on the national territory. Because demilitarised weapons in Afghanistan were not tracked after reaching an arms depot, it was only belatedly apparent that poor stockpile management and inadequate protection of transports and storage areas were facilitating their illicit sale or seizure. In many cases, weapons obtained through the disarmament process found their way back into the hands of illegal armed groups.

Afghanistan and Burundi highlight the complex, non-linear relationship between disarmament, SSR and development gains. In Afghanistan, offering development incentives for disarmament is only a viable approach if targeted armed groups are interested in the future of communities. If groups are guided by 'hard' criminal or political agendas, enforced disarmament will be the more effective option. The two cases also show the interrelated dynamics of arms availability across armed groups and civilians. SSR is critical to a reduction in the availability of arms because both civilian and military processes require a level of confidence in order to succeed.

Development implications of demobilisation without SSR

DDR programmes often cite the numbers of ex-combatants going through demobilisation as a marker of success. However, a quantitative approach can obscure the developmental implications of demobilisation. In Afghanistan, some 'demobilised' armed groups retained both internal staff structures and their small arms and light weapons. In short, incentives to integrate within communities proved less strong than the link between commanders and their troops. As a consequence, armed groups in certain regions continued to exert authority over local populations via the collection of taxes and exploitation of natural resources following formal demobilisation. These illegitimate governance structures have undermined prospects for social and economic development at the community level while seriously eroding the credibility of the central government.[38]

The lack of commitment to demobilisation shown by certain armed groups has been compounded by narrow approaches to reforming state security providers in Afghanistan. On the one hand, the process to reform the Afghan National Army ruled out an alternative career option by rebuilding a new army from scratch rather than seeking to integrate militias. On the other hand, concerns over police officers retaining loyalties to armed groups have not resulted in thorough efforts to review the composition of the ANP.

Rather than attempting to identify and demobilise those with a direct affiliation to illegal armed groups, international support has focused on additional training as a means to plug capacity gaps within the police.

In Burundi the army was instrumentalised from early post-colonial times to maintain the authority of the Tutsi elite over the majority Hutu population. Defence reform to rebalance the army's ethnic composition has therefore been a key requirement for Hutu groups negotiating terms for DDR within the peace process. The multiplicity of agreements binding the various armed factions required a juggling act which fixed quotas for different groups within the security sector. For this reason, certain ex-combatants electing to join the police or army were subsequently demobilised so that a given group would not be 'over quota'.

Up until the recent development of a more holistic national security sector development programme supported by the Netherlands, army and police reform in Burundi has followed a similar pattern to Afghanistan, focusing on professionalisation initiatives such as the construction of barracks, logistical support and training. This approach reflects an absence of political will to engage in an in-depth vetting process that addresses suitability for service and past conduct issues. Instead, the presence of war criminals and rights abusers compounds capacity gaps, undermining citizens' trust in the security sector.

Rumin offers an example of the unintended consequences of demobilisation without penal reform in Burundi. As a result of weak management and oversight capacities within the prison system, common law prisoners mistakenly considered as war or political prisoners were released under the terms of the demobilisation process. Although the lack of evidence-based analysis obscures the extent of the problem, communities were certainly exposed to an influx of criminals who should have remained in prison. Improving the records management on the status of prison inmates to address this issue is a non-politically sensitive measure that could have been introduced as an early SSR priority.[39]

Awareness of the potential cumulative impact of demobilisation and SSR initiatives is missing across these cases. The lack of analysis for either Afghanistan or Burundi linking programme decisions to the security of individuals and communities reflects an absence of engagement with local stakeholders. For the former in particular, very limited analysis is available on the composition and characteristics of different armed groups. This would provide a point of departure for developing compelling demobilisation options and linking these to appropriate development incentives.

Perceptions and realities in reintegration and integration choices

Recognising that reintegration is a two-way trade-off between ex-combatants and communities can help mitigate tensions and ensure that programme goals are matched by local realities. Ex-combatants are in many cases unlikely to be welcomed into communities. They are often perceived as having been rewarded (through a reintegration package) for their violent pasts. They are also more likely than other groups to resort to violence to resolve disputes. This reality provides a strong argument for including community representatives in a negotiated process over the design and implementation of reintegration programmes. Only through a process of programme development that closely draws on local knowledge can viable reintegration options be identified that will contribute to longer-term recovery.

The frustrations felt by ex-combatants and receiving communities illustrate the need to understand reintegration dynamics from an individual's point of view. For example, ex-combatants have in certain cases been reluctant to participate in labour-intensive community development projects because this is felt to be beneath their dignity or outside their experience. Even when willing, a former rebel fighter recruited at a very early age may lack the requisite social and income-generating skills to contribute meaningfully to communal life. Moreover, although not necessarily evident when making the choice, on selecting reintegration the individual loses the sense of group identity associated with combatant status. An ordered way of life is swapped for an uncertain future once the short-term financial guarantees of the reintegration package expire.[40] If these social factors are not acknowledged, it will be impossible to build bridges between ex-combatants and receiving communities.

One DDR/SSR issue with significant consequences for social and economic development relates to housing, land and property (HLP) rights. Clarity on property ownership is an important enabler of economic development. In particular, access to land is a prominent issue in Burundi because of the combination of extremely high population density with a predominantly rural population dependent on agricultural subsidies. The significant levels of displacement caused by the civil war and the subsequent influx of returnees following the end of the conflict have put additional pressure on the already weak economy and exacerbated existing land disputes. Although ex-combatants form only a small part of the much larger group of returning refugees and internally displaced people, the heightened risk of violence associated with ex-combatants makes them a special category. A focus on HLP issues for ex-combatants through the deployment of appropriate law enforcement capacities in sensitive areas could resolve conflicts before they become violent. Equally, data on land availability generated through HLP schemes could be used to verify the viability of farming as a reintegration option in specific localities.

The National Commission for Demobilisation, Reinsertion and Reintegration (NCDRR) is the national body responsible for DDR decision-making in Burundi. It has been supported by the donor community precisely because it embodies national ownership of the process. The balancing act between acknowledging different political interests and delivering on DDR and SSR is conducted within this body. In reality, the NCDRR is composed of predominantly military staff within the Ministry of Defence as well as ex-combatants. Community voices are absent, so experience is lacking in the practical challenges faced by ex-combatants and receiving communities. The mono-ministerial profile also prevents participation by other government departments with key responsibilities for socio-economic development, notably the ministries of interior (responsible for civilian disarmament), labour and social affairs. As a result, the decision-making base is narrow and militarily oriented. Sidelining these ministries delinks reintegration from wider national processes to support socio-economic recovery. At the same time, the military profile of the NCDRR has meant that less attention is given to the role of the police and other internal security actors that can provide security to communities.

Ex-combatants in Afghanistan and Burundi were offered a seemingly voluntary choice between community reintegration and integration into the police or army. In Afghanistan a high proportion of individuals opted for reintegration. This bias was caused by two factors: the seemingly more attractive benefits of the reintegration package when weighed against the reputedly low pay and bad conditions associated with army life; and a restrictive age requirement for recruitment of 18–28 years of age. In practice, given the absence of labour market surveys, the lack of viable jobs for ex-combatants was not flagged, creating unrealistic expectations that turned to anger once the reality of non-combatant status became apparent. An influx of disillusioned ex-combatants thus swiftly became a negative factor in community life. Moreover, in both Afghanistan and Burundi soldiers' pay was increased some time after demobilisation processes had taken place, to facilitate recruitment and retention. This compounded the frustration felt by many who had recently demobilised before improved benefits were offered.[41]

In 2008 the Burundian police increased by a factor of ten as a result of the integration process, with very limited data gathered on the background and skills of these new entrants. An unintended consequence of the quota system is that in some cases new recruits were admitted at the expense of experienced police and gendarmes. This points to a central programming challenge. On the one hand, DDR design acknowledged the underlying ethnic tensions critical to navigating a politically viable process. Indeed, according to Rumin, a more nationally representative integrated army and police force raised public confidence in these bodies.[42] On the other, from an SSR perspective the programme caused unrest among ex-combatants as well as undermining the ability of the army and police to deliver security. The fragmented nature of the political process thus resulted in poorer service delivery at the operational level.

Changing the game: A security sector governance approach to the DDR/SSR nexus

DDR/SSR dynamics in Afghanistan and Burundi highlight the difficulties of implementing coherent programmes that can consolidate security and enable development. This section builds on these insights to consider the utility of an SSG-driven approach to DDR and SSR as a means to realise common security and development goals. Three sets of issues are addressed. First, ways to address critical knowledge gaps are considered. This is intended to move beyond narrow, technical initiatives that fail to exploit their development potential through ignoring the needs of ex-combatants and communities. A second set of issues surrounds mainstreaming approaches to DDR and SSR that link state and human security concerns. Finally, the relationship is considered between building more effective security sector governance institutions and the resulting developmental pay-offs.

Filling critical knowledge gaps

If DDR/SSR programmes are to address the different causes of insecurity that constrain development, they must be built on deep knowledge of context and prevailing political/security dynamics. Yet the case studies point to a reliance on quantitative statistics that measure outputs – soldiers trained, weapons surrendered, etc. – rather than a combination of quantitative and qualitative analysis that could shed light on the developmental impact of DDR and SSR. The positive effects of DIAG compliance were undermined by the lack of rapid follow-up development initiatives.[43] This shows that if you do not monitor disarmament and related development activities, their cumulative effects on socio-economic conditions in a given area will be unclear.[44] An alternative approach would have been to combine data on weapons collected or numbers of ex-combatants reintegrated with conflict-sensitive considerations, such as whether and how militias have been broken up or commander patronage networks disrupted.

It is particularly important that information flows are sustained. The affiliations of ex-combatants to their former armed groups, the nature of employment opportunities and reintegration impacts on community security can all shift significantly over time. This is demonstrated by the effects of gaps between demobilisation and the provision of reintegration packages. As seen in Afghanistan, delays between demilitarisation and socio-economic incentives may result in negative tendencies, such as the remobilisation of armed groups. Beyond the evident problems associated with slow disbursement, Rumin highlights that in Burundi the development benefits of reintegration were undermined because activities focused narrowly on the ex-combatant rather than her/his social and economic links to the community.[45] Without local involvement in programme planning, communities will be unprepared for returnees in terms of shaping viable livelihood options or supporting the social reintegration of ex-combatants and their families.

The lack of knowledge on these issues displays a generalised absence of public surveys and focus group analysis as a basis for programme design and implementation. The relapse of DIAG-compliant districts in Afghanistan shows that development incentives and security enforcement activities were not sufficiently tailored to the distinct motivations and agendas driving the various armed groups. These gaps can only be overcome by engaging with a broad range of national stakeholders, and at the same time providing an entry point to increase agency for disfavoured and underrepresented groups within DDR and SSR decision-making. Nuanced analysis would enable DDR and SSR activities to respond to individual and community needs and link these to targeted development incentives.

Knowledge gaps are apparent at both international and national levels. While much attention is given to the shortcomings of inappropriately designed external interventions, the example of the monochrome national commission in Burundi highlights how national elites are frequently far removed from internal dynamics within their own countries. Security and development goals will not be realised if DDR and SSR decisions at the national level are simply agreed top-down with the unquestioning support of international partners.

Prioritising state and human security

Reorienting the security sector towards safeguarding the security of citizens should be a central concern for DDR and SSR.[46] Security institutions that remain focused purely on state or regime security will not provide an enabling environment for development. However, in many cases security institutions remain locked into these ingrained modes of behaviour. By failing to reflect the differentiated security needs of individuals and communities, programmes ignore actual threats to physical security while at the same time missing an opportunity to build confidence through demonstrating the ability of the state to provide security.

The reorientation of security sector values, mission, priorities and objectives is clearly a long-term project that is intimately linked to wider governance issues. At the same time, shifting mindsets and cultures from external to domestic security imperatives is something that can be introduced early on. Rather than attempting to force through specific policies or strategies where political space is lacking, supporting inclusive national dialogues can provide a vehicle for the elaboration of a shared vision of SSG that can subsequently mould institutional and policy frameworks for the security sector.[47] Part of this discussion will be on the benefits of and modalities for cooperation between state and non-state actors with a stake in security provision, management and oversight.

What are the potential developmental pay-offs that can flow from taking human and state security concerns more seriously in DDR and SSR programmes? Collectively identifying and responding to challenges relating to the DDR process and its relationship with SSR can provide a step forward in redefining relations between the security sector and the community. To do so, the mandate and capacities of local security forces need to be sensitive to citizens' security needs. Local security plans linked to DDR activities should therefore be developed jointly with community stakeholders in order to identify requirements and ensure adequate law enforcement capacity to support this process. Greater dialogue can also point to specific challenges to development within communities, such as security vacuums caused by demobilisation, the vulnerability of ex-combatants to re-recruitment or the continued existence of patronage networks linked to former commanders.

Acknowledging the intrinsic link between ex-combatants and receiving communities can make the difference between *pro forma* DDR and a genuine shift away from combatant behaviour and status. Without recognising this distinction, demobilisation may actually reverse potential openings for individual and community development. To avoid this, the needs and vulnerabilities of ex-combatants must be acknowledged, including special groups such as former child soldiers, women fighters and dependants. With such an approach, DDR and SSR can jointly contribute to healing cleavages between groups. Perceptions of DDR as 'rewarding' ex-combatants can be nuanced through engagement with communities. Rebuilding the police and other state security actors can also generate trust in these bodies through recasting their image as a positive force capable of assuring the protection of the entire population. Thus, beyond their direct security benefits, community security initiatives should be considered as a mechanism for encouraging communities to accept ex-combatants and at the same time enhancing the status of local law enforcement actors.

If wide participation in decision-making enhances the legitimacy of DDR and SSR, it also permits informed decision-making. Reintegration 'success' from a development perspective needs to be qualified in relation to a refined set of indicators. Crime rates, prison populations and the flows of ex-combatants are all variables that can inform how SSR initiatives can best combine with DDR. Through demonstrating these development benefits, clarifying synergies between DDR and SSR can make a strong case for more joined-up approaches to programming.

Building effective institutions

While short-term security priorities often seem to prevail over longer-term institutional development, the two imperatives nevertheless need to be addressed in parallel. The security sector cannot provide security or support development if decision-making is narrowly informed or undermined by illegitimate governance structures. In the language of the WDR, shortcomings can be addressed through enhancing institutional integrity. From an SSG perspective, good governance of the security sector can be usefully understood in terms of the legitimacy and sustainability of security institutions.

Building a deeper sense of national ownership in DDR and SSR processes is essential to the establishment of legitimate security institutions. Where trust is low, cooperation will not take place within and between institutions. The political and economic stakes mean that resistance to change is inevitable. However, visible efforts to move beyond political exclusion and support institutions that transcend narrow political or ethnic affiliations can be as important for development prospects over the longer term as structural changes in the short and medium term. This can help to create the kind of multilevel partnerships that feature in the WDR's roadmap to transform state institutions and reinforce capacities to break out of cycles of violence.[48]

One reason for the delinking of DDR and SSR from social and economic development is that many national actors with a stake in these areas are not represented in decision-making bodies. SSG-driven reforms that draw in a wide range of national stakeholders from across government as well as parliamentarians, civil society actors and local community leaders can help to redefine security needs and priorities. An important part of this process is the lowering of barriers between different security and development actors. Promoting greater transparency through widening participation can make these bodies less partisan and encourage collaboration.

Enhancing management and oversight functions can mitigate debilitating cultures of donor dependency. The development of cross-cutting skills such as line management and human resource and financial management that can reinforce the overall integrity of security institutions is often ignored. Yet these are the very skills that will make national programmes more effective while building capacities with wider application. This is the best response to externally generated approaches that seek to make 'their' institutions look more like 'ours'. Reinforcing national commissions and related oversight bodies can therefore increase the effectiveness of DDR support while delivering on the key SSR goal of building representative institutions that promote sustainability through tailoring programmes to national needs. It can also mitigate risks experienced during transition periods when international support draws down.

A prominent part of the dependency culture is the reliance on external funding. The consequences of unsustainable assistance are often felt when support is scaled down (e.g. when security forces have to be rapidly downsized). Yet investing in SSG institutions can help match resources to local needs and limitations. It can also support fiscal sustainability by addressing the corruption risks that often accompany large volumes of combatants and significant flows of money. At the same time, SSR initiatives that strengthen national control of conflict-driving resources – from arms to drugs or precious materials – can increase the likelihood of armed groups entering the DDR process by closing down alternative options.

Conclusion

Policy debates across security and development communities have received a new impetus as a result of the publication of the WDR. However, discussions in donor capitals and the headquarters of international organisations have not necessarily been reflected by changes in the ways actors interact on the ground. Counterintuitively, one useful outcome of these ongoing discussions may be a more realistic assessment of not only the opportunities offered by but also the limitations of DDR and SSR in supporting development. Additional clarity at the conceptual level on synergies between DDR and SSR should not only highlight potential development contributions, but also inform when and how these activities can be folded into longer-term processes of recovery and development.

This chapter argues that a focus on security sector governance concerns can improve the impact of DDR and SSR programmes in their own right while making a more meaningful contribution to wider security and development goals. At the heart of an SSG approach is the need to 'change the game' by basing institutional reforms on the development of new cultures of cooperation. For both national stakeholders and international development partners, therefore, the DDR-SSR nexus is about bridging divides and demonstrating the benefits of collaborative approaches.

The daunting political and security-framing conditions present in contexts such as Afghanistan or Burundi make it difficult to identify ways to support sustainable development. This is particularly apparent where a baseline of security is lacking. However, early opportunities do exist, even in non-enabling environments, that can widen the optic of DDR/SSR beyond 'security first' approaches. The absence of national political will for reform is an argument frequently deployed to justify narrow international approaches to supporting these activities. But this is nonetheless insufficient, ignoring the ways that citizens in even the most constrained environments are demanding greater transparency, accountability and responsiveness from political and security elites. Steps that facilitate a gradual shift from regime to state and human security logic can be put in place, and are feasible and non-threatening when embedded within a process-driven approach. Strengthening national security sector governance actors and mechanisms can improve state responsiveness to citizens' security needs while building more sustainable, legitimate institutions that are better integrated within the national governance architecture and thus better placed to contribute to security and development.

This chapter has highlighted a number of gaps in knowledge that would merit further empirically grounded research as a means to influence practice. A first area relates to the need to bridge divides across issues and communities. There have been significant advances in thinking about the linkages between different elements of the post-conflict peace-building agenda. The new IDDRS modules addressing the relationships between DDR, SSR and transitional justice provide positive examples of this evolution. However, far fewer policy-relevant insights have emerged on how these agendas can contribute to longer-term recovery and development. Concerns over the securitisation of development notwithstanding, this represents a critical knowledge gap.

A second focus should be on the empirical base that needs to inform strategies for linking state and human security concerns. In many contexts, knowledge of who actually delivers security and justice remains patchy and superficial. Contextually nuanced insights are essential as a basis for developing complementary relationships between state and non-state actors.

Finally, there is a need to identify lessons that can contribute to maximising development benefits through more effectively operationalising the DDR-SSR nexus. Positive examples can be found of states that have transitioned from conflict through stabilisation to peace-building and then developmental orientations. While these cases may be more the exception than the rule, an important feature has been the transformation of security institutions in both practice and public perception. There is a need to tap into processes that have broken down barriers between different constituencies and built new cultures of security based on consensus across national, regional and community levels. Strong, legitimate security institutions that are integrated within wider national governance frameworks and development plans can be powerful champions of national development.

Notes

1 See Alan Bryden, 'Understanding the DDR-SSR Nexus: Building Sustainable Peace in Africa', DCAF Issue Paper (Geneva: DCAF, 2007), available at www.dcaf.ch; W. Andy Knight, 'Linking DDR and SSR in Post-conflict Peacebuilding: An Overview', *African Journal of Political Science and International Relations* 4, no. 1 (2010): 29–54; Michael Brzoska, 'Embedding DDR Programmes in Security Sector Reconstruction', in *Security Governance in Post-conflict Peacebuilding*, eds Alan Bryden and Heiner Hänggi, (Münster: LIT Verlag, 2005): 95–115. The UK Department for International Development also commissioned a study on DDR and SSR linkages within the context of a broader project on DDR and Human Security: Post-Conflict Security Building in the Interests of the Poor. See the project page at www.ddr-humansecurity.org.uk/.

2 This chapter draws on the findings of a multi-year project developed to understand better policy and programming issues related to the DDR-SSR nexus. The project, mandated by the UN Development Programme and the UN Department for Peacekeeping Operations, has resulted in a new module within the UN Integrated Standards for Disarmament, Demobilization and Reintegration (IDDRS) – Module 6.10, 'Disarmament, Demobilization and Reintegration and Security Sector Reform'. The module is available at www.unddr.org. The project has also generated an edited volume which includes field-based case studies undertaken to inform the drafting of the module: Alan Bryden and Vincenza Scherrer, eds, *Disarmament, Demobilisation and Reintegration and Security Sector Reform – Lessons from Afghanistan, Burundi, the Central African Republic and the Democratic Republic of the Congo* (Münster: LIT Verlag, forthcoming in 2012).

3 World Bank, 'Overview', in *World Development Report 2011* (Washington, DC: World Bank, 2011): 2.

4 Many of the key messages in the *World Development Report 2011* can be found in Robert B. Zoellick, 'Fragile States: Securing Development', *Survival* 50, no. 6 (2008/2009): 67–84.

5 Dylan Hendrickson, 'Key Challenges for Security Sector Reform: A Case for Reframing the Donor Policy Debate', GFN-SSR Working Paper (Birmingham: University of Birmingham, 2009).

6 The development of a series of UN DDR standards was initiated in 2004. The three main aims of the IDDRS are to set out clear, flexible and in-depth guidance for DDR practitioners; to establish a shared basis for integrated operational planning; and to provide a training resource for the DDR community. An inter-agency working group on DDR (IAWG-DDR) was subsequently established by the Executive Committee on Peace and Security in March 2005 with the mandate to improve UN performance in DDR. The 26 modules comprising the first edition of the IDDRS were jointly developed and approved by the IAWG-DDR in July 2006.

7 IDDRS 'Operational Guide', available at www.unddr.org/iddrs/iddrs_guide.php: 2.

8 IDDRS 1.20 Glossary and Definitions.

9 For the evolution of DDR approaches see Mats Berdal and David Ucko, eds, *Reintegration of Armed Groups after Conflict: War to Peace Transitions* (London: Routledge, 2009); Robert Muggah, 'No Magic Bullet: A Critical Perspective on Disarmament, Demobilisation and Reintegration and Weapons Reduction during Post-conflict', *The Round Table: The Commonwealth Journal of International Affairs* 94, no. 379 (April 2005): 239–252.

[10] The Stockholm Initiative on Disarmament, Demobilisation and Reintegration was a multi-stakeholder process initiated by Sweden to improve international understanding of the political, economic and social implications of DDR. A series of meetings and the commissioning of background papers in 2004–2005 culminated in the presentation of a final report to the UN Secretary-General in March 2006. See www.regeringen.se/sb/d/4890.

[11] Stockholm Initiative on Disarmament, Demobilisation and Reintegration, 'Final Report' (Stockholm: SIDDR, 2006): 27–28.

[12] For an overview of the security sector transformation discourse see Alan Bryden and 'Funmi Olonisakin, 'Conceptualising Security Sector Transformation in Africa', in *Security Sector Transformation in Africa*, eds Alan Bryden and 'Funmi Olonisakin (Münster: LIT Verlag, 2010): 3–23, available at www.dcaf.ch/publications.

[13] UNDP, 'Practice Note: Disarmament, Demobilization and Reintegration of Ex-combatants' (New York: UNDP), available at www.undp.org/bcpr/whats_new/ddr_practice_note.pdf.

[14] See Kathleen M. Jennings, 'Seeing DDR from Below: Challenges and Dilemmas Raised by the Experiences of Ex-Combatants in Liberia', FAFO Report no. 3 (Oslo: FAFO, 2008); Muggah, note 9 above.

[15] The importance of contextual determinants is explored in Nat J. Colletta and Robert Muggah, 'Context Matters: Interim Stabilisation and Second Generation Approaches to Security Promotion', *Conflict, Security and Development* 9, no. 4 (December 2009): 425–453.

[16] See Nat J. Colletta, Jens Samuelsson Schjorlien and Hannes Berts, *Interim Stabilisation: Balancing Security and Development in Post-conflict Peacebuilding* (Stockholm: Folke Bernadotte Academy, 2008); UN DPKO, *Second Generation Disarmament, Demobilisation and Reintegration Practices in Peace Operations* (New York: UN DPKO, 2010).

[17] See Hans Born and Albrecht Schnabel, eds, *Security Sector Reform in Challenging Environments* (Münster: LIT Verlag, 2009).

[18] OECD/DAC, *Handbook on Security System Reform: Supporting Security and Justice* (Paris: OECD/DAC, 2007): 105, available at www.oecd.org/dac/conflict/if-ssr.

[19] Alan Bryden and 'Funmi Olonisakin, 'Enabling Security Sector Transformation in Africa', in *Security Sector Transformation in Africa*, eds Alan Bryden and 'Funmi Olonisakin (Münster: LIT Verlag, 2010): 225–227.

[20] Note of the Secretary-General to the General Assembly on the administrative and budgetary aspects of the financing of the United Nations peacekeeping operations, UN Doc. A/C.5/59/31 (2005).

[21] This section draws on case studies focusing on the DDR-SSR nexus in Afghanistan and Burundi authored by Mark Sedra and Serge Rumin in support of a DCAF project to develop a module for the IDDRS on DDR and SSR. For the full case studies, see Bryden and Scherrer, note 2 above.

[22] Islamic Republic of Afghanistan, 'Afghanistan Millennium Development Goals Report 2010', available at www.undg.org/docs/11924/MDG-2010-Report-Final-Draft-25Nov2010.pdf: 8.

[23] Burundi, 'Rapport Burundi 2010 sur les Objectifs du Millénaire pour le Développement', available at www.undp.org/africa/documents/mdg/burundi_july2010.pdf.

[24] Michael Brzoska, 'Development Donors and the Concept of Security Sector Reform', Occasional Paper no. 4 (Geneva: DCAF, 2003): 10.
[25] Mark Sedra, 'Afghanistan', in *Disarmament, Demobilisation and Reintegration and Security Sector Reform – Lessons from Afghanistan, Burundi, the Central African Republic and the Democratic Republic of the Congo*, eds Alan Bryden and Vincenza Scherrer (Münster: LIT Verlag, forthcoming in 2012).
[26] Stéphanie Pezard and Nicolas Florquin, *Small Arms in Burundi: Disarming the Civilian Population in Peacetime* (Geneva: Small Arms Survey, 2007): 41–44.
[27] World Bank, note 3 above: 7.
[28] DIAG Project, 'First Quarter Report 2010', UNDP Afghanistan, available at www.undp.org.af/projects/Q1.ProgRep.2010/ANBP_DIAG_QPR_Q1_2010.pdf: 6.
[29] Sedra, note 25 above.
[30] UNDP, 'United Nations Development Programme Afghanistan Disbandment of Illegal Armed Groups (DIAG) Annual Project Report 2010', available at www.undp.org.af: 19
[31] Islamic Republic of Afghanistan, note 22 above: 44.
[32] US Government Accountability Office, 'Afghanistan Security. U.S. Programs to Further Reform Ministry of Interior and National Police Challenged by Lack of Military Personnel and Afghan Cooperation', Report to the Committee on Foreign Affairs, House of Representatives (Washington, DC: GAO, 2009): 12.
[33] Ibid.: 11.
[34] Sedra, note 25 above.
[35] Islamic Republic of Afghanistan, note 22 above: 44.
[36] US Government Accountability Office, note 32 above: 13–14.
[37] Sedra, note 25 above.
[38] International Crisis Group, 'The Insurgency in Afghanistan's Heartland', Asia Report no. 207 (Brussels: ICG, 2011): 25–28.
[39] Serge Rumin, 'Burundi', in *Disarmament, Demobilisation and Reintegration and Security Sector Reform – Lessons from Afghanistan, Burundi, the Central African Republic and the Democratic Republic of the Congo*, eds Alan Bryden and Vincenza Scherrer (Münster: LIT Verlag, forthcoming in 2012).
[40] Rumin (ibid.) provides a valuable comparison of the different implications of integration and reintegration from the perspective of the individual ex-combatant.
[41] Ibid.
[42] Ibid.
[43] UNDP, note 30 above: 20.
[44] Ibid.
[45] Rumin, note 39 above.
[46] UNDP, 'Community Security and Social Cohesion: Towards a UNDP Approach' (New York: UNDP Bureau for Crisis Prevention and Recovery, 2009), available at www.securitytransformation.org.
[47] Although the level of international support was much higher in the former case, this approach was a feature of SSR in Sierra Leone and South Africa – two reform processes generally held as examples of SSR good practice.
[48] World Bank, note 3 above: 13.

Chapter 8

Guerrillas, Gangsters and Contractors: Integrating Former Combatants and Its Impact on SSR and Development in Post-conflict Societies

Henri Myrttinen

Introduction

One of the many challenges faced by societies emerging from violent conflict is the disarmament, demobilisation and reintegration (DDR) of former combatants and others associated with fighting forces. Especially the third stage, successful reintegration into society, is a crucial but difficult element of post-conflict settlements. Given their experiences and allegiances to networks of loyalty from the conflict years, political clout, possible access to weapons and often a lack of skills needed on the labour market, former combatants and others previously associated with fighting forces have often formed a volatile demographic which can derail post-conflict reconstruction and development. The key question is whether and how security sector reform (SSR) – of which DDR processes are a part – can achieve the goal of integrating veterans, and whether this improves security and furthers positive political, social and economic development. Can DDR and SSR processes simultaneously help to integrate potential spoilers and potentially marginalised ex-combatants, and provide security environments conducive to development?

This chapter charts some of the different paths taken by veterans in post-conflict settings and analyses what impacts these have for the relevant SSR processes and post-conflict development in general. It draws mainly on field research I conducted in Aceh (Indonesia) and East Timor as well as secondary sources. The chapter begins with a brief outline of the conflict and post-conflict dynamics in Aceh and East Timor, followed by an outline of four common paths taken by former combatants: integration into security sector institutions, successful integration into post-conflict society, sliding into illegality and drifting into socio-economic marginality. In all four paths – which are not mutually exclusive – the importance of gender role expectations will be considered, as these can often determine which path the demobilised ex-combatant goes down. Men are, for example, often more likely than women to be seen as 'real' (i.e. armed) ex-combatants and suitable material for recruitment into police or armed forces, but also more likely to be members of networks of patronage. Men's participation in struggles is also more likely to be either valorised or excused as inevitable in hindsight (depending on whether they fought on the side of the victors or not), while women's contributions are often forgotten, belittled or vilified. The chapter concludes with a discussion of the interplay between DDR, SSR and development in the two case studies.

Aceh and East Timor: A brief overview of conflict and post-conflict dynamics

Situated at opposite ends of the Indonesian archipelago, both Aceh and East Timor went through decades-long independence struggles against the Indonesian central government from the mid-1970s onwards, followed by internationally brokered peace deals, DDR processes and complex international support operations.[1] In both cases, reintegration efforts were characterised by *ad hoc* approaches and marred by allegations of favouritism. In both cases, women associated with the fighting forces were, at least initially, neglected in the reintegration efforts. In neither case were DDR and broader SSR processes explicitly linked to post-conflict development agendas, but rather seen more narrowly as technical, security-sector-related issues.

In spite of numerous similarities, both processes also differ in stark ways – while Aceh remained a province of Indonesia, East Timor became an independent state, and arguably, for reasons discussed below, the post-conflict settlement in Aceh has been the more stable of the two with higher rates of socio-economic development.[2]

Aceh – Peace in the wake of the tsunami

The separatist uprising of GAM (Gerakan Aceh Merdeka – Free Aceh Movement) began in 1976 and the subsequent almost 30-year-long struggle resulted in an estimated 12,000–15,000 deaths, mostly civilians.[3] GAM was divided into a civilian and a military wing, the Tentara Neugara Aceh (TNA), which included a 'battalion' of female combatants, called the *inong balee*.[4] By the end of the struggle, GAM claimed in the peace negotiations to have a strength of 3,000 combatants, though this was more or less accepted by both sides as an 'artificial figure', lower than the actual number of active and former TNA, let alone other supporters of the struggle or pro-government militias.[5] Pressed by donors to give a precise figure on female GAM members who were to receive demobilisation benefits, the leadership of the movement gave the figure of 844, although this clearly does not include all women who supported GAM by performing different tasks during the struggle.[6]

While previous attempts to solve the conflict between GAM and the government of Indonesia had failed, the Southeast Asian tsunami on 26 December 2004 changed the situation radically. Though the government of President Susilo Bambang Yudhoyono had already made initial contact with GAM, the disaster, which killed an estimated 170,000 people in Aceh alone, gave both sides an additional pressing impetus to reach a settlement, along with local and outside pressure to find a peaceful solution to the conflict – unlike, for example, the case of Sri Lanka.

The memorandum of understanding signed in Helsinki by the representatives of the Indonesian government and GAM in 2005 covered political reforms in Aceh, human rights, amnesty and reintegration of GAM members into society, security arrangements (including DDR of 3,000 ex-GAM combatants, allocation of farmland or social security benefits to former combatants and pull-back of all non-locally recruited Indonesian troops), increased Acehnese control over natural resources, the establishment of the Association of Southeast Asian Nations/European Union Aceh Monitoring Mission and a dispute settlement mechanism.[7] No explicit link was made between DDR and post-conflict economic rehabilitation or the ongoing post-tsunami reconstruction efforts.

The settlement gave Aceh wide-ranging special autonomy in exchange for GAM dropping demands for independence. The package allowed for local political parties to be created – which GAM duly did with the establishment of Partai Aceh (the Aceh Party), the dominant party at local and provincial levels.[8] Due to the special autonomy provisions, this political power also equals vastly greater fiscal power than previously. According to the World Bank, 'special autonomy boosted natural resource revenues kept within Aceh by more than 150 times, from Rp 26 billion (US$2.7 million) in 1999 (or 1.4 per cent of the revenue) to Rp 4 trillion (US$421 million) in 2004 (or 40 per cent)'.[9] The combination of post-tsunami reconstruction and special autonomy funds has led to a construction boom, especially around the capital, Banda Aceh.

While the disarmament part of the DDR process ran relatively smoothly (though not without minor hiccups), the demobilisation and reintegration process were hampered by squabbles over the lists of the names of former GAM members.[10] The Indonesian government insisted upon a verifiable list of names for reasons of transparency, which GAM refused to give, citing security concerns. In the end the veterans' organisation Komite Perahilan Aceh (Aceh Transitional Committee), which more or less replicated the territorial structure of GAM, had its way and was allowed to distribute the US$1.8 million of reintegration money with little oversight.[11] In the distribution of the funds, those best integrated into male-dominated networks of patronage dating back to the conflict years fared best.

Practical reintegration efforts such as vocational training were to be carried out by Badan Reintegrasi Aceh (Aceh Reintegration Board), with technical and financial support from Japan, the World Bank and the International Organization for Migration (IOM). The work of Badan Reintegrasi Aceh was generally seen as lacklustre at best, hobbled by a lack of institutional capacity, accused of a lack of transparency and often seen by the supposed beneficiaries as not having delivered much beyond promises.[12]

Both the peace deal and the post-conflict development of the province need to be seen in the context of the 2004 tsunami and its aftermath. Though tentative peace talks had been under way before the catastrophe, the massive devastation and the pressing needs of reconstruction opened a window of opportunity for the peace deal. The amount of aid money which flowed into the post-tsunami relief efforts far outstripped the financial, material and technical support for the post-conflict reconstruction efforts – by a factor of 20 according to a World Bank estimate.[13] Though there has been some degree of discontent with this imbalance, the post-disaster relief industry created new economic opportunities for demobilised GAM members, especially as security guards or in transport, logistics and construction. By default rather than by design, the DDR efforts and post-disaster rehabilitation efforts became enmeshed as well-connected former GAM members secured their share of the reconstruction pie.[14]

Though peace has held in Aceh, internal political and economic turf battles between former GAM members (and possibly involving members of the Indonesian security forces) have led to several murders and attacks between 2009 and 2011.[15] While many of the lower-ranking former GAM members have not gained substantially from the economic and political possibilities of the post-conflict settlement, a return to a separatist conflict looks unlikely.[16] In contrast to the rank and file, many in the former leadership of GAM have benefited economically and politically from the settlement, at times reaping sizeable personal 'peace dividends' which are in part passed down to former comrades through networks of patronage.[17] Thus in spite of formal demobilisation, the formal structures from the time of the struggle have been transformed into informal networks which often can determine the pathways open to former combatants in the post-conflict settlement.

East Timor: Independence and instability

While the Acehnese independence struggle ended in the compromise of a special autonomy package, the East Timorese independence struggle, which was shorter by six years (1975–1999) but far more costly in lives, ended with independence for East Timor.[18] The Indonesian involvement began with the collapse of the Salazar/Caetano dictatorship in Portugal in 1974, which triggered a rapid decolonisation process in Portugal's African colonies – and in what was then known as Portuguese Timor. The left-wing Fretilin party emerged victorious from a brief civil war. Fearful of the spread of leftist influence in the region, neighbouring Indonesia, with Western approval, invaded on 7 December 1975.[19]

Though the Indonesian armed forces had established *de facto* control of the territory by 1980, Falintil (Forças Armadas de Libertação Nacional de Timor-Leste), the armed wing of the resistance, continued a small-scale guerrilla struggle with the support of a clandestine civilian network, the *clandestinos* and *clandestinas*.[20] The fall of the Suharto regime in 1998 opened a window of opportunity for conflict settlement, and a UN-organised referendum on independence was held in 1999. The run-up to the referendum and its aftermath were marked by a massive campaign of violence by pro-Indonesian militias and the Indonesian security forces. The international outcry over the violence after the August 1999 referendum led to the deployment of a UN peacekeeping force (INTERFET) in September 1999 and the establishment of a temporary UN administration (UN Transitional Administration in East Timor – UNTAET). UNTAET administered the territory until its independence in 2002, during which time the future structures of the state administration were established.

Prior to the independence referendum, the remaining approximately 1,900 Falintil voluntarily withdrew to cantonment areas to await demobilisation.[21] The DDR process of the former Falintil combatants fell indirectly under the mandate of UNTAET. Crucially, however, the DDR process was not in the mission's original mandate and this lack of leadership had important ramifications for the process. The DDR process thus relied heavily on *ad hoc* initiatives by various donors and peacekeeping contingents.[22] Several international agencies (including the IOM, Canadian International Development Agency, USAID and UNDP) were involved in the initial FRAP (Falintil Reinsertion Assistance Program) and subsequent UNDP-led RESPECT (Recovery, Employment and Stability Programme for Ex-Combatants and Communities in East Timor) programmes for ex-Falintil members not recruited by the new armed forces or police. While FRAP was aimed at individual ex-combatants, RESPECT had a more community-based approach aimed at vulnerable groups in general (in addition to ex-combatants, widows, conflict victims and unemployed youth were among the targets), with a more explicit development agenda. These programmes were less effective than had been hoped for, however, and were more of a stop-gap measure for ex-combatants than a real bridge into civilian life. FRAP failed to have any longer-term impact beyond the immediate reinsertion of ex-Falintil, while RESPECT was broadly seen as not having reached its goals due to structural problems with the programme, haphazard planning, lack of political support and poor community outreach.[23]

Demobilisation of pro-Indonesian militias, on the other hand, was by default the responsibility of the Indonesian authorities, as the militia groups had fled to Indonesian West Timor *en bloc*, forcibly deporting several hundred thousand civilians in the process. No comprehensive demobilisation was implemented for these groups. As Indonesian state interest in supporting them waned rapidly after 1999, the groups dissolved, though an association of former militia members was established to lobby for more financial support, at times resorting to violence.[24]

Two main career opportunities which opened up for former members of the resistance were inclusion into the new armed forces F-FDTL (Falintil-Forças de Defesa de Timor-Leste) and new police force PNTL (Polícia Nacional de Timor-Leste). Approximately 650 ex-Falintil were integrated into the former and some 150 into the latter, out of a total force size of approximately 1,500 and 3,000 respectively, based on a plan drawn up by foreign consultants.[25] As far as I am aware, no female ex-Falintil joined the

F-FDTL or PNTL, although both forces have relatively high percentages of women by international standards. This may be due to a preference given in the national recruitment process to those ex-Falintil who 'carried a gun' during the struggle, which tended to favour male ex-guerrillas.[26] Police and military training was given by the UN missions and through bilateral support programmes.

One of the main problems with the DDR process was that the terms 'veteran' and 'ex-combatant' were ill-defined from the outset. In the context of East Timorese society, the term 'veteran' is often seen in a broad perspective. It is not only the former weapons-bearing combatants who see themselves as veterans of the struggle, but also women and children associated with fighting forces and the urban network of *clandestinas* and *clandestinos* who supported the Falintil guerrilla force logistically as well as organising civilian protests in both occupied East Timor and Indonesia proper.[27] Several commissions have been established to register veterans, and at the time of writing the number of applicants for veteran status and possible financial compensation has swollen to over 200,000 persons. From the outset there were allegations that the commissions favoured those politically close to the president and former guerrilla commander, Xanana Gusmão.[28] Female ex-combatants were at least initially heavily sidelined, though pressure from local civil society organisations and international donors has changed this to a degree.[29]

The frustration of younger male veterans, especially former *clandestinos*, at the perceived lack of proper acknowledgement and of social and economic benefits led to a sizeable number of them joining violence-prone pressure groups, including veterans' associations.[30] The largest of the veterans' associations are Sagrada Familia and the Committee for the Popular Defence of the Democratic Republic of Timor-Leste. These have formed links to gangs and martial arts groups as well as to political parties.[31] Thus especially the young male veterans in East Timor have been far more vocal – and violent – than their Acehnese or female counterparts in demanding what they perceive as being their dues.[32]

The new security forces also proved internally fragmented and antagonistic to each other. Tensions came to a head in April–May 2006 when around 500 members of the F-FDTL – mostly from the western part of the country and mostly new members rather than ex-combatants – petitioned the political leadership alleging discrimination based on their regional background.[33] As demonstrations turned violent, the police and armed forces imploded, with armed civilians, police and military fighting street battles that left 37 people dead.

The violence led to a renewed and more robust UN mission, UNMIT (UN Integrated Mission in Timor-Leste), which had a limited mandate to assist in the reform of the PNTL. The reform process has been widely criticised and has fallen well short of its goals.[34] Nonetheless, full policing responsibility was handed back to the PNTL in March 2011, although by far not all criteria for reform had been met.[35] Due to reservations on the East Timorese side with regard to perceived outside interference in sensitive internal matters, neither a more comprehensive SSR process for the PNTL nor any kind of SSR beyond increased training for the F-FDTL was politically feasible.

Post-conflict trajectories: Four possibilities

Looking at the Acehnese and East Timorese case studies, there were four possible trajectories which veterans could follow in the post-conflict period: integration into security institutions (SIs); integration into the post-conflict/post-disaster reconstruction industry; integration into semi-legal or illegal structures; or fading into the margins of society.

These four options, as I discuss below, are not mutually exclusive.[36] Due to the dynamics of the processes in both Aceh and East Timor, in which the actual number of former combatants remains disputed, there are possibilities of multiple trajectories and there are statistical difficulties in tracking illegality and social exclusion, it is not possible to place exact figures on the sizes of the various groups.

Integration into security institutions

A favoured option for the reintegration of former members of fighting forces (and, usually to a lesser degree, those associated with them) is their incorporation into security institutions. Often this has meant a complete restructuring of the respective security forces to allow the incorporation of former adversaries into one force and turning regular and irregular combatants into members of SIs which are, ideally, accountable and transparent.[37] These SIs are mainly the police and armed forces but include other institutions, such as border guards, the penal service and private security companies (PSCs). Former combatants may also continue to be involved with security sector issues through civil society organisations, research institutes, media work or parliamentary oversight committees. In both cases, social norms of gender-appropriate behaviour, a tendency to see work in the security sector as being a male prerogative, the benefits of membership in male-dominated networks and preference given to those who 'carried a gun' tend to make involvement with SIs much more of an option for male rather than female ex-combatants.

In East Timor the integration of former Falintil into the new armed forces and, less prestigiously, the police was seen as the favoured option by many veterans.[38] In addition to a steady income, especially the F-FDTL as the mantle-bearer of Falintil brought with it the aura of the independence struggle. The image of the PNTL, at least initially, was tainted in the public eye by the inclusion of former members of the Indonesian police force POLRI.[39] A third SI option is the private security sector, either through PSCs or employment as an individual private security guard.[40] The work of all three of these SI options, however – armed forces, police and PSCs – has been undermined by problematic relationships of individual (mostly male) members to criminal and violence-prone groups.[41] This was especially visible during the crisis years of 2006–2008, though the problems have continued to undermine the professionalism of the SIs.[42]

More seriously, unresolved tensions between the F-FDTL and PNTL, and between competing networks of patronage within the two forces, led to an implosion of both forces and the near-collapse of the country in 2006. Though the PNTL has officially been undergoing a reform process, institutional resistance in the force and poor working relations between the UN mission and the East Timorese security sector and political elite have meant that many of the underlying problems have not been seriously addressed.[43]

Given the political history of the conflict and continued mutual animosity between former GAM members and the Indonesian security forces, incorporation into the state security sector did not play a major role for former combatants in the case of Aceh. Tensions between former GAM members, police, military and intelligence run high, often intermingling political differences with turf battles and shady dealings in which both members of the security forces and former GAM members are involved.[44] Rather than integrating into the Indonesian police or armed forces, a more common option for male ex-GAM members has been to join the forestry police force (*polisi hutan*), municipal security services or private security outfits. As in the case of East Timor, in the latter option it is not always possible to draw a clear line between *bona fide* security provision and direct or indirect extortion.[45]

Joining the gold rush

A second group of veterans in both Aceh and East Timor has successfully become involved in post-conflict/post-disaster reconstruction and development efforts. Those among the veterans who are not integrated into the SIs but are well connected and entrepreneurial, i.e. especially those formerly in command positions, have been able to use their wartime connections to gain from the opportunities presented by the post-conflict environment. In both cases, leading former resistance fighters have taken up senior political positions (including the governor of Aceh and the prime minister of East Timor) and have at times used their positions to further the interests of former combatants over other social groups. In the case of East Timor, this has for example included the blocking of both individual and community-based reparations to conflict victims by ex-Falintil parliamentarians who demand that veterans' needs be addressed first.[46]

A number of well-connected former combatants have integrated into the reconstruction economy in both Aceh and East Timor by becoming contractors. The term 'contractor' as it is used in both cases can be somewhat misleading. Being a *kontraktor*, as Indonesians term it, does not necessarily mean one or one's company actually carries out reconstruction efforts or provides real services, but rather that one is in a position to facilitate deals and receive commissions.[47] These dealings can be fully legitimate, but often take place in a legally murky system where the extra coercive edge which veterans can have over other business partners is their real or perceived access to violence, thus giving the activities a rent-seeking nature which is not uncommon in the construction industry.[48]

In both the Acehnese and East Timorese cases, field commanders or leaders of the clandestine support movement, some with legally questionable backgrounds, have been able to establish themselves as construction, logistics and transport contractors.[49] In Aceh a key role was played by the Komite Perahilan Aceh, which accommodated former members and kept old command structures and networks of loyalties in place which could then be translated into political and economic power, both legally (through positions in administration and the political machinery of Partai Aceh or legitimate jobs) and in illegal activities, as discussed below.[50] In East Timor it has allegedly been connections to key government figures, veterans' organisations and politically well-connected gangs led by former *clandestinos* which have opened up paths to employment for veterans.[51] By gaining access to state contracts, former commanders have been able to secure both financial benefits and 'jobs for the boys', such as working in construction or as drivers. Given the nature of much of the employment in the reconstruction industry, which is seen as being 'men's work', and the male-dominated nature of the networks of patronage in both Aceh and East Timor, it is male ex-combatants who have on the whole been more successful in utilising the economic and political opportunities opened up by post-conflict development processes.

Drifting into illegality

Post-conflict societies are often characterised by unclear legal and political situations, meaning that the business practices used may willingly or unwillingly extend into illegality, undermining goals of more transparent and legal processes and leading to increased criminality and thus insecurity. Individually or as a group, non-integrated veterans may end up operating in full illegality, e.g. through criminal gangs, often posing a serious threat to stability. These groups may also, through old ties dating back to the conflict years, extend into the new SIs, leading to conflicting loyalties and a possible undermining of SSR processes.

In both Aceh and East Timor allegations abound of former insurgents becoming involved in criminal economic activities, including theft, extortion, smuggling schemes and illegal logging. In part these are allegedly carried out in cooperation with active members of SIs, such as the police, forest rangers, employees of PSCs and border guards.[52] In part the criminal activities can build upon *modus operandi* and connections to black-market networks from the time of the struggle. Armed insurgents are by definition dependent on illegal activities to some degree for financing the insurgency, and these often lucrative habits can be hard to break – especially if legal options are either not available or economically less promising.

In Aceh, part of the income of GAM during the conflict years was assured by levying a 'revolutionary tax' from civilians and companies operating in the province, plus diaspora donations and criminal activities.[53] GAM field commanders in West Aceh were reportedly especially notorious for their propensity to resort to crime.[54] Other criminal fundraising activities for which GAM members were deemed responsible were kidnapping businesspeople for ransom, involvement in the marijuana trade, illegal logging and engaging in acts of piracy in the Malacca Straits.[55] Though comprehensive and reliable crime statistics are not available for the whole post-conflict period, conflict monitoring by the World Bank in Aceh reported 588 outbreaks of violence between October 2006 and September 2008, many of which were related to armed robbery and were mostly attributed to former combatants.[56]

Given in part the geographical location of East Timor, Falintil was far less diverse in its fundraising activities and relied mainly on public donations for its supplies and money with which to buy weapons and ammunition off Indonesian soldiers.[57] The degree to which coercion was used to obtain 'revolutionary taxes' from East Timorese and Indonesian businesspeople is not known, but that this did happen is probable. Members of the East Timorese diaspora occasionally also dabbled with illegal fundraising schemes in order to buy weapons, most famously perhaps the future minister of the interior, Rogério Lobato, who was involved in diamond smuggling in Angola, for which he was jailed for four years in 1983.[58]

Following the end of the conflict, former combatants – especially *clandestinos* associated with gangs and martial arts groups – have been linked to a range of crimes, including murder, arson, sexual assault, extortion and the drug trade (mostly methamphetamines).[59] Several veterans' organisations have also been linked to cross-border smuggling operations (e.g. car tyres), coercion and extortion (e.g. from small-scale salt miners), although all have denied this and claimed political motivations behind the allegations.[60]

The gendered nature of this option is in both cases rather striking. In my research, I did not come across any cases of female ex-combatants or supporters of the fighting forces becoming involved with violent crime. This has in both cases remained very much a male-dominated space.[61]

Fading into the margins

A fourth group of veterans is those who are neither incorporated into the new SIs nor fully integrated into the new opportunities of the post-conflict economy – be it legally or illegally. These are above all non-combatant members and supporters of the fighting forces, and this category therefore tends to include many women. These individuals are those who return to civilian life without joining pressure groups and often without reintegration packages. While from a political and security point of view they are arguably the least problematic, they can easily become socio-economically marginalised and be denied benefits enjoyed by other veterans. A range of factors often reduces their chances of successfully participating in post-conflict development processes, including lack of education, lack of political clout and connections and a lack of understanding of the workings of official DDR processes. Due to the fragmented and politically marginalised nature of this group and the domination of the discourse by more powerful ex-combatants' associations, they have mostly not been able to form power bases from which to lobby for their needs. Alliance-building with civil society organisations (e.g. women's organisations) in both Aceh and East Timor has often remained at the rhetorical level.[62]

Given the dynamics of the reintegration processes in Aceh and East Timor, several thousand potentially eligible persons were not initially included in the processes. Even among those who did enter the DDR process in one way or another, many, especially in Aceh, did not receive the full compensation and reintegration package, or quickly used up the limited funds from the packages. Lacking necessary education, experience and skills, former combatants have struggled with finding new sources of livelihood, leading to widespread sentiments of a failure of the process. The sense of disappointment can in many cases have been augmented by unrealistic expectations of the size of reinsertion packages and the 'peace dividend' at the end of the conflict.[63] In East Timor, disaffected younger male *clandestinos* have tended to gravitate towards ritual and martial arts groups and veterans' associations, and older, less well-connected male ex-combatants and women have tended to drift back into the socio-economic margins. In Aceh there has been no similar phenomenon of veterans' associations emerging – in part due to the more centralised nature of GAM and in part because this would in all likelihood not be tolerated by the Indonesian security forces. As in East Timor, though, less well-connected ex-combatants and women have mostly faded into political invisibility.[64]

In spite of UN Security Council Resolution 1325 (2000), women were sidelined throughout much of the political process leading to the post-conflict settlements in both Aceh and East Timor. Women associated with the respective fighting forces were only belatedly and partially taken into account in reintegration packages, and often only after pressure from local civil society organisations and outside actors.[65] For women, additional obstacles can come from prevailing societal attitudes disapproving of independent women and a labelling of the returnees as 'damaged goods', reducing their chances of marrying, which especially in the rural areas of Aceh and East Timor is a key element in the social integration of women.[66] In spite of their contributions to the struggles and the new social, political and economic opportunities (in addition to burdens) which women had during the conflict, the majority of women involved with and in Falintil and GAM have returned to traditional roles as subsistence farmers and/or small-scale vendors.[67]

Discussion

The examination of the case studies of the post-conflict settlements in Aceh and East Timor through the trajectories of former combatants and others associated with the former fighting forces highlights some of the linkages between SSR (and especially DDR) processes and development. In neither case, however, was the nexus explicitly reflected in policy or implementation, in part due to the relatively *ad hoc* manner in which the processes were carried out. Beyond the narrow scope of the provision of one-off reinsertion packages in the form of cash and other benefits and very limited vocational training, SSR and DDR were in both cases mainly regarded as being technical, strictly security sector processes separate from broader socio-economic issues. The clearest attempt to link reintegration with broader community needs and development goals, the RESPECT programme in East Timor, largely failed to meet its goals.

In both Aceh and East Timor, former combatants and those demanding recognition for their role in the conflict years have in part remained a restive demographic, threatening stability. In Aceh this has mostly taken the form of criminal activity by former combatants (and occasional political violence), while East Timor has seen the emergence of violence-prone organisations which tap into the disaffection of *clandestinos* and former combatants. A key reason for this disenchantment with the post-conflict settlement can be found in the way the DDR processes were run in both cases. The lack of a proper definition of eligibility for reintegration benefits and a lack of accountability (especially in the case of Aceh) have allowed politically better-connected ex-guerrillas to benefit at the expense of other social groups.

The breakdown in 2006 of the East Timorese police and armed forces, coupled with the emergence of violent groups that involved ex-combatants, underlined in the starkest possible way the risks involved with unsuccessful SSR processes – and in terms of post-conflict development. The crisis laid bare the problems caused by a lack of democratic oversight, abuse of power, corruption and links to criminal groups within the SIs. The violence which emanated from the implosion of the new armed forces and police paralysed the country between 2006 and 2008, led to 10–15 per cent of the population permanently living in refugee camps for up to two years, brought to a halt commercial activity in the capital city whenever street fighting erupted and saw important infrastructure going up in flames, setting the fragile economy back by years.

In both Aceh and post-2006 crisis East Timor, potentially troublesome ex-insurgents have been 'bought out' strategically, either through integration into the respective SIs or by allowing them to gain preferential access to post-conflict development funds as 'contractors' and in the process allowing some degree of semi-legal or illegal business practices to blossom. In both cases, these partially incorporated ex-combatants have benefited from mini-economic booms caused by the influx of outside funds into public coffers controlled by former comrades – post-tsunami rehabilitation and special autonomy funds in the case of Aceh and oil/gas revenues in East Timor. At the same time, many ex-combatants and others associated with the former fighting forces, especially women, who neither had the connections and power to cash in on the new political and economic opportunities nor joined pressure groups have been left to fade into socio-economic marginality. While this has contributed in both cases to political stability, it also raises two troubling questions. Is the price of allowing some degree of corruption, collusion and nepotism, as well as other illegal economic activities which would help smooth over the process of transition, worthwhile and better than a return to war?[68] And given the limited funds, project timeframes and donor attention spans, is it better to focus on the potential troublemakers than on those who are unlikely to raise their voices, let alone take up arms?

Affirmative answers to both these questions go against the grain of the accepted spirit (and, arguably, legal frameworks) of donor-supported SSR processes. Yet in the practical implementation of the processes in both Aceh and East Timor, this has been implicitly the case. Short- and medium-term corruption, criminal activities and impunity by veterans and/or members of SIs have been officially decried but accepted in practice for the sake of keeping the peace.

In both in Aceh and East Timor the DDR/SSR processes have tended to favour those among the protagonists in the struggle who were part of the armed resistance and had access to influential, mostly male-dominated and in part militarised networks of loyalty and patronage. Jobs created tend to be in fields which are considered male domains, such as drivers, construction workers or security guards. In both cases, however, these networks have also shown their potential to become factors of instability, be it in the political or criminal arena. Ominous grumblings of a return to violence by these groups if their demands are not met continue to be heard in Aceh and East Timor – and in both the demands continue to grow. As in the past, open and veiled threats have helped other would-be spoilers in reaping benefits, as there is little incentive to refrain from making these demands.

Women directly associated with the fighting forces, members of the civilian resistance and those men and women who did not have access to the right networks have been sidelined, though in both cases local civil society organisations and parts of the donor community have lobbied for more comprehensive approaches. In spite of the increased rhetorical commitment, however, no further efforts were made (beyond the ill-fated RESPECT programme) to target more vulnerable groups specifically.

In the short and medium terms, this implicit, expedient favouring of the strong, the savvy and those with access to political and coercive capital among the veterans can help to buy peace. Quite apart from ethical concerns, however, this tacit acceptance of predatory behaviour – be it by politically powerful veterans, rent-seeking contractors or police officers connected to criminal gangs – encourages in the longer term the development of institutional cultures within the SIs and society in general which can be detrimental to political and economic development. The possibility of rent-seeking which has been opened up can be a disincentive to finding more strenuous but socially and economically more productive livelihoods – why toil away in a coffee cooperative for a pittance when vastly greater sums can be made by facilitating fake construction contracts? While the influx of money in Aceh and East Timor has allowed a 'buying of the peace', this has bred a sense of entitlement among those now receiving a share of the pie and increasing demands – demands which may one day no longer be met once the influx of funds dries up, making a return to criminal or political violence a distinct possibility, with detrimental impacts on development.

Post-conflict settlements inevitably require some degree of 'buying off' potential spoilers as well as processes of recognition, compensation and reconciliation for former combatants within the frameworks of SSR and DDR. In the two cases of Aceh and East Timor, these processes were seen mostly as being separate from broader developmental agendas, and most lacked any serious consideration of gender issues beyond rhetoric. Interestingly, this occurred in spite of major post-conflict/post-disaster rehabilitation efforts occurring in parallel to the DDR/SSR processes and a commitment (especially by international actors) to gender mainstreaming. While the buying-off strategy has been successful in the short term in ensuring the security necessary for development, its long-term costs for society may outweigh the benefits. Marginalisation of women and subordinate men has increased socio-economic disparities.

Arguing counterfactually, would SSR/DDR processes with a broader outreach and a more explicit link to development have been more successful? The potential benefits of such an approach might have been a more equitable post-conflict settlement, in which access to political and economic power would not be determined by belonging to a male-dominated 'in group' of ex-combatants. This could possibly have avoided the emergence of ex-combatants' groups displaying predatory behaviour, better socio-economic integration across the board, increased gender equality and avoiding socially damaging competition between various groups, e.g. weighing victims' claims against veterans' claims. As Alan Bryden argues (Chapter 7), a more community-based approach to reintegration has the potential to reduce tensions between former combatants and the rest of society. Also, a more locally rooted and bottom-up approach can increase the chances that combatants are actually reintegrated rather than maintaining combatant identity and behaviour in another form. Had the conceptual link between SSR/DDR and development been greater from the outset, more of a focus would have been on the reintegration part of DDR, and on what is needed from security sector institutions in order for them to be conducive to development. Echoing Bryden, a more community-centred approach allows critical local knowledge to be tapped into in order to build the links between development and security goals.

A positive outcome of a more inclusive and community-based approach is, however, far from guaranteed. As the example of the RESPECT programme shows, proper implementation needs political backing and local buy-in and ownership as well as sound planning and execution.

Notes

1 In both cases only the respective insurgent group underwent DDR processes, though pro-Indonesian militia groups were demobilised and disarmed outside official DDR processes. The other conflict parties, the Indonesian police and armed forces, have obviously not entered a DDR process, although there have been slow-moving efforts at SSR since 1998. See for example Fabio Scarpello, 'SSR and Hybrid Democracies: A Critique, Post-authoritarian Democratisation and Military Reform in Indonesia', Master's dissertation, Murdoch University, Perth (2010).

2 Though only a partial and imperfect indicator, the respective GDPs per capita for Aceh (US$1,962.75 in 2006) and Timor-Leste (US$518.20 in 2008) give a rough idea of the levels of economic development. Figure for Aceh from World Bank, Aceh Economic Development, available at http://web.worldbank.org/WBSITE/EXTERNAL/COUNTRIES/EASTASIAPACIFICEXT/INDONESIAEXTN/0,,contentMDK:21653738~pagePK:141137~piPK:141127~theSitePK:226309,00.html; data for Timor-Leste from UN Data, available at http://data.un.org/CountryProfile.aspx?crName=Timor-Leste.

3 Figure from Anthony Reid, ed., *Verandah of Violence – The Background to the Aceh Problem* (Singapore: Singapore University Press, 2006), which also gives a comprehensive overview of Acehnese history.

4 Kirsten E. Schulze, 'Insurgency and Counter-insurgency: Strategy and the Aceh Conflict, October 1976–May 2004', in *Verandah of Violence – The Background to the Aceh Problem*, ed. Anthony Reid (Singapore: Singapore University Press, 2006): 228. Whether or not the *inong balee* 'battalion' was an actual combat unit or set up mainly for propaganda purposes remains an issue of debate among scholars.

5 International Crisis Group, 'Aceh: Post-conflict Complications', Asia Report no. 139 (Brussels/Jakarta: International Crisis Group, 4 October 2007): 10. The low number has been in part explained by security concerns by GAM, but can also have been due to a lackadaisical approach to the issues by the negotiators of the memorandum of understanding, which was being agreed with a very tight schedule. Later Badan Reintegrasi Aceh estimates include 6,200 non-TNA GAM members; 3,204 GAM who surrendered before the memorandum; 6,500 anti-GAM militias; approximately 6,000 former political prisoners; approximately 62,000 other conflict victims; and 14,932 disabled. Figures from Roman Patock, 'Reintegration in Aceh – Continuation of War with Other Means?', discussion paper presented at ICAS/AAS Conference, Honolulu, 31 March–3 April 2011.

6 International Crisis Group, ibid.: 10.

7 For a full text of the 'Memorandum of Understanding between the Government of the Republic of Indonesia and the Free Aceh Movement' see www.aceh-mm.org/download/English/Helsinki%20MoU.pdf.

8 This is a constitutional anomaly in Indonesia, where otherwise only parties registered nationally, with a presence in all provinces and a party headquarters in Jakarta, are allowed to contest elections at any level. Despite a rhetorical commitment to increasing women's representation, all Partai Aceh deputies in the provincial parliament are men.

9 Patrick Barron and Samuel Clark, 'Decentralising Inequality? Center-Periphery Relations, Local Governance and Conflict in Aceh', World Bank Social Development Papers no. 39 (Washington, DC: World Bank, 2006): 7.

10 These included GAM members handing in fewer weapons than agreed or weapons in an unusable condition which Indonesian authorities refused to include in the final count. See

for example Katri Merikallio, *Miten rauha tehdään – Ahtisaari ja Aceh* (Helsinki: WSOY, 2006).

11 International Crisis Group, note 5 above: 9.

12 Interviews with former GAM members and Acehnese civil society organisations quoted in Henri Myrttinen and Nicole Stolze, 'Ignore at Your Own Peril? – Notes on the Lack of a Gender Perspective in, and the Implications for, the DDR Processes in Aceh and Timor-Leste', paper presented at Fifth EuroSEAS Conference, Naples, 12–15 September 2007. For similar and more damning estimates see Edward Aspinall, 'Combatants to Contractors: The Political Economy of Peace in Aceh', *Indonesia* 87 (April 2009): 1–34; International Crisis Group, note 5 above: i–iii.

13 The World Bank estimate is cited in Patrick Barron, 'The Limits of Reintegration Programming: Lessons for DDR Theory and Practice from Aceh', in *Small Arms Survey 2009* (Geneva: Small Arms Survey, 2009).

14 Author's interviews in Aceh, May–June 2011; Aspinall, note 12 above; Patock, note 5 above.

15 These shady dealings led to several unresolved murders of Partai Aceh members in the run-up to the 2009 and 2011 elections, with fingers being pointed at criminal elements, former GAM members or members of the Indonesian security forces. International Crisis Group, 'Indonesia: Deep Distrust in Aceh as Elections Approach', Asia Briefing no. 90 (Brussels: International Crisis Group, 2009); International Crisis Group, 'Indonesia: GAM vs GAM in the Aceh Elections', Asia Briefing no. 123 (Brussels: International Crisis Group, 2011).

16 Author's interviews in Aceh and Jakarta, May–June 2011.

17 The introduction to Aspinall has a very vivid description of this. Aspinall, note 12 above: 1–2.

18 The conservative estimate of the East Timorese Truth and Reconciliation Commission is 102,800 deaths during the occupation, the majority of them civilians killed by disease and starvation during the initial years of the Indonesian occupation. CAVR, *Chega! Relatório da Comissão de Acolhimento, Verdade e Reconcilição de Timor-Leste – Executive Summary* (Dili: Comissão de Acolhimento, Verdade e Reconcilição de Timor-Leste, 2005): 44.

19 Don Greenless and Robert Garran, *Deliverance – The Inside Story of East Timor's Fight for Freedom* (Crow's Nest, NSW: Allen & Unwin, 2002): 12–15. It was the year in which Cambodia, Laos and South Vietnam had been taken over by leftist forces.

20 Reliable figures on membership in the armed and unarmed resistance, as discussed in the section on registering veterans, are contested and there are no gender-aggregated data available. Women did, however, join the struggle at all levels, including in combat units.

21 IOM, 'Falintil Reinsertion Assistant Program (FRAP) Final Evaluation Report' (Dili: International Organization for Migration, 2002): 8.

22 Gordon Peake, *What the Veterans Say: Unpacking Disarmament, Demobilisation and Reintegration (DDR) Programmes in Timor-Leste* (Bradford: University of Bradford, 2008); Edward Rees, *Under Pressure: FALINTIL Forças de Armadas de Timor-Leste – Three Decades of Defence Development in Timor-Leste* (Geneva: DCAF, 2004): 6.

23 FRAP provided former Falintil with transport to their home communities from the cantonment areas, monthly payments of US$100 for five months and a one-off reintegration package of US$550. These funds were generally quickly spent and their actual meaning was not clear to all demobilised Falintil. In spite of the explicit mention of

supporting vulnerable women, RESPECT has been criticised for its lack of actually engaging with women in the implementation phase. See La'o Hamutuk, 'Observations Regarding the RESPECT Program in East Timor', *La'o Hamutuk Bulletin* 5, no. 5/6 (Dili: La'o Hamutuk, 2004): 1–3; Myrttinen and Stolze, note 12 above; Peake, ibid.: 11–12; Rees, ibid.: 47.

[24] Henri Myrttinen, 'Histories of Violence, States of Denial – Militias, Martial Arts and Masculinities in Timor-Leste', PhD thesis, University of KwaZulu-Natal, Durban (2010): 214–215, 306.

[25] King's College London, *Independent Study on Security Force Options and Security Sector Reform for East Timor* (London: Centre for Defence Studies, King's College London, 2000).

[26] This bias in favour of armed former members of the resistance *vis-à-vis* others tends to be reflected more generally in the definition and treatment of veterans in Timor-Leste. See for example Kate Roll, 'Defining Veterans, Capturing History: State-led Post-conflict Reintegration Programmes in Timor-Leste', MPhil thesis, University of Oxford, Oxford (2011).

[27] Peake, note 22 above; Roll, ibid.

[28] Author's interviews in Dili, June 2011. See Roll, note 26 above: 60–66 for a comprehensive overview of the various registration processes.

[29] Myrttinen and Stolze, note 12 above; Elisabeth Rehn and Ellen Johnson Sirleaf, *Women, War and Peace: The Independent Experts' Assessment on the Impact of Armed Conflict on Women and Women's Role in Peace-building* (New York: UNIFEM, 2002): 117.

[30] José Sousa-Santos, '"The Last Resistance Generation": The Reintegration and Transformation of Freedom Fighters to Civilians in Timor-Leste', in *Nation-building across the Urban and Rural in Timor-Leste, Conference Report*, eds Damian Grenfell, Mayra Walsh, Januario Soares, Sofie Anselmie, Annie Sloman, Victoria Stead and Anna Trembath (Melbourne: RMIT, 2010): 67–71; James Scambary, 'Anatomy of a Conflict: The 2006–2007 Violence in East Timor', *Conflict, Security & Development* 9, no. 2 (2009): 265–288.

[31] Scambary, ibid.: 269–271.

[32] Recognition of the contribution of younger members of the resistance and *clandestinos* remains a contentious issue. To date, veterans' benefits are linked to age and seniority, with a strong bias towards members of the armed resistance. The government, civil society organisations, the Catholic Church and the donor community have sought to address 'the youth issue' (as the phenomenon of young, violent men is often referred to) through classic social work (e.g. setting up youth centres and organising sports activities) and limited vocational training programmes (Myrttinen, note 24 above; author's interviews in Dili and Gleno, July 2010 and June 2011).

[33] One of the underlying issues was the question of who has been more deserving of recognition for contributions to the independence struggle. The younger F-FDTL members from the western regions felt that older ex-Falintil were getting a better deal in the armed forces, that 'westerners'' contribution to the struggle was not being sufficiently recognised and that 'easterners' were generally being given preferential treatment. Prior to the petitioners' crisis, it had tended to be groups from the eastern areas (such as Sagrada Familia) which claimed socio-economic discrimination. In terms of post-conflict economic development, however, there is little difference between the eastern and

western districts and they are marked by equality in poverty, with economic activity being heavily concentrated in the capital, Dili.

[34] See for example Jim Della-Giacoma, 'The UN's Lame Security Review for Timor-Leste', *The Interpreter*, Weblog of the Lowy Institute for International Policy (17 February 2009), available at www.lowyinterpreter.org/post/2009/02/17/The-UNs-tame-security-review-for-Timor-Leste.aspx; Yoshino Funaki, *The UN and Security Sector Reform in Timor-Leste: A Widening Credibility Gap* (New York: Center on International Cooperation, 2009); International Crisis Group, 'Timor-Leste: Time for the UN to Step Back', Asia Briefing no. 116 (Brussels: International Crisis Group, 2010); Elisabeth Lothe and Gordon Peake, 'Addressing Symptoms but not Causes: Stabilisation and Humanitarian Action in Timor-Leste', *Disasters* 34, suppl. 3 (October 2010): S427–S443; Jùlio Pinto, 'Reforming the Security Sector: Facing Challenges, Achieving Progress in Timor-Leste', *Tempo Semanal* (18 August 2009); Bu Wilson, 'Crime Fiction: Regulatory Ritualism and the Failure to Develop the East Timorese Police', presentation at Conflict, Interventionism and State-building: Lessons from the Melanesian Pacific and Timor-Leste Workshop, Canberra, 7–8 December 2010.

[35] This fact was noted rather prominently in UN Security Council Resolution 1969 (2011) mandating the extension of the UNMIT mission until February 2012.

[36] An ex-combatant turned police officer may well, for example, moonlight as a security guard for a legal construction site and occasionally be hired as extra muscle for extortion – not unlikely prospects in other post-conflict societies, either.

[37] This is more common in the case of armed forces (e.g. in Nepal or post-apartheid South Africa), but in the case of Timor-Leste former Falintil guerrillas and former Indonesian police officers were recruited into the PNTL while the author knows of no former members of the Indonesian military joining the F-FDTL.

[38] Henri Myrttinen, Monika Schlicher and Maria Tschanz, eds, *'Die Freiheit, für die wir kämpften...' Osttimor: Facetten eines Wandels* (Berlin: Regiospectra Verlag, 2011): 155–168; Peake, note 22 above; Roll, note 26 above.

[39] Eirin Mobekk, *Law Enforcement: Creating and Maintaining a Police Service in a Post-conflict Society – Problems and Pitfalls* (Geneva: DCAF, 2003).

[40] These positions tend to be highly male-dominated, with the exception of clerical staff – who tend to be female but younger and better educated than former female Falintil members. Author's observations.

[41] Henri Myrttinen, 'Poster Boys No More – Gender and SSR in Timor-Leste', DCAF Policy Paper no. 31 (Geneva: DCAF, 2009).

[42] This was also noted in UNSCR 1969 (2011) on the extension of the UNMIT mission.

[43] See for example International Crisis Group, note 34 above; Lothe and Peake, note 34 above; Wilson, note 34 above.

[44] International Crisis Group, 2009, note 15 above.

[45] Author's interviews in Aceh, May–June 2011.

[46] La'o Hamutuk, 'Proposed Laws on Reparations and Memory Institute', *La'o Hamutuk Bulletin* (18 February 2011), available at www.laohamutuk.org/Justice/Reparations/10ReparIndex.htm.

[47] Aspinall, note 12 above: 2.

[48] This is not merely the case in post-conflict Aceh and Timor-Leste. As the annual report of Transparency International noted in 2005, the construction industry is worldwide one of the least transparent – and often one of the most dynamic in post-conflict societies where

reconstruction needs are great and legal oversight minimal. Transparency International, '2005 Annual Report' (Berlin: Transparency International, 2005).

49 In Timor-Leste, several veterans who were key figures in the 2006 crimes and named as criminal suspects by the UN Commission of Inquiry report have since been given lucrative positions in government infrastructure development and construction projects which have been financed by increased public revenue from the Petroleum Fund. The impact of these projects on the ground has, however, based on the author's observations, been mostly minimal to non-existent. A further lucrative field for ex-combatants has been controlling rice distribution. Author's interviews in Aceh and Timor-Leste, June 2011. For Aceh see also Aspinall, note 12 above; International Crisis Group, note 5 above; International Crisis Group, 2009, note 15 above.

50 Aspinall, note 12 above: 14–17; International Crisis Group, note 5 above: 5.

51 Myrttinen et al., note 38 above: 155–168; author's interviews in Timor-Leste, 2009–2010.

52 Author's interviews in Aceh, Jakarta and Timor-Leste, 2007–2010. Similar claims are also raised for example in Jacqueline Siapno, 'Human Safety, Security and Resilience: Making Narrative Spaces for Dissent in Timor-Leste', in *East-Timor: How to Build a New Nation in Southeast Asia in the 21st Century?*, eds Christine Cabasset-Semedo and Frédéric Durand (Bangkok: IRASEC, 2009); and more discreetly in Fundasaun Mahein, *Operasaun Krime Organizadu iha Timor Leste* (Dili: Fundasaun Mahein, 2010).

53 The line between collecting a revolutionary tax and extortion is often a very thin one. The tax was also levied upon transnational companies active in the province, such as ExxonMobil. Kirsten E. Schulze, 'Indonesia, GAM and the Acehnese Population in a Zero-Sum Trap', in *Terror, Insurgency, and the State: Ending Protracted Conflicts*, eds Marianne Heiberg, Brendan O'Leary and John Tirman (Philadelphia, PA: University of Pennsylvania Press, 2007): 95; Aspinall, note 12 above.

54 Patrick Barron, Samuel Clark and Muslahuddin Daud, *Conflict and Recovery in Aceh – An Assessment of Conflict Dynamics and Options for Supporting the Peace Process* (Jakarta: World Bank, 2005); Merikallio, note 10 above: 178.

55 Schulze, note 4 above: 228–229; Schulze, note 53 above: 92–93. Incidentally, members of the Indonesian security forces also routinely engaged in illegal economic activities in Aceh, including extortion, illegal logging, smuggling and the drugs trade. Edward Aspinall, 'Violence and Identity Formation in Aceh under Indonesian Rule', in *Verandah of Violence – The Background to the Aceh Problem*, ed. Anthony Reid (Singapore: Singapore University Press, 2006): 164–165; Damien Kingsbury and Lesley McCulloch, 'Military Business in Aceh', in *Verandah of Violence – The Background to the Aceh Problem*, ed. Anthony Reid (Singapore: Singapore University Press, 2006): 199–224.

56 Patrick Barron, Enrique Blanco Armas, David Elmaleh and Harry Masyrafah, 'Aceh's Growth Diagnostics: Identifying the Binding Constraints to Growth in a Post-conflict and Post-disaster Environment', paper presented at Second ICAIOS Conference, Banda Aceh, 23–24 February 2009.

57 Constancio Pinto and Matthew Jardine, *East Timor's Unfinished Struggle – Inside the Timorese Resistance* (Boston, MA: South End Press, 1997): 102.

58 Loro Horta, 'East Timor: A Nation Divided', *Open Democracy*, available at www.opendemocracy.net/democracy-protest/easttimor_3629.jsp.

59 Robert Muggah, ed., 'Urban Violence in an Urban Village: A Case Study of Dili, Timor-Leste', Geneva Declaration Working Paper (Geneva: Geneva Declaration Secretariat, 2010); James Scambary, Hipolito da Gama and João Barreto, 'A Survey of Gangs and

Youth Groups in Dili, Timor-Leste', report commissioned by AusAID Timor-Leste, Dili (2006).

[60] Kate McGeown, 'Aceh Rebels Blamed for Piracy Monday', BBC News Online, (8 September 2003), available at http://news.bbc.co.uk/2/hi/asia-pacific/3090136.stm; 'Court Decides L-7 as Suspect of Magnesium Case', *Suara Timor Loro Sa'e* (1 October 2010); 'Ex-guerilla Fighters Threaten Belo and STL', *Timor Post* (16 July 2010); 'Manatuto District Police Detain Four People Due to Illegal Activity', Radio Timor-Leste (23 February 2011). The cases involving the former Falintil fighters were not settled yet at the time of writing.

[61] Women and girls have been involved in gang violence in Timor-Leste to a very small degree, but at least from the information I have been able to collect they have not been former combatants. On possible links between concepts of masculinities and gang violence in Timor-Leste, see for example Myrttinen, note 24 above.

[62] Author's interviews in Aceh and Timor-Leste, November 2010 and May 2011; Myrttinen et al., note 38 above: 155–168.

[63] Author's interviews in Aceh and Timor-Leste, May–June, 2011; International Crisis Group, note 5 above; Myrttinen and Stolze, note 12 above; Peake, note 22 above: 12; Roll, note 26 above.

[64] Nonetheless, disaffection in Aceh has led to grumblings among ex-combatants about returning to more violent ways to make their voices heard. Patock, note 5 above. On marginalisation: author's interviews in Aceh and Timor-Leste, October 2010 and May–June 2011.

[65] Irene Cristalis and Catherine Scott, *Independent Women – The Story of Women's Activism in East Timor* (London: CIIR, 2005): 40–41; Myrttinen et al., note 38 above: 157–159.

[66] Vanessa Farr, *Gendering Demilitarization as a Peacebuilding Tool* (Bonn: BICC, 2002): 15–17.

[67] Author's interviews, Aceh and Timor-Leste, October–November 2010 and May–June 2011.

[68] The point is raised also by Aspinall, note 12 above, and to a degree echoes lines of thinking outlined in Christopher Cramer, *Civil War Is Not a Stupid Thing: Accounting for Violence in Developing Countries* (London: Hurst & Company, 2006); Philippe Le Billon, 'Buying Peace or Fuelling War: The Role of Corruption in Armed Conflicts', *Journal of International Development* 15, no. 4 (2003): 413–426.

PART V

TOWARDS 'DEVELOPING' SSR POLICY?

Chapter 9

Security Sector Reform and State-building: Lessons Learned

Paul Jackson

Introduction

Research shows that the number of wars and their lethality have been declining since 1992, and over the same time the worst conflicts declined by over 80 per cent.[1] However, research also shows that the improvements result from more wars ending: the onset of new wars, regrettably, remains constant.[2] 'Failed', 'weak' or 'fragile' states, home to the poorest billion of people living in fewer than 60 countries, 70 per cent of which are located in Africa,[3] are still most at risk of falling into conflict.

Many of these states may also have a dysfunctional security sector that is either politically compromised, chronically underfunded or subject to conflict and unable to control sovereign territory or criminal activity. From an international donor perspective, ignoring such states risks furthering their decline, while carefully designed interventions, including the reform of their security apparatus, may help them develop. There is a danger, however: adding a security component to overseas development aid could affect strategic decisions about aid allocation and shift objectives to meet Western security concerns. This would amount to a full securitisation of aid. Given scarce resources and global political realities, difficult decisions must be made and a clear agenda set to ensure that development and SSR overlap and support each other.

By highlighting the conflict-development link, donors like the UK may be in a better position to show that aid money not only helps prevent poor countries from declining into conflict, but contributes to keeping the West safe. The assumption is that the recurring cycle of violence that derails development and human security in general could be broken by a more strategic use of international funding aimed at developing opportunities for those in conflict-affected areas to make a living other than by resorting to violence to survive. In this approach, a post-conflict agenda based on a broader definition of security and its relationship to development could set out a new strategic logic for development aid that may make sense for both the West and the poorest and most vulnerable.

This approach, however, raises the question of what or who development is for. Are development and support for failed states intended to maintain the status quo of existing governance systems and the interests of the donors, or do they aim to assist the people on the ground in the affected countries? The history of interventions that attempt to construct governance systems that deliver development outcomes to the general population, as opposed to primarily security outcomes for the general community of states, is not necessarily a good one, although such interventions continue, as in the international efforts in Iraq and Afghanistan.

This chapter outlines a series of challenges to post-conflict security sector governance, understood as management of the national security sector. It is written within a conceptual framework that emphasises 'governance' rather than 'government', and recognises the large diversity of actors and processes and the multiplicity of contexts in which security sector reform (SSR) takes place. Making the post-conflict environment more secure involves managing, demobilising and integrating militias, establishing the rule of law (and justice more broadly), ensuring that past crimes are redressed and constructing a security governance system that prevents future threats to the general population.[4] The security governance perspective facilitates a comprehensive approach to delivering legitimate, accountable and publicly owned security. This goes to the heart of what it means to govern well.

The post-conflict environment places extreme pressure on the relationships within the national security sector, incorporating both uniformed and non-uniformed security services (military, police, intelligence) and the state institutions and government oversight mechanisms that monitor those organisations authorised to use force. Functioning oversight mechanisms create a useful pressure to govern the security sector accountably, particularly where the military has a history of brutality. Delivering appropriate security remains critical to the core functioning of governance more broadly.

This chapter works within a framework that moves beyond institution-building as exclusively Westphalian. It attempts to place current approaches to state-building within a broader historical process and also show that the reconstruction of governance following conflict is best understood as a function of political networks rooted in substate and regional networks.

The post-conflict environment

In post-conflict environments, security sector governance is frequently seen as part of the broader development of public administration and governance. However, ministries of defence are not always part of unified governance reform agendas. In Sierra Leone, for instance, Ministry of Defence reform was an integral part of SSR programming, but was completely excluded from the more general public sector reform programme within core ministries.[5] As another example, the post-conflict environment within Nepal is dominated by military tension between the Maoist Army and the National Army, and a political situation in which the core political parties find it extremely difficult to agree. The Ministry of Defence, as far as it exists at all, is not capable of policy formulation and the political impasse effectively prevents it from developing governance powers. What this means in practice is that the discipline of the two forces is achieved by informal political agreements and a general commitment to the Comprehensive Peace Agreement. In Nepal's mistrustful post-conflict atmosphere, the security governance system remains fragile and risky, notwithstanding the fact that the peace has held for some five years and there has been very little violence by international standards.[6]

Security governance itself has been seen as an integral element of SSR programming within a number of countries. In fact, the development of SSR itself (and security governance) has been shaped by engagement in post-conflict situations. The UK's experience in Sierra Leone coincided with its leadership of the OECD/DAC group that produced the guidelines on SSR, for example. As discussed in Chapter 10, while this initially reflected a security-driven view of post-conflict intervention, it also incorporated a number of broader governance and development objectives, including recognition that economic and political development is necessary to support security more broadly.[7]

However, security sector governance did not start with the OECD/DAC, and as the early example of Zimbabwe shows clearly, poorly executed security policies aiming, for instance, to reintegrate former combatants following civil wars can have political consequences later on. In Zimbabwe the political allegiance of the security services has steadily undermined the possibility that development gains can be achieved.[8] To prevent further situations like this, holistically designed post-conflict SSR is important in setting the future political agendas of the state and ensuring that development trajectories do actually contribute to lasting peace.

The study of post-conflict states is blessed with a wide and varied lexicon of terms that overlap, contradict and confuse while trying to describe varying forms of state collapse. Whether fragile, weak, collapsed or neopatrimonial, dysfunctional states all suffer from vulnerability to external shocks, internal conflict, competing economic and political structures and an inability to exercise effective legal control within state borders. A post-conflict state may exhibit all these features and be subject to continuing, cyclical violence, making the prospect of lasting SSR all the more difficult.

For an inexperienced designer of SSR, the challenge may be that dire conditions create the illusion of a 'blank slate', which may appear attractive for reconstruction and SSR. However, this notion is dangerous and illusory, as it leads those designing SSR interventions to ignore existing norms, structures and the country's previous history. This may result in a 'one-size-fits-all' approach that can undermine long-term security and development sustainability. While SSR donors should be cautious of treating post-conflict states as a 'blank slate', there remains nonetheless a window of opportunity for reform through the provision of a series of entry points. For instance, there may be a national will to accept some forms of external support, even in sensitive areas like security. This may be complicated when the environment is not actually 'post-conflict' at all, as in Afghanistan and Iraq, where SSR is taking place under combat conditions. However, when closely examined, in many ways the current process within Afghanistan is not fully SSR, but rather comprises various SSR-related elements (e.g. security sector training, development of a national security strategy) that when combined with a broader and more holistic approach could then more closely resemble SSR.

There are usually four core areas identified as central for assessing the moment for appropriate intervention:[9] context, politics and socio-economic position of the population; political will and commitment of international actors; local ownership and tension with external interventions; and integrated and coherent sequencing. However, given that post-conflict interventions are so contextual, it is likely that there is no one set methodology or timing, and these four will not be the same in each intervention. This means that any international intervention needs to be essentially political in terms of picking the right moment to intervene, intervening in a sensitive and diplomatic way and taking into account domestic political sensitivities within a heightened political situation.

What has tended to happen is that many interventions have been fundamentally technically focused rather than politically aware. The US-led SSR intervention in Liberia, for example, was driven partly by technical approaches to efficiency and capacity within the armed forces through a private contract between DynCorp and the US government.[10] In other interventions there has been a tendency to carry out the 'easier' technical tasks of training police and military while neglecting the more difficult governance aspects.[11] From this perspective, it is all too easy to overlook the political environment in which the intervention occurs, which may be a serious obstacle to it progressing effectively.

SSR undertaken in a post-conflict state always needs to deal with the legacy of the past, which often includes a long authoritarian regime. In such cases both the governance structure and the institutional framework will need to be reformed. In many African contexts, for example, armed conflict resulted from an authoritarian, individualised, political structure that excluded specific members of the population (Sierra Leone, Liberia) or involved the replacement of a colonial-authoritarian regime with an indigenous-authoritarian state (Zimbabwe). The main distinguishing features of such post-conflict environments are usually the need to provide immediate security, to demobilise and reintegrate combatants, to manage post-conflict increases in violence, particularly against women, and to downsize security institutions while instituting civilian oversight mechanisms that will hopefully prevent the security forces from taking over too much authority again in the future.

Additionally, political considerations come into play due to the variety of actors involved in post-conflict reform and governance processes. These include international agencies, international militaries, private companies and non-statutory security actors, encompassing parties such as insurgent groups, religious transnational actors and warlords, as well as civil society and government itself.

State-building as the practical face of the security-development nexus

The debates on the security-development nexus are vast, and are set out in Chapter 2 of this volume. However, what do they mean in practice? The World Bank identifies a number of different reasons why security should be incorporated into poverty reduction strategies.[12] Importantly, the betterment of their security is identified as a major issue by poor states themselves. Clearly there may be ulterior interests in declaring security as an issue for a government caught up in an armed conflict, particularly, in the current global context, if a terrorist threat can be defined. However, the importance of security at a community level is demonstrated in the World Bank's Voices of the Poor survey, which shows that poor people also identify insecurity and access to justice as two core concerns.[13] It is not made clear, however, exactly what is included in their definition of security. Understandings of what it means to be secure can also, of course, shift. In Sierra Leone there was a very noticeable change in local views of security in the post-conflict period, from an immediate desire to stop the killing and re-establish order to more development-oriented concerns, including reducing crime (particularly drug smuggling), economic insecurity (particularly employment opportunities) and domestic and sexual violence.[14]

The World Bank goes on to cite studies from Paul Collier that show the extent to which conflict affects the economy, but then, perhaps unsurprisingly, moves on to identify security as a core government issue, a public good and an issue of service delivery. It thus returns to the idea of security being defined by the capability of the state to provide a service to its citizens in a very Hobbesian way.[15] This view demonstrates the strong link between SSR, security sector governance and state-building as a global project.

Unsurprisingly, state-building has become a focus of much international aid, but unfortunately attempts at realising its goals in practice have frequently been problematic. A core reason for this is the methodology of state-building. As argued earlier, the vast majority of states that have been subject to contemporary state-building approaches have received interventions that concentrate very much on technical issues, especially effectiveness and functionality, rather than on the idea of what a state actually is and should deliver to its citizens. There is a clear difference between constructing a state apparatus and building a state that delivers rights to its citizens, including the right to live free from harm, not least in separating the technical process of what states do from the political processes involved in what states actually are.

In Iraq, for example, the United States attempted to construct a Western-style state armed with an entire range of neoliberal theories that view the institutions of the state as being technocratic and separate from politics. As a result of this thinking, the United States dismantled the existing state and started all over again, constructing a new set of ahistorical institutions alien to the local population.[16] Similarly, examples such as East Timor (see Chapters 6 and 8) and Kosovo point to the limitations of an externally led UN approach that incorporated local elites but marginalised the majority of the population, effectively producing states that exist legally and are managed by an elite, but remain hollow because they are unrelated to local political processes or representation and may lack legitimacy beyond the ruling elite or the United Nations (see Chapter 4 by Kunz and Valasek, who argue this point through a gender analysis).[17] Both these examples show that externally led, technocratic solutions do not necessarily result in a successful state.[18]

Much state-building is dominated by the construction of exit strategies for the intervening party, which often designates a 'democratic election' as the end point. However, holding an election does not mark the successful conclusion of state formation, even though technocrats might argue that democracies can be created in this way. Apart from the problems in establishing a multiparty democracy in a post-conflict situation, there may be a fundamental misunderstanding of what the project of state-building actually means in practice. This has important implications for security governance, because security institutions are a core element of the state and are often identified by poor people as a major threat to their security. Constructing security institutions that are representative is therefore critical to the future stability of the state and the human security of the population.

There is much literature on state-building, but it is useful to look at representative illustrations of some main approaches.[19] Fukuyama, for instance, outlines a set of approaches posited on a completely ahistorical and technocratic view of states.[20] One of the initial points he makes in his analysis concerns the lack of institutional memory about state-building within policy bodies such as the United Nations. This is complemented by the point that state-building takes a long time – it is a long-term commitment and requires sustained investment in time and resources.

Other analysts add to these ideas, but many of these generalised comments do not really provide a comprehensive theoretical framework for state-building. For example, Hippler outlines a three-point plan based on improving living conditions, structural reform of ministries and integration of the political system.[21] Again, this is a depoliticised version of reality that takes the politics out of state-building. In addition, such interventions are frequently carried out by bureaucrats, or in the case of security governance by military officers from the international community whose concerns are primarily technical rather than political.[22]

What does this actually mean in practice? Into what is the political system being integrated? If it means (as it usually does) integration of the political system into the international order, then who owns this process? Is it something that enjoys some form of local ownership among those who are supposed to benefit, or does it benefit international states relying on a state system? A significant silence in Hippler's analysis is that no attention is given to the role of a functioning security sector capable of maintaining a safe environment in which state-building can actually flourish.

While virtually all current analysts accept that there are problems with the nation-state in many of the contexts in which states are failing, there is still a tendency to accept the technocratic parameters of state-building as laid out by Fukuyama. This casts the nation-state as the norm in international relations, ignoring the broadening and deepening of security at international and subnational levels, particularly the intra-state nature of much conflict, international conflict actors and also the role of the state itself as an actor in non-state conflict. There remains an assumption that if we can develop the right mixture of policies, then we can create a healthy nation-state that can exist in the international order. Rebuilding states on paper does not mean that they exist in reality. All states rely on people to make them work, and this means that states need to be political structures as well as institutional bodies. The implications of this begin with people needing to buy in to the state at some level. Commonly related to ideas of legitimacy, there has to be some level of support for the state as an institution that represents something its populace recognise as a state. In a liberal sense this is realised by multiparty democracy, but in reality this type of democratic structure may not deliver representation in conflict environments, partly because nascent democratic institutions take time to bed down. Somalia is the archetypal collapsed state, but this is not simply a function of its own history but also a problem of contemporary international relations, particularly the universalisation of one model of the nation-state.[23] UN-sponsored external state-building in East Timor, as mentioned earlier and argued in Chapter 6 in this volume, is another example of a failure to embed legitimacy within government beyond local elites; and, as the example of Zimbabwe shows, replacing one autocracy with another can have dire consequences for the population more generally.[24]

This raises the second main point, namely that the construction of a new state requires a significant cultural change in terms of how people relate to that state as well as how they conduct everyday business. In Iraq, for example, attempts by the United States to construct a Western state, and its initial emphasis on deconstructing Saddam's state and political party, effectively superimposed an artificial state over subnational political systems. That state existed solely because the United States supported it, and not because there was an underlying belief in it in Iraqi society.[25] The risk now is that the new Iraqi state will effectively become another faction rather than an oversight mechanism for controlling warring factions at subnational level.

Thirdly, state-building is extremely 'capacity hungry'. In Sierra Leone, for example, the UK provided a lot of technical support for the security institutions without giving many resources to building the corresponding political support – mainly because it would have been difficult to secure. The technical support offered resulted in many UK officials taking decisions because those inside Sierra Leone lacked the capacity to do so. Ten years of SSR in Sierra Leone have effectively created an overdeveloped security force, including intelligence, but without the culture of civil oversight to control it.[26] This problem is also discussed in Chapter 6 on Australian technical capacity-building in the South Pacific.

Fourthly, given the fact that modern state-building is so resource intensive, it is usually externally funded. Because of the degree of financial investment, on a political level the process becomes externally driven. This creates significant problems with regard to funding and funding priorities, particularly when considering local ownership – or lack of it – and, most recently, the more limited availability of funds from countries affected by the current financial crisis. It raises serious questions about the long-term sustainability of reform and security, and also the relative balance between different activities; for instance, should donors fund the military more than development activities? This remains a core dilemma of international intervention. The example of the shifting definitions of insecurity over time within Sierra Leone, cited above, shows that the balance of donor intervention also needs to change over time to account for changes in the security situation, but entrenched interests and the inflexibility of many donor planning systems effectively mean that states may be locked into set trajectories for some time.

Fifthly, the creation of functioning state institutions can be very uneven. Even where states have had a functioning core before, during or after conflict, this core rarely penetrates into the rural areas.[27] As a result, many people simply do not receive services directly from the state. In the area of justice provision, for example, the majority of the population may receive justice from customary authorities such as chiefs or village headmen, legitimised because a local leader controls local security by controlling the local police, militias or 'vigilantes'.[28] At best this can produce a functioning governance system in which local people have both a say and a choice in terms of accessing services, including security. However, there is a risk that such hybrid systems, relying on both traditional approaches and modern systems of governance, will also reinforce the position of local elites and shore up the kleptocratic tendencies of neopatrimonial rule to the detriment of the population.[29]

Lastly, there are inconsistencies between state-building, security and development. There is an (unwritten) assumption that human security can be best served by creating a functioning state that will, it is theorised, provide security as a public good. Then, it is conjectured, development will provide benefits to the general population. However, there is a problem with exactly how diverse individuals fit into this picture. It is clear that the history of institutional development within state-building has not been a happy one for many people in terms of guaranteeing their security, and access to security has a sad tendency to remain uneven between states, groups and individuals. Human security, or 'freedom from fear', which implies an entitlement to protection by the state in which they are citizens, remains elusive for many people. Moreover, states' (and by extension the international community's) responsibility to protect citizens is yet to be realised in many places. This sets up a vicious cycle that justifies or legitimises international intervention in failed states.[30]

State-building, SSR and security governance

The development of SSR has been closely intertwined with the growth of state-building as a set of activities that coalesced following the collapse of many states in the post-Cold War era. In recent years, building the capacity of civil servants to provide oversight of defence ministries in particular has become more entwined with the development of civil service reform programmes as a whole, while security in general has remained central to the entire state-building approach from the point of view of both individual citizens and the international community, however that may be defined. Furthermore, SSR is now understood as an integral part of the international community's approach to conflict management. The reconstruction and reform of security institutions following conflict have become central elements of international intervention, bolstered by the belief that 'relatively cheap investments in civilian security through police, judicial and rule of law reform … can greatly benefit long-term peacebuilding'.[31]

SSR is intended to improve the performance and accountability of police, military and intelligence organisations, among others, with the aim of improving the basic elements of security for individuals. As a process, SSR should ideally move far beyond narrow technical definitions of setting up functioning security institutions and follow a more ambitious agenda of reconstructing or strengthening a state's ability to govern the security sector in a way that serves the population as a whole rather than the narrow political elite. As argued by Hudson in Chapter 3, this involves a radical restructuring of values and cultures within usually secretive and insular institutions that are inaccessible to particular subgroups within the population, particularly women and youth. The process usually takes place in contexts where the general population are mistrustful of security services and hostile to organisations that may be viewed as a direct threat to their individual security. An SSR process must therefore encompass an ambitious set of approaches that can contribute to restoring the social contract.

Despite obvious difficulties resulting from the political nature of these interventions, many international actors are currently involved in SSR programmes, including the UK, the United States, the United Nations and the European Union. The programmes they deliver employ an array of approaches and involve a complex mixture of international organisations, governments, non-state actors and private companies. While there are significant differences between the US approach in employing DynCorp to carry out 'SSR' in Liberia and the UN intervention in security and police reform in East Timor, there is a family resemblance in terms of the general approaches adopted. Some of the challenges of this 'one-size-fits-all approach' are discussed in Chapters 3, 6, 8 and 11 in this volume.

There has been much written about SSR, but, as mentioned above, it has been subject to what Peake et al. refer to as 'benign analytical neglect'.[32] This neglect has emerged despite the concept having been developed partly from an academic pre-history of civil-military relations. However, much of what has been written on SSR has tended to focus on practical policy-related analysis rather than being rooted in conceptual or theoretical approaches.[33] Particular activities have received attention rather than looking at wider interventions as an expression of and in relation to broader social and economic reform.[34] In particular, specifics of case studies have been used as gateways into discussions surrounding security without really reflecting on broader implications.

Governance, development and security

In a recent article on the macro-history of the security-development nexus, Björn Hettne posits three possible futures: neo-Westphalian, neo-medieval and post-national.[35]

In a neo-Westphalian scenario the current system would effectively continue to function through a state-based structure (with gaps), greatly enhanced by stronger multinational organisations with greater and more securitised powers. Such a structure could be multipolar, and might involve the inputs of the BRICS nations (Brazil, Russia, India, China and South Africa) as active security hegemons in their respective regions. Such a system may be violent, and create revolution and reaction within non-core areas of the global economy.

Neo-medievalism, on the other hand, represents a less violent option in terms of scale, but offers no solution for those areas that are outside organised nation-states. With neo-medievalism there is a loosening of the state to allow smaller units based on primitive accumulation or warlord economics in the short term, leaving those who live in localised pockets of violence to suffer that violence.

Lastly, Hettne posits the idea of a post-national future based on global development, which in turn is built on the inter-regional approach proposed by the European Union, among others. In this scenario regional governments act as vehicles to promote human rights, democracy and conflict prevention, and such arrangements are (at least in theory) cooperative and voluntary.

However, none of these offers a practical solution to developing a security-development nexus that provides freedom from fear. Clearly the first two scenarios are linked, with the first being both more aggressive and perhaps less certain to protect individuals from violence. In the first scenario one may be subject to international violence, and in the second to localised 'low-scale' violence (of course, it is not low scale to those suffering the violence!). The third scenario may offer some way forward, but there is a real problem with an EU-inspired solution, namely that EU decisions are based on an arrangement between functioning states that share a great deal of common ground, including the collective experience of a European war that no one wishes to repeat. This is not the case in, for example, Africa, where the experience of regional organisations has been woeful, partly because the states that sign up to regional agreements are frequently the first to break them. Prospects for the development of comprehensive regional actors remain bleak precisely in those areas where conflict is greatest.[36]

Regional approaches may offer some way forward in terms of renegotiating the colonial boundaries that have contributed to conflict (in the Horn of Africa, possibly in the Middle East and clearly in Sudan), but the fundamental issue is the nature of the state and the close ties between the state, the regime and the individual at the head of the regime.[37] Failed states incorporate varied political orders, some more legitimate than others. A failed state typically lacks a monopoly of force and is unable to extend its authority across its entire sovereign territory. It may also suffer from a lack of legitimacy, be fragmented by alternative sources of power and face continual threats to its authority. 'Traditional' and state functions coexist, but may form avenues to political power that the existing regime is concerned about. When faced with regimes that have a tendency to creeping authoritarianism, the construction of alternative sources of security (paramilitaries rather than militaries) and use of the security services to protect regimes rather than protecting the state or the population are often a problem.[38]

All these scenarios offer diverse sets of challenges for SSR approaches to tackle if they are to contribute to development and security. If SSR is to work, it has to derive from the political structures and history of the place it is working in. This is frequently acknowledged in donor documentation but not carried out in practice. I argue that the SSR intervention in Sierra Leone, despite its shortcomings, was more successful than that in Liberia because the Liberian/US approach was effectively to contract SSR out to a technical provider and not to engage with the government. This echoes the approach taken in East Timor and Kosovo, where failure to understand and then engage with the population (as opposed to receptive elites) has resulted in states that are not representative and may perhaps provide security for the elite/regime but questionable results for the population.[39] In the case of Kosovo this may be alleviated by accession to the European Union, but in East Timor, as in Sierra Leone and Liberia, the long-term survival of the state is at least partially dependent on the international community.

Given this set of problems, at least in the short term, we are left with the state as the basic building block of any international approach to security and development and also as the main means of delivering both security and development to national populations. A more nuanced, patient and flexible approach to constructing states – a development approach – is therefore necessary, as outlined below.[40]

Firstly, there should be proper recognition that security is a political entitlement of citizens as part of a social contract with the state. It is an obligation of the state to provide security for its citizens, not to protect personal regimes. It needs to be recognised that this will require substantial change on the part of security services, including individual security actors committing to not becoming agents of insecurity themselves.

Secondly, interventions need to be rooted in the specific historical-cultural-political situation of the country itself, and not just derived from the international experience of donors or non-governmental organisations. State-building has become problematic partly because it does not take into account the specific contexts of its application, and the emphasis on multiparty elections as an indicator of the legitimacy of states (or as an exit point for donors) is a mistake that may become dangerous, since it may worsen civil conflict and entrench it for years to come.[41]

Thirdly, it is important to provide a voice to those who are subject to violence and support access to justice for victims of state and other forms of violence. Poverty imprisons people in situations of extreme vulnerability, as do the social and economic roles assigned to those with a lack of employment opportunities. Development in the form of functioning delivery of justice must be combined with access to income-generating opportunities. Both would open a route to emancipation for those trapped in vulnerable situations.

Fourthly, it is important to ensure that security from below is grounded in evidence, not idealism or ideology. This applies to the 'off-the-shelf' interventions of some development agents, but also the highly romanticised view of some grassroots organisations. Warlords may provide a degree of governance, but only in so far as it benefits them and only to the limits of state power. Traditional authorities and chiefdom systems may be cheap and easily understood, but traditional systems usually discriminate against some loser groups at a local level. Not everything at local levels is positive or enjoys universal support.[42]

The state itself may also be seen as complicit in either making people more insecure, through using security services or militias to oppress people directly, as in Zimbabwe, using violent organisations to enforce political power and patronage, as in Sudan, or through links between criminal gangs and state security organisations such as terrorist groups, as in the case of the Pakistan secret services.

In short, 'smarter' and more targeted interventions are needed and, above all, a far deeper understanding of the politics of intervention over and above the technical expertise required to design an SSR intervention. Security is an integral element of governance more generally, and the provision of security is a key element of legitimacy. Those subject to poverty identify security as a key need. In essence, it does not matter what the academic debate says about the separation between security and development: those who are beneficiaries of development at the lowliest levels have already made that decision and accept security as a core need.

Conclusion

Contemporary state structures, this chapter argues, are not always the best models to deliver security to their citizens. The only way forward, then, is to realise the expected connections between the social contract and inclusive security. Current neoliberal state-building models are creating more poverty and exclusion. If we concede that state-building as social engineering has failed, then a discussion of the alternatives is overdue. Just leaving states to evolve themselves through some form of 'historical logic' is clearly not an option if the immediate security of the population is a concern. Politically, economically and ethically, it would be extremely difficult to cordon off an area of the world and label it 'failed'. This calls for a way forward that relies on pluralistic solutions to different contexts and an understanding of the state that does not merely rehash medieval Europe. However, this is typically left unsaid in contemporary development and security approaches.

Shifting colonial boundaries is not the only solution, although that may make a difference in specific circumstances like Sudan. In particular, there must be an acknowledgement of the pluralism of institutions at local level within areas labelled as 'states'. Politically hybrid institutions, combining traditional approaches with modern notions of successful governance, exist across most failed states and provide services to populations, including security and justice. The question is how can the provision of services to the population be delivered without simply generating power for local elites?

Western political theory finds it difficult to engage with failed states in which governance institutions continue to function at some level. There is a reality of political order that exists with or without the state. Surely non-state providers offer an alternative approach that may accommodate heterogeneous polities and social organisation and therefore strengthen peace-building?[43] It is clear that governance does exist beyond the formal state sector in many areas, and it is the incorporation of these social institutions into security management that remains important. For example, intelligence organisations existed right down to the village level in places as diverse as Sierra Leone and Nepal. These locally based organisations functioned far better than the state versions.[44]

At the same time, there is a sometimes uneasy coexistence between state and 'traditional' authorities in the security area.[45] The delivery of security and justice at the local level can be dominated by local leaders, including tribal chiefs, who generally exercise considerable power.[46] They might be able to appoint a customary court, be involved in social regulation through membership of a secret society, have at their disposal a range of actions they might take against non-conformists and see the dispensation of justice as an exercise of power.[47] It is important to note that local authorities such as chiefs see the provision of security as a means to maintain their power, and they therefore need to be consulted closely when local-level SSR is envisioned. The idea of hybrid political orders and the incorporation of non-state institutions into SSR and security governance overall rests on a number of key assumptions about those institutions. In particular, there is a critical question of seeing local institutions as far more legitimate than an externally imposed state-building solution. One solution may be to incorporate competing claims to legitimacy and authority, and recognise that 'traditional' and 'modern' institutions may coexist. However, it is also necessary to recognise that those forms of hybrid governance have differing dynamics, and may not only be coexisting but mutually influential or even mutually reinforcing. As von Trotha points out, this concept of a hybrid order is too frequently seen as being static, downplaying the continuing conflicts within such systems to produce variable outcomes as part of an ongoing political process.[48] These existing social and power structures are usually seen as obstacles to the successful implementation of SSR programmes, rather than sources of energy that can be assimilated into security governance or development programmes. Indeed, without the incorporation of some of these networks it may be impossible to achieve many desired development outcomes or to construct a sustainable structure of security governance.

A negative view of such actors tends to ignore what security apparatuses look like in those areas beyond effective state control. Whenever states abandon an area, other actors step in to fill the vacuum, ranging from predatory warlords to traditional authorities and 'other non-state actors'.[49] Consequently, alternative (to the state) sources of violence emerge and develop as proto-states. Contemporary wisdom argues that intervention is necessary in such cases, and should centre on state-building since failed states have largely failed through succumbing to continual conflict. SSR itself, taking security governance as being central, has a tendency to follow particular blueprints based on assumptions of what states are. In addition, SSR programmes are usually guided by service personnel of donor countries, who bring their own experience to bear but usually have no experience of the local politics and history in the area where they are operating. As a consequence, many officers tend to be naive in their assessment of local partners. They also tend to take command themselves, creating internal weaknesses in capacity once they return home.

Poorly regulated governance systems are open to abuse, vulnerable as they are to developing neopatrimonial tendencies which benefit the local elite and maintain patterns of social exclusion. Such structures offer little distinction between public and private, state and non-state and public and secret organisations. In particular, such clientalist systems tend to undermine security governance, replacing 'security for all' with security for the 'regime' at a local level. This is usually reinforced by control over local power encompassing security, justice and also development decisions in the local area. Many ordinary people in the countryside may not be in favour of a hybrid solution that just replicates a neopatrimonial system.[50] Indeed, many people want a just outcome rather than a particular system, and the usual claims of local systems being cheap, easy to access and easy to understand might be neither true nor a guarantee of justice for groups outside local elites.

A genuinely hybrid system needs to provide security to both state and non-state actors. Such a system will differ from place to place. The question arises as to what balance needs to be struck when a hybrid system of security governance is encouraged in order to maximise the security and development opportunities of the population.

Clearly, this question opens a Pandora's box. Nevertheless, I have identified a number of potential ways forward, all of them pragmatic. I would suggest that interventions by external actors need to be carefully contextualised and, in particular, take into account the politics surrounding security. Secondly, there has to be some realism regulating how we work with hybrid institutions. There is no simple dichotomy between 'formal' and 'informal' security systems (however these are defined), and in practice these two systems are closely intertwined. In accepting just one or the other there is a risk of leaving significant groups of people isolated from services, including access to justice. In addition, acceptance of traditional or customary systems implies acceptance of a number of elements that may not conform to desired development outcomes, including the enforcement of human rights. There is no reason why a local community should not provide local security (and many do), but there is a thin line between local security and thuggish vigilantism. The answer may not be to sweep away systems that are imperfect and replace them with another imperfect system based on formal law, but to make the existing systems work better so they provide more security for more people, more reliably.

Notes

[1] Human Security Centre, *Human Security Report, 2005: War and Peace in the 21st Century* (New York: Oxford University Press, 2005).

[2] Joseph Hewitt, Jonathan Wilkenfeld and Ted Robert Gurr, *Peace and Conflict 2008* (College Park, MD: Center for International Development and Conflict Management, 2008).

[3] Paul Collier, *The Bottom Billion: Why the Poorest Countries Are Failing and What Can Be Done About It* (Oxford: Oxford University Press, 2007). At the same time, eight out of ten of the countries ranked lowest on the Human Development Index have been recently or are at war. All of the top ten failed states in the world are experiencing conflict and eight of them are in Africa. Major causes include a heady cocktail of dysfunctional governance; political, economic and social inequalities; extreme poverty; economic stagnation; poor government services; high unemployment; and environmental degradation.

[4] Alan Bryden, Timothy Donais and Heiner Hänggi, 'Shaping a Security Governance Agenda in Post-conflict Peacebuilding', DCAF Policy Paper no. 11 (Geneva: DCAF, 2005).

[5] Peter Albrecht and Paul Jackson, *Security System Transformation in Sierra Leone, 1997–2007* (Birmingham: University of Birmingham, 2009).

[6] The author is currently an international adviser to the Nepali Parliament on the military integration of the Maoist combatants.

[7] OECD/DAC, *Handbook on Security System Reform: Supporting Security and Justice* (Paris: OECD, 2007). The author was part of the discussions on the handbook and Sierra Leone was frequently cited as an example of 'good practice', even if this was not reproduced within the text. Rather, it forms a subtext of the original version of the handbook.

[8] See, for example, Paul Jackson, 'Military Integration from Rhodesia to Zimbabwe and Beyond', in *Military Integration*, ed. Roy Licklider (Oxford: Routledge, 2011), which is a collection of historical and contemporary examples of integration, security governance and the consequences of intervention.

[9] See Bryden et al., note 4 above.

[10] Adedeji Ebo, *The Challenges and Opportunities of Security Sector Reform in Post-conflict Liberia* (Geneva: DCAF, 2005).

[11] See, for example, Albrecht and Jackson, note 5 above.

[12] Donata Garrasi, Stephanie Kuttner and Per Egil Wam, *The Security Sector and Poverty Reduction Strategies* (Washington, DC: World Bank, 2009). Further discussion on the World Bank can be found in Chapter 10 in this volume.

[13] See World Bank 'Voices of the Poor' project, available at http://web.worldbank.org/WBSITE/EXTERNAL/TOPICS/EXTPOVERTY/0,,contentMDK:20622514~menuPK:336998~pagePK:148956~piPK:216618~theSitePK:336992,00.html.

[14] Albrecht and Jackson, note 5 above.

[15] Hobbes's *Leviathan* is credited as the model for an all-powerful sovereign state. However, Hobbes himself, having just experienced the English Civil War, understood that governments could be dangerous and grounded his philosophy in a social contract whereby loyalty of citizens was repaid by a state guarantee of the safety of citizens, including from itself.

[16] Martina Fischer and Beatrix Schmelzle, eds, *Building Peace in the Absence of States: Challenging the Discourse on State Failure*, Berghof Dialogue Series no. 8 (Berlin: Berghof Research Center, 2009).

[17] Nicolas Lemay-Hebert, 'The "Empty-Shell" Approach: The Setup Process of International Administrations in Timor-Leste and Kosovo, Its Consequences and Lessons', *International Studies Perspectives* 12, no. 2 (2011): 190–211.

[18] Susan L. Woodward, 'A Case for Shifting the Focus: Some Lessons from the Balkans' (Berlin: Berghof Research Center, 2009), available at www.berghof-handbook.net/documents/publications/dialogue8_woodward_comm.pdf.

[19] See Mark T. Berger, *From Nation-building to State-building* (Oxford: Routledge, 2007) for a very good discussion of these issues.

[20] Francis Fukuyama, *State-building: Governance and World Order in the 21st Century* (Ithaca, NY: Cornell University Press, 2004); Francis Fukuyama, *Nation-building: Beyond Afghanistan and Iraq* (Baltimore, MD: Johns Hopkins University Press, 2006).

[21] Jochen Hippler, ed., *Nation-building: A Key Concept for Peaceful Conflict Transformation?* (London: Pluto Press, 2005).

[22] See Albrecht and Jackson, note 5 above; Andreas Mehler, 'Hybrid Regimes and Oligopolies of Violence in Africa: Expectations on Security Provision "From Below"' (Berlin: Berghof Research Center, 2009), available at www.berghof-handbook.net/documents/publications/dialogue8_mehler_comm.pdf.

[23] Wolfgang Heinrich and Manfred Kulessa, 'Deconstruction of States as an Opportunity for New Statism: The Example of Somalia and Somaliland', in *Nation-building: A Key Concept for Peaceful Conflict Transformation?*, ed. Jochen Hippler (London: Pluto Press, 2005): 57–67.

[24] Lemay-Hebert, note 17 above; Jackson, note 8 above.

[25] Similar comments could be made of Afghanistan. This fundamental tension was also evident in later disagreements between the nascent, emerging Iraq state and US authorities.

[26] Albrecht and Jackson, note 5 above.

[27] See Paul Jackson, 'Reshuffling an Old Deck of Cards? The Politics of Decentralisation in Sierra Leone', *African Affairs* 106, no. 422 (2007): 95–111.

[28] See Bruce Baker, 'Beyond the Tarmac Road: Local Forms of Policing in Sierra Leone and Rwanda', *Review of African Political Economy* 35, no. 118 (2008): 555–570.

[29] This is not a new argument. It stems from work by Olsen on the difference between static and mobile bandits, the theory being that one wishes to be ruled by a static bandit since they have an interest in keeping you alive – basic feudalism. Mancur Olsen, 'Dictatorship, Democracy, and Development', *American Political Science Review* 87, no. 3 (1993): 567–576. See also Paul Jackson, 'Warlords as Alternative Forms of Governance System', *Small Wars and Insurgencies* 14, no. 2 (2003): 131–150; Mehler, note 22 above.

[30] See Robin Luckham, 'Introduction: Transforming Security and Development in an Unequal World', *IDS Bulletin* 40, no. 2 (2009): 1–10.

[31] UN High Level Panel on Threats, Challenges and Change, *A More Secure World: Our Shared Responsibility* (New York: United Nations, 2004): 74.

[32] Gordon Peake, Eric Scheye and Alice Hills, *Managing Insecurity: Field Experiences of Security Sector Reform* (New York: Routledge, 2007).

[33] See for example, Gavin Cawthra and Robin Luckham, eds, *Governing Insecurity: Democratic Control of Military and Security Establishments in Transitional Democracies*

(New York: Zed Books, 2003); Michael Brzoska, 'Development Donors and the Concept of Security Sector Reform', DCAF Occasional Paper no. 4 (Geneva: DCAF, 2003), available at http://se2.dcaf.ch/serviceengine/Files/DCAF/18353/ipublicationdocument_singledocument/8ac3f8ba-cce6-43c3-be56-e14e8549152e/en/op04_development-donors.pdf.

[34] See Alan Bryden and Heiner Hänggi, eds, *Security Governance in Post-conflict Peacebuilding* (Münster: LIT Verlag, 2005).

[35] Björn Hettne, 'Development and Security: Origins and Future', *Security Dialogue* 41, no. 1 (2010): 31–52.

[36] James Hentz, 'The Southern African Security Order: Regional Economic Integration and Security among Developing States', *Review of International Studies* 35, supplement s1 (2009): 189–214.

[37] Trutz von Trotha, 'The "Andersen Principle": On the Difficulty of Truly Moving Beyond State-centrism' (Berlin: Berghof Research Center, 2009), available at www.berghof-handbook.net/documents/publications/dialogue8_trotha_comm.pdf.

[38] Laurie Nathan, 'Domestic Instability and Security Communities', *European Journal of International Relations* 12, no. 2 (2006): 275–299.

[39] Lemay-Hebert, note 17 above.

[40] See Luckham, note 30 above, who constructs an argument for rethinking security around four main points, which are included here as part of the analysis. Luckham's four points are unequal distribution of security; recognition of agency; empirical evidence; and complicity of the state.

[41] Burcu Savon and Daniel Tirone, 'Foreign Aid, Democratization and Civil Conflict: How Does Democracy Aid Affect Civil Conflict?', *American Journal of Political Science* 55, no. 2 (2011): 233–246.

[42] See Jackson, note 27 above; Ulrich Schneckener, 'Spoilers or Governance Actors? Engaging Armed Non-state Groups in Areas of Limited Statehood', SFB Governance Working Paper Series no. 21 (Berlin: SFB Research Centre, 2009).

[43] See Fischer and Schmelzle, note 16 above, for an articulate approach to these issues. The initial premise of this Berghof Dialogue is precisely to put the case for the incorporation of hybrid political institutions.

[44] Albrecht and Jackson, note 5 above; author's own interviews in Nepal, 2010–2011.

[45] See, for example, Peter Albrecht and Lars Buur, 'An Uneasy Marriage: Non-state Actors and Police Reform', *Policing and Society* 19, no. 4 (2009): 390–405.

[46] See, for example, Richard Crook, Kojo Asante and Victor Brobbey, 'Popular Concepts of Justice and Fairness in Ghana: Testing the Legitimacy of New or Hybrid Forms of State Justice', African Power and Politics Working Paper no. 14 (London: ODI, 2010): 1–31; Bruce Baker and Eric Scheye, 'Access to Justice in a Post-conflict State: Donor-supported Multidimensional Peacekeeping in Southern Sudan', *International Peacekeeping* 16, no. 2 (2009): 171–185.

[47] Richard Fanthorpe, 'On the Limits of the Liberal Peace: Chiefs and Democratic Decentralization in Sierra Leone', *African Affairs* 105, no. 408 (2006): 27–49; Jackson, note 27 above.

[48] von Trotha, note 37 above.

[49] See Jackson, note 29 above; Crook et al., note 46 above; Baker and Scheye, note 46 above; Schneckener, note 42 above.

[50] Crook et al., note 46 above.

Chapter 10

Opportunities to Support Security and Development: International Organisations' Evolving SSR Approaches

Willem F. van Eekelen

Introduction

This century began with agreement on the Millennium Development Goals (MDGs), which aim by 2015 to decrease drastically poverty and hunger, provide basic education for everybody, increase equality between men and women, reduce HIV/AIDS and tropical diseases, protect the environment and establish a worldwide partnership for development. All these are laudable goals, but remain rather disconnected from another critical, cross-cutting and fundamental objective, namely the need for good and transparent governance in accordance with the rule of law as a basis for sustainable development.

The lack of a comprehensive and integrated analysis weakens the design and potential realisation of the MDGs. For example, in the adoption of the MDGs, the elimination of poverty was erroneously made a goal in itself, while in fact it could only be the outcome of a comprehensive policy integrating many separate development objectives. The European Commission, in its EU Strategy for Africa, took a more helpful approach by combining the first six MDGs into 'making health education and basic social services available for the poorest people in Africa, contributing to the establishment of a social safety net for the most vulnerable: women, elderly, children and disabled people'.[1] Even there, it could be argued that the list (which in fact includes all people save perhaps for able-bodied, well-to-do, middle-aged men) does not describe how societies work and how such a social safety net would actually protect people. Further, much like the MDGs, the EU strategy is not conflict-sensitive. Indeed, neither the MDGs nor the EU strategy reflects an appreciation of the security-development nexus, although both are intended to be used in contexts where long-term violence impedes long-term development.

To be fair, some progress has been made in some countries in the last decade. In many African countries the middle class is growing and incomes are rising as a result of more stable and predictable governments and political and economic systems.[2] There are indications that the number of people living below the poverty line and the number of casualties in violent conflict have decreased as well. While many more people die from other causes than armed conflict and statistics say little about real living conditions, the reduction of armed conflict has arguably resulted in fewer refugees and improved chances for better governance in fragile environments.

As the 2011 *World Development Report* (WDR) shows, despite this progress there is still a strong link between conflict and development challenges. According to the WDR, people living in countries affected by violence are twice as likely to be undernourished and 50 per cent more likely to be impoverished, while 42 million are displaced today as a result of conflict, violence or human rights abuses, of whom 17 million are refugees. In addition, low-income, fragile or conflict-affected countries face daunting challenging in achieving the MDGs.[3] Certainly, political and economic instability tends to be seen as an obstacle to development. There seems to be an interaction between weak government and violent conflict, while civil unrest and revolts are most likely to take place in countries with a large, young and unemployed male population.[4]

If one assumes that unless development agendas are pursued in tandem with the provision of broad national and human security considerations, they will likely not be successful, those two agendas need to be linked more effectively. International organisations play an important role in supporting the provision of both security and development in many transition societies, partly through the support of security sector reform (SSR) initiatives. This chapter thus focuses on the efforts of a number of those organisations in coordinating their security and development objectives – with an emphasis on how this is accomplished through their evolving SSR strategies and activities.

The Organisation for Economic Co-operation and Development (OECD) was instrumental in stressing the link between security and development. It argued that without a basic level of security, development aid would be ineffective, while lasting security could not be achieved without development. Conceptually this is important, as it provides a bridge between security and the development communities.[5] However, despite this apparent nexus, there is resistance to putting such conceptual links into practice. For example, because of its emphasis on improving security, transparency and justice, among other issues, SSR should serve as a link between security and development. However, while SSR holds much promise, few countries wish to be reformed. Thus instead of maintaining an emphasis on the 'reform' component of SSR, a focus on security sector governance through security sector 'transformation' sometimes serves as a more acceptable approach.[6]

Additionally, although a security-development nexus approach recognises that without a minimum level of security development objectives are less likely to be achieved, at the field level it is often difficult for development actors to implement such linkages due to their reticence to be associated and cooperate with military missions.[7] Similar sentiments are expressed by non-governmental organisations (NGOs). Still, development and security communities are becoming more integrated, as is evident in a review of international organisations' engagement with SSR and UN security mandates. However, the degree to which international organisations are involved in SSR and the scope of their interaction vary considerably and are made more challenging by bureaucratic competition and difficulties in coordinating activities with member states, NGOs and local partners. The pursuit of effective cooperation and coordination therefore remains a crucial challenge for achieving both security and development objectives.

The remainder of this chapter focuses on a review of key international organisations and their efforts to achieve better coordination in furthering security and development objectives, chiefly through their particular SSR strategies. The conceptual role of the OECD in defining the security sector and its links with development will be examined. I then discuss the role of the United Nations, particularly the Security Council, followed by SSR approaches of financial institutions, namely the World Bank and the International Monetary Fund. Finally I focus on NATO and the European Union and their roles in implementing UN security objectives and mandates. Due to its broad spectrum of instruments conducive to the provision of both security and development, the main emphasis will be on the European Union as a model for other international organisations.

The Organisation for Economic Co-operation and Development

In 1997 the OECD Development Assistance Committee (DAC) published 'Conflict, Peace and Development Cooperation on the Threshold of the 21st

Century', a set of guidelines that emphasise the strong linkages between security and development, the need for donor coordination and a shift from crisis response to prevention.[8] Although they had considerable impact on the policies of donor countries, they did not influence the definition of the MDGs as the United Nations was not yet ready to follow this donor-driven initiative. A 2001 supplement to these guidelines, 'Helping Prevent Violent Conflict', views security as an all-encompassing condition in which people and communities live in freedom, peace and safety, participate fully in the governance of their countries, enjoy the protection of fundamental rights, have access to resources and the basic necessities of life and inhabit an environment which is not detrimental to their health and well-being.[9] This understanding of security was seen as consistent with the notion of 'human security' promoted by the UN Development Programme (UNDP) and other development actors. It was given practical importance in the debate regarding which security-related assistance could be considered official development assistance (ODA), the leading indicator of official development financing flows. The DAC agreed to extend ODA eligibility to six new activities: strengthening the role of NGOs in the security field; support for legislation against recruiting child soldiers; non-military activities in peace-building, conflict prevention and conflict resolution; the control, prevention and reduction of small arms/light weapons; and finally improving democratic governance in the security sector. The DAC decided that SSR-related activities to improve democratic governance and civilian control of security institutions would be ODA eligible.

The OECD/DAC defines the security sector – or 'security system', the preferred term used by the DAC – along four main categories, which also informed the concepts developed subsequently by the European Union:[10]

- *The core security actors:* armed forces, police, gendarmerie, paramilitary forces, presidential guards, intelligence and security services (both military and civilian), coastguards, border guards, customs authorities and reserve or local security units (civil defence forces, national guards, militias).
- *Management and oversight bodies:* the executive, national security advisory bodies, legislature and legislative select committees; ministries of defence, internal affairs and foreign affairs; customary and traditional authorities, financial management bodies (finance ministry, budget office, financial audit and planning units); and civil society organisations[11] (civilian review boards and public complaints commissions).

- *Justice and law enforcement institutions:* judiciary, justice ministry, prisons, criminal investigation and prosecution services, human rights commissions and ombudsmen and customary and traditional justice systems.
- *Non-statutory security forces, with which donors rarely engage:* liberation armies, guerrilla armies, private bodyguard units, private security companies, political party militias.

The guidelines were followed in 2005 by a reference document on 'Security System Reform and Governance' containing a policy statement and an extensive analysis of SSR in general and its regional dimensions in particular.[12] The DAC proposed ten recommendations for action to promote peace and security as the fundamental pillars of development and poverty reduction. The recommendations stressed the importance of applying a 'whole-of-government' approach to SSR and the manner in which OECD governments address security-related issues such as international corruption; money laundering; organised crime; perpetuation of militia-linked private security forces, including support from multinational enterprises; human trafficking; the proliferation of weapons of mass destruction; terrorism prevention; and illicit trade in small arms and light weapons.[13]

As this comprehensive list of issues shows, as well as the previous list of security actors, the DAC approach to the security sector is broader than just the uniformed services and the civil authorities responsible for their oversight. It also encompasses judicial and penal institutes and civil society organisations. While such a wide scope of security-relevant actors and issues could have easily compromised the effectiveness of a concept that should have concise policy relevance, the DAC drafted specific policy-relevant recommendations that emerged from extensive discussions held over the course of several years. As suggested by its title – 'Security System Reform and Governance' – the primary objective of the reference document was to aid in the provision of better governance. The DAC achieved this by broadening the focus of security policy from state stability and regime security to the well-being of populations and respect for human rights, paving the way to mainstream security as a public policy and governance issue. Traditional security providers such as the military and police were seen as being among numerous instruments of security policy, as greater attention was given to legal, social and economic instruments. The reference document pointed to the necessity that a security system should be managed according to the same principles of accountability and transparency that apply across the public sector, in particular through greater civil oversight of security processes.[14] It moreover defines three core requirements of a well-functioning security system, calling upon national political authorities to:

- Develop a nationally owned concept of security.
- Strengthen governance of the security institutions responsible for formulating, executing, managing and monitoring security policy.
- Build institutional mechanisms for implementation and capacity throughout the security system, which should include building up civil control and supervision bodies in order to avoid any increase in the power and influence gap between military and civil bodies.

The principles included in the reference document were meant to be more than merely a declaratory statement of desirable aims, and thus contained examples of good practice among member countries. Nonetheless, the document was not without critics. For example, it remained donor-driven and donor-oriented; thus it was criticised for not drawing on input from the receiving countries, which, after all, were supposed to be in the driving seat of reform. Additionally, some recommendations were disliked by parts of the development community. For instance, the African Peace Facility was criticised for spending more on short-term peacekeeping than on preventive action and long-term capacity-building.[15] This particular debate is likely to continue, but is increasingly leading to the conclusion that all three of these elements (peacekeeping, prevention and capacity-building) are necessary when analysed through the security-development nexus. Lastly, significant gaps remain between the OECD's approach to SSR theory, its translation into policy and the subsequent implementation in SSR 'theatres'.

In 2007 the OECD published its *Handbook on Security System Reform*, which made a shift from promoting guidelines to developing practical tools to facilitate better SSR implementation. It offered flowcharts from structure to objectives, key issues and finally the desired outcome. It furthermore treated SSR as a 'multi-layered service delivery', addressing the political nature of SSR and focusing on outcomes rather than outputs (i.e. on results rather than effort), with a key role for local ownership, non-state actors and long-term sustainability.[16] As Bryden argues, the OECD/DAC must continue to act as both a facilitator and a watchdog entity to monitor progress, and capacity-building among member states requires significant further effort.[17] As no single actor alone could possibly cover the entire SSR agenda, complementarity of efforts is crucial.[18] Yet the handbook was not meant to be a tool primarily for security actors – development partners could now judge their donor programmes in accordance with the good practices it defined. If used as a resource on good and bad SSR practices, the handbook allows security and development actors to be more sensitive to each other's agendas and operating procedures. Given its broad base of some of the most powerful donors worldwide, the OECD's embracing of SSR has been a most significant step in putting it on the agenda of the donor community, while pointing to the relevance of development agendas for SSR actors.

The United Nations

The United Nations was relatively slow in moving towards an operational concept of SSR that would be relevant UN-wide, across the entire range of the organisation's work. In 2007, during its membership of the Security Council, Slovakia issued a presidential statement on the role of the Security Council in supporting SSR and created the 'Group of Friends of Security Sector Reform' with some 30 members.[19] Together with the Republic of South Africa it organised an international workshop in Cape Town, which called for a report by the UN Secretary-General on SSR. This request was endorsed by the General Assembly, which in 2008 led to a report by Secretary-General Ban Ki-moon to the General Assembly and the Security Council.[20] The report admitted that, despite extensive experience, support for SSR remained largely *ad hoc* and lacked elaborated principles and standards. It emphasised that 'security is a precondition for sustainable peace, development and human rights' and announced the need for an 'integrated approach' with the following requirements:

- A shared vision of the UN's strategic objectives.
- Closely aligned or integrated planning.
- A set of agreed results, timelines and responsibilities for the delivery of tasks critical to consolidating peace.
- Agreed mechanisms for monitoring and evaluation.

As the Secretary-General's report argues, 'SSR describes a process of assessment, review and implementation as well as monitoring and evaluation'.[21] The major institutional consequence of the report was the creation of the UN Inter-Agency Security Sector Reform Unit, located within the Office of Rule of Law and Security Institutions of the Department of Peacekeeping Operations. This places SSR firmly within the UN's activities in armed peace support, rather than its development and political programmes. Yet an SSR taskforce was created with the objective to broaden ownership and include all relevant UN actors (such as, for instance, the UNDP, the Department of Political Affairs, UN Women, the UN Children's Fund and numerous others who are active in supporting peace-building in transition and post-conflict societies).

While the Secretary-General's report provides a comprehensive narrative of the entry points into SSR processes open to various UN bodies, it lacks practical guidance in the form of guidelines or the flowchart precision of the OECD handbook. Guidelines are currently being developed by the SSR taskforce and its members to assist in mainstreaming and implementing the report's operational consequences throughout the UN system. This is a challenging task, and the outcome will likely be difficult to monitor. Yet the operationalisation of SSR within the UN's work is an important and long-overdue step towards standardising the international community's approach to SSR in at least those countries in which the United Nations is actively supporting peace-building processes. Within UN peace support operations, for instance, there has so far been little capacity and staff dedicated to SSR, although in a rudimentary form such capacity has been included in some cases. For instance, the UN Operation in Burundi (ONUB) listed a disarmament, demobilisation and reintegration (DDR)/SSR section in its organisational chart of 2005; and the UN Organization Mission in the Democratic Republic of the Congo (MONUC) established dedicated offices for DDR, rule of law and human rights, yet was criticised for not ensuring effective links between its governance and rule-of-law activities early in its mandate.[22]

Major bilateral actors: Lessons learned from UN and OECD practice

The call to move from mere coexistence to cooperation, coordination and integration of security and development policies was echoed by international engagements of many UN and OECD member states. As a result, increasingly a whole-of-government approach was applied to both sides of the security-development nexus. For instance, the United Kingdom departed from its practice of regular defence reviews and in 2008 introduced a national security strategy that very specifically adopted a whole-of-government approach. As early as 2001 the Foreign and Commonwealth Office, the Ministry of Defence and the Department for International Development combined resources and jurisdiction in a conflict prevention pool. Civilian and military specialists were integrated in a post-conflict reconstruction unit, which in 2007 was renamed a 'stabilisation unit'. A further example of more effective and flexible funding was the Netherlands Stability Fund, which pursued an integrated approach to peace, security and development.

Austria, Finland, Sweden and Germany are also working on improved coherence of their policies and actions in crisis management and peace-building. Canada, the Netherlands and the United States coined the '3D concept' of defence, diplomacy and development, but recently switched to the notion of an 'integrated policy'. Although still commonly referred to, some actors feel uncomfortable about the 3D concept when in most cases it is security instead of defence, and good governance instead of diplomacy, which, along with development, lie at the heart of peace-building. Nevertheless, these approaches have greatly contributed to increased awareness about the linkages between security and development.

Improved cooperation and coordination are necessary at various levels: between government ministries, between different states, between bilateral and multilateral actors, and between all of those and civil society. While there is a well-established history and practice of civil society engagement in development or humanitarian assistance, similar collaboration on SSR acquires expertise that cuts across multiple, well-established state domains, such as defence intelligence, policing and judicial and penal systems. Integrating insights from these domains is a major challenge for civil society actors working on SSR issues.[23]

In January 2009 the US Department of State, Department of Defense and US Agency for International Development published an official document outlining the aims of SSR programmes (including the 'management of the legacies and sources of past or present conflict or insecurity'). Its guiding principles were support of host-nation ownership; incorporation of principles of good governance and respect for human rights; balance of operational support; linking security and justice; fostering transparency; and taking a 'do-no-harm' approach by avoiding donor assistance becoming part of the conflict dynamic.[24]

These aims, amplified by lessons learned in linking military operations and the subsequent post-conflict stabilisation and reconstruction phases in Iraq and Afghanistan, undoubtedly contributed to building a consensus on SSR within the donor community. On the other hand, they also ran the danger of becoming mere slogans. For example, local ownership, an important and repeatedly discussed principle of sustainable SSR, still remains an imprecise concept in both security and development. There is little clarity on whose ownership over which issues seems to be central to reform efforts.[25] Insistence on local ownership is not always a helpful approach to take, as it might for instance translate into further entrenchment of authoritarian warlords or corrupt politicians. Here lies the link with the need for transparency and accountability in a parliamentary system based on practices of 'reveal, explain and justify' in terms of government actions.

The World Bank and the International Monetary Fund

The World Bank group was created in 1944 to promote 'a smooth transition from wartime to a peacetime economy' by facilitating capital investment for productive purposes, promoting balance of payments stability and the balanced growth of international trade. The main focus of the World Bank's projects was shifted in the 1980s to equitable economic growth and development, and again in the 1990s to poverty reduction and good governance. The Bank conducted the Multi-Country Demobilization and Reintegration Program in Central Africa, but refrained from describing its activities as SSR, possibly through fear of being drawn further into military and traditional security activities. In 2002 a new taskforce was established on 'low income countries under stress', in line with the principle that 'state building is the central objective in fragile states, and that effective donor programs require integrated approaches across the political-security-development nexus'.[26] Most recently, in its 2011 *World Development Report* the Bank argued that the building of government institutions – which can mediate political and communal violence – is more important than the short-term goal of simply stopping conflict.[27]

The International Monetary Fund (IMF) contributes indirectly to joint security and development objectives. In the first instance the IMF was established to promote international monetary cooperation, exchange stability and orderly exchange arrangements; to foster economic growth and high levels of employment; and to provide temporary financial assistance to countries to help ease balance of payments adjustments.[28] While having developed a 'Code on Good Practices and Fiscal Transparency', the IMF does not address off-budget expenditure in the security sector unless it has an active programme in the country which would be placed at risk. Nevertheless, it has played an important indirect role through donor efforts to agree to limits on the amount that governments spend on their military budgets. Still, the IMF does not formally institutionalise SSR or indeed its development implications.[29]

NATO and the European Union

NATO's Partnership for Peace has had a considerable impact on restructuring armed forces and parliamentary oversight, particularly in the countries which have joined the alliance since the end of the Cold War. At its September 2010 summit meeting in Lisbon, NATO adopted a new strategic concept; this mentions the importance of conflict prevention and crisis management, integrated civilian-military planning throughout the crisis spectrum and the capability to train and develop local forces in crisis zones. On the whole, however, it still views NATO as a threat-oriented defensive alliance focusing on military capabilities for expeditionary operations. Paradoxically, compared to the Cold War years when political consultations among its members were very intensive, NATO has become even more military in nature.[30] However, components of NATO are beginning to engage with development efforts, which complicate their roles and missions. Most notably, the provincial reconstruction teams in Afghanistan, which initially focused on security provision, acquired an important development dimension. In addition to infrastructure development, justice sector reform activities aimed at police, judges and prisons have become a priority. Nonetheless, an explicit development-oriented strategy is absent from NATO's official strategic concept.

The European Union (EU) is the world's largest donor of development assistance and emergency aid, contributing €45 billion annually (of which 80 per cent are direct contributions by its member states). In contrast to NATO, for a long time the European Union remained a very reluctant security actor. The Treaty of Maastricht of 1991 assigned the military aspects of security to the Western European Union. However, in 1997 the Treaty of Amsterdam shifted those functions to the European Union, whereas the Western European Union continued only as a parliamentary assembly and a treaty obligation to render automatic military assistance in case of aggression.[31] The Treaty of Lisbon, which came into force in 2009, merged the three 'pillars' of the European Union and agreed on a common security and defence policy (CSDP) and the European External Action Service to represent all aspects of the European Union abroad.

Important groundwork was laid for this development in 2003 by High Representative Javier Solana, who launched a major initiative to formulate a European security strategy as a conceptual underpinning of the common foreign and security policy (CFSP).[32] The main points were the comprehensive listing of the current challenges and threats, the emphasis on 'effective multi-nationalism', with the UN Charter as its fundamental framework, and the recognition that crises cannot be solved by military means alone. The strategy concluded that 'security is the first condition for development' and contained a brief reference to SSR:

> As we increase capabilities in different areas, we should think in terms of a wider spectrum of missions. This might include joint disarmament operations, support for third countries in combating terrorism and security sector reform. The last of these would be part of broader institution building.[33]

A full SSR document was adopted by the Council of the European Union in November 2005, complementing existing mission concepts in the areas of rule of law and civilian administration in crisis management.[34] The preceding working document states that support for SSR in partner countries will contribute to:

> an accountable, effective and efficient security system, operating under civilian control consistent with democratic norms and principles of good governance, transparency and the rule of law, and acting according to international standards and respecting human rights, which can be a force for peace and stability, fostering democracy and promoting local and regional stability.[35]

To achieve these goals, EU support must be based on respect for local ownership and coherence with other areas of EU external action. This support is defined in close consultation with the partner government and adapted to the particular situation. In terms of specific activity, the document notes that DDR initiatives constitute a significant pillar of SSR and are regarded as central to conflict resolution and internal stability. Yet the point is made that SSR goes well beyond DDR, and should be considered as the umbrella concept.[36]

Additionally, the document lists an extensive range of measures in support of reforming the defence sector and the police, as well as strengthening the justice sector and the rule of law. It emphasises that EU action would require as a legal basis a UN Security Council resolution or an invitation by a host partner state or international, regional or subregional organisation.

The EU SSR concept certainly has its merits, but as is the case in many other documents produced in Brussels, it suffers from undue length and numerous statements of the obvious. Its main weakness, however, is neglecting to address the links between the various security sectors and the measures that are necessary to meet the ambition of a truly holistic approach. Moreover, most of the institutions mentioned might function in relatively mature states, but will only be able to play a significant role in fragile or transition states (where reforms tend to take place) when stability has been restored. For example, especially in post-conflict contexts it is difficult to talk about local ownership as long as legitimate governing structures are not yet in place.

On 24 May 2006 the European Commission sent a communication to the Council and the European Parliament, 'A Concept for European Community Support for Security Sector Reform.'[37] This communication emphasised the non-military aspects of SSR. The Commission's preference for the term 'security system' (also used by the OECD) was based on the wish to underline that its reform went beyond enhancing the effectiveness of individual services. The communication also emphasised that SSR should be seen as a 'holistic process'.[38]

The Commission defined the security system as 'all state institutions and other entities with a role in ensuring the security of the state *and* its people'.[39] In the paragraph on 'security management and oversight bodies' it changed the order used in the Council document and placed parliament first; it also reverted to the OECD list, and added 'media, academia and NGOs' as elements of civil society. On the other hand it deleted the civilian review boards and public complaint commissions, presumably because these would belong to the ombudsman or other functions listed under justice institutions.[40]

The Commission listed a large number of challenges that face countries receiving EU assistance: oversized and underpaid regular forces; irregular forces and security firms operating outside the law; lack of judicial independence, status and resources; lack of capacity, legal competence and sometimes political will by parliaments to ensure accountability of security services; human rights abuses by police and defence forces; a culture of state impunity; and the inability to protect the population against terrorist acts. In helping to meet these challenges the European Community (EC, after the Treaty of Lisbon the European Union) was engaged in over 70 countries through both geographical and thematic programmes and political dialogues. Among the factors crucial for successful reform was its 'commitment to policy coherence for development, in particular where EC policies have a significant impact on developing countries'.[41] The European Community should take 'into account the close inter-linkages between security, development and governance, including democratic principles, rule of law, human rights and institutional capacity building'.[42] The Commission emphasised the need to focus more clearly on the governance aspects of SSR, including the strengthening of parliamentary oversight, judicial independence and media freedom. In addition, it advocated more short- to medium-term engagement that could kick-start and complement long-term instruments. This included, for example, a rapid reaction mechanism, which had been set up in 2001 to enhance the EU's ability to intervene quickly in a (potential) crisis within six months. It was followed in 2007 by the 'instrument for stability', which covers a longer time span (18 months) and finances a considerable number of 'crisis response projects', particularly in the Great Lakes region of Africa. Examples are support for mediation, confidence-building, interim administration, rule of law and traditional justice, and, as a very specific activity, an analysis of the role of natural resources in conflict situations. Its peace-building partnership aims at strengthening the expertise of civil society organisations and their involvement in EU programmes and projects.

The Commission's views, which contained a broader security-development linkage than the Council's document, were barely discussed in the Council, ostensibly because it was only a 'communication' and not a 'directive', but most probably because of perpetual disputes over competencies. Nevertheless, the EU General Affairs Council welcomed the Commission's views on 12 June 2006. It considered them as complementary to the CFSP concept, as together they could constitute a policy framework for an EU role in SSR, supported by a cross-pillar approach to this long-term process.[43]

After the Lisbon Treaty entered into force, merging the functions of the high representative for the CFSP and vice president of the Commission for external relations, no attempt was made to integrate the two SSR documents – which, taken together, are among the most visible expressions by international organisations of the link between security, development and SSR. Renewed interest in and commitment to SSR was demonstrated in May 2011 by a joint communication of the Commission and the high representative, 'A New Response to a Changing Neighbourhood.'[44] This communication stated that 'We shall adapt levels of EU support to partners according to progress on political reforms and building deep democracy', to include free elections, freedom of expression and assembly, rule of law administered by an independent judiciary, right to a fair trial, the fight against corruption and SSR. The latter was defined as 'security and law enforcement sector reform (including the police) and the establishment of democratic control over armed and security forces'. By expressly linking EU support to political reforms and 'deep democracy', which include reforms to the security sector, the document articulated a new element of conditionality within the large array of EU instruments.[45] SSR is thus considered to be a key ingredient of political, economic and security reform processes worthy of EU support. Given the political clout of the European Union, such commitment will likely add significant political momentum to prioritising SSR in overall reform processes.

However, while concepts, frameworks, strategies or even merely rhetorical statements are important steps in maintaining and strengthening commitment to the significance of SSR in the context of broader reform processes, ultimately only results matter. The common pursuit of shared political, diplomatic, economic and security objectives by a variety of local, national and international actors requires joint and cooperative planning and implementation, which is of course a demanding task. Such common approaches require common assessments, which in turn must be applied to concrete joint as well as separate efforts in policy planning and implementation. Elements of conditionality in the application of the spectrum of available instruments, as suggested by the European Union in the joint communication cited above, will increase chances for successful cooperative approaches to joined-up reform efforts. Early efforts to facilitate joint efforts tend to hold the greatest promise for sustained cooperation. Assessments are the ideal entry points for subsequent joint programmes.

Moreover, not only front-end assessments during the planning phase but increasingly assessments of programme results – evaluations – are considered central elements in controlling the relevance of particular reform and assistance programmes. As with evaluating the outcome of any intervention, assessing the results of development efforts is a two-way street between donors and recipients, and also increasingly attracts the interest of the taxpayer in donor countries.[46] Local ownership implies the willingness, initiative and capacity to assess and define one's own reform and development needs. It also requires joint, whole-of-government initiatives in collaboration with all major stakeholders in society in harmonising assessment processes and results at country level, thus laying the foundation for policy dialogues that promote context-relevant reforms. On the other hand, multiple and uncoordinated donor assessments, possibly without suitable involvement of local and national actors, may do more harm than good. In that spirit, an OECD/GovNet-sponsored conference in 2008 highlighted growing donor interest in helping partner countries diagnose their own governance challenges.[47]

Particularly within the European Union, an evolving pattern of linking security and development concerns and reform requirements is an expression of the ambition to create stable conditions for economic and social activity. Security, rule of law, democratic governance and development are considered to be vital parts of an integrated approach. Assessment frameworks help judge their potential and actual effectiveness and formulate criteria for effective and successful implementation. Ultimately, they also assist donors in deciding upon where, how, with whom and for how long a helping hand should be extended to countries requiring assistance in their efforts to build peace and stability along a sustainable development path.

So far, activities in the context of the EU's CSDP have largely been demand-driven, primarily as a reaction to eruptions of violence. In the future, however, such involvements should be increasingly determined in a strategic context and include preventive measures. Development planners tend to focus their activities, such as poverty reduction, on countries that are not in conflict or have reached a post-conflict stabilisation phase, thus assuring reasonable expectations of the sustainability of their development activities. The CSDP, on the other hand, is more acutely concerned with immediate operational efforts in restoring a minimum level of law and order. In both cases peace and stability are not assured, and a relapse into violence can offset all ongoing peace-building and development efforts. A most effective EU operation was its mediation role in resolving the long-running conflict in Aceh, Sumatra. Just when all attempts at settling the conflict had failed, the December 2004 Southeast Asian tsunami struck.[48] The European Union rendered substantial economic assistance, but also contributed a multinational force of approximately 300 personnel, acting with the consent of both parties and in cooperation with ASEAN (the Association of Southeast Asian Nations), which was able to change the context by supervising both the collection of arms surrendered by guerrillas and the phased withdrawal of the Indonesian army.

Several factors support the evolution of increasingly integrated approaches by the European Union. The various financial crises beleaguering it inevitably put greater pressure on development budgets. While shrinking budgets are of course worrisome, by default they might lead to better coordination and a more critical eye on programme performance. The EU Financial Perspectives 2014–2020 offers an opportunity to emphasise both the security-development nexus and the need for transparency, efficiency and accountability. This could influence its development activities, such as its most important framework of development cooperation, the Cotonou Treaty of 2003. With 79 partner countries in Africa, the Caribbean and the Pacific, the treaty covers not only development aid and trade, but also conflict resolution, good governance and respect for human rights.

Both the European Union and its member states are called upon to improve the coordination of their SSR interventions.[49] There will be greater demand for involvement and investment in security provision in the context of peace and stabilisation operations, while development activities will play a significant role prior to the outbreak of armed conflict and after it has ended – supporting prevention and reconstruction. While post-conflict reconstruction is a broader agenda than SSR, in places such as the Democratic Republic of the Congo both depend on successful DDR efforts. Coordinating armed interventions, SSR support and longer-term development assistance is a major challenge, not only among official institutions, both internal and external, but also with and among the dozens and sometimes hundreds of non-governmental agencies actively involved in a broad range of peace-building activities before, during and after an outbreak of armed violence. In the case of all those activities, long-term commitment is crucial, while quick successes are essential in ensuring growing confidence and continuing commitment of the population to security and development measures.

In the context of SSR activities, special attention should be paid to assisting the creation of responsible and honest police forces and judges, both of which have deep roots in and are closely watched by the communities in which they serve. In Afghanistan this has become a major objective of the International Security Assistance Force. The European Union also possesses a unique capability in the grey area between the military and the police in the form of the European Gendarmerie Force, created in 2004, which, as an added value, tends to be more readily available for deployment into conflict zones than regular police officers.[50]

The growing willingness among member states to envisage joint action of the ministries of foreign affairs, development assistance and defence, as mentioned above, is a major step forward. Under the Treaty of Lisbon, decision-making and implementation of the EU's security and development roles will likely be more pronounced, making each possibly more effective.

Conclusion

In 2012 the European Union should amend its security strategy in light of the new challenges and approaches discussed above.[51] NATO has a new strategic concept, which should make it easier for the European Union to develop and pursue a complementary conceptual approach. NATO has experience with its provincial reconstruction teams in Afghanistan and the ensuing difficulties of inter-institutional cooperation, while on the whole it views itself as a threat-oriented defensive alliance with a main focus on military capabilities for expeditionary operations. Its SSR and development activities are limited to the duration of a specific operation, and for the most part focus on assistance in the restructuring of the armed forces. While the European Union cannot match these capabilities, NATO will not muster the civilian instruments available to the European Union.

The European Union would do well in balancing its approach by creating a joint military-civilian headquarters for the CFSP/CSDP, ready to cooperate with other regional organisations, open to participation of other countries on a case-by-case basis and with an appropriate section on SSR to communicate with its counterparts at UN headquarters and in UN missions. The current structure of an EU military committee and a military staff should be integrated with the Civilian Planning and Conduct Capacity created in 2008, as well as with the Crisis Management and Planning Directorate put in place in 2009. This needs to be operationalised with some urgency, as the need for SSR will continue to figure prominently in post-conflict settings and international organisations are well placed to coordinate and integrate their efforts in jointly assisting national actors to meet both development and security challenges, based on their respective comparative competencies and experiences.

Notes

[1] Communication from the Commission to the Council, the European Parliament and the European Economic and Social Committee, 'EU Strategy for Africa: Towards a Euro-African Pact to Accelerate Africa's Development', SEC (2005) 1255 (Brussels, 12 October 2005), COM 92005, 489 final, available at http://eur-lex.europa.eu/LexUriServ/LexUriServ.do?uri=COM:2005:0489:FIN:EN:PDF: 23. Under the heading 'Sustain peace in post-conflict situations' it recommended implementing 'linking relief, rehabilitation and development' efforts.

[2] Africa's Growing Middle Class', *The Economist*, 12 May 2011.

[3] World Bank, *The 2011 World Development Report*, 'Facts and Figures', available at http://wdr2011.worldbank.org/early-findings.

[4] Joris Voorhoeve, *Negen plagen tegelijk. Hoe overleven wij de toekomst?* (Amsterdam/Antwerp: Uitgeverij Contact, 2010). See also Henrik Urdal, 'A Clash of Generations? Youth Bulges and Political Violence', *International Studies Quarterly* 50, no. 3 (2006): 607–630.

[5] Neclâ Tschirgi, Michael S. Lund and Francesco Mancini, 'The Security-Development Nexus', in *Security and Development: Searching for Critical Connections*, eds Neclâ Tschirgi, Michael S. Lund and Francesco Mancini (Boulder, CO: Lynne Rienner Publishers, 2010): 1–16. They concluded that political uncertainty and instability emerge as causes rather than consequences of development failures. Moreover, countries with very youthful age structures are most at risk of conflict.

[6] Alan Bryden and 'Funmi Olonisakin, eds, *Security Sector Transformation in Africa* (Münster: LIT Verlag, 2010): 6–10.

[7] For recent UN experiences with integrated missions, see Heiner Hänggi and Vincenza Scherrer, eds *Security Sector Reform and Integrated Missions* (Münster: LIT Verlag, 2008).

[8] OECD/DAC, 'Conflict, Peace and Development Co-operation on the Threshold of the 21st Century', policy statement (Paris: OECD, 5–6 May 1997).

[9] OECD/DAC, *The DAC Guidelines: Helping Prevent Violent Conflict* (Paris: OECD, 2001). See also OECD/DAC, 'Security System Reform and Governance', DAC Guidelines and Reference Series, DAC Reference Document (Paris: OECD, 2005): 20.

[10] See OECD/DAC, *OECD Handbook on Security Sector Reform: Supporting Security and Justice* (Paris: OECD, 2007): 22–23. This mentions a possible alternative categorisation in criminal justice, intelligence and state security systems.

[11] Although not mentioned in detail by the DAC, civil society groups extend beyond such specialised oversight bodies to encompass the media, professional and research groups, religious and community groups, and other NGOs. See Albrecht Schnabel, 'Ideal Requirements versus Real Environments in Security Sector Reform', in *Security Sector Reform in Challenging Environments*, eds Hans Born and Albrecht Schnabel (Münster: LIT Verlag, 2009): 9. As we shall see later, the significance of this category of civil society actors was also emphasised by the European Commission.

[12] OECD/DAC (2005), note 9 above.

[13] Another follow-up was the decision of the OECD/DAC High-Level Meeting of Development Ministers and Heads of Agencies on 3 March 2005, which agreed that the principles drafted by the Fragile States Group on 'good international engagement in fragile states', by both development and security stakeholders, should be piloted in nine

fragile states until the end of 2006 and adopted by the High-Level Meeting in 2007. See www.oecd.org/site/0,3407,en_21571361_34391787_1_1_1_1_1,00.html.

[14] The DAC, like the European Commission later, chose to use the term 'security system' instead of 'security sector', for in its view the word 'sector' gave the impression of a closed and tightly defined area of responsibility. Similarly the reference to 'security processes' indicates a variety of security-related issues and actors.

[15] Damien Helly, 'Assessing the Impact of the European Union as an International Actor', in *Securing Europe? Implementing the European Security Strategy*, eds Anne Deighton and Victor Mauer, Zürcher Beiträge zur Sicherheitspolitik no. 77 (Zürich: ETH Zürich, 2006): 86.

[16] OECD/DAC, note 10 above.

[17] Alan Bryden, 'From Policy to Practice: Gauging the OECD's Evolving Role in SSR', in *Intergovernmental Organisations and Security Sector Reform*, ed. David Law (Münster: LIT Verlag, 2007): 65–83.

[18] DCAF followed this up in a project with the UNDP, which resulted in the following publication: Eden Cole, Kerstin Eppert and Katrin Kinzelbach, *Public Oversight of the Security Sector: A Handbook for Civil Society Organisations* (Geneva: UNDP, 2008).

[19] Statement by the President of the Security Council at the 5632nd meeting of the Security Council, UN Doc. S/PRST/2007/3*, 20 February 2007. See also *Enhancing United Nations Support for SSR in Africa: Towards an African Perspective*, proceedings of international workshop, Cape Town, 7–8 November 2007 (prepared by DCAF on behalf of the Ministry of Foreign Affairs of the Slovak Republic and the Department of Foreign Affairs of South Africa); United Nations, annex to letter dated 20 November 2007 from the Permanent Representatives of Slovakia and South Africa to the United Nations addressed to the Secretary-General, 'Statement of the Co-Chairs of the International Workshop on Enhancing United Nations Support for Security Sector Reform in Africa: Towards an African Perspective', 29 November 2007, UN Doc. S/2007/687. For a review and assessment of pre-2008 SSR activities within the UN system, see Vincenza Scherrer, 'Challenges of Integration: Cooperation on SSR within the UN System and Beyond', in *Intergovernmental Organisations and Security Sector Reform*, ed. David M. Law (Münster: LIT Verlag, 2007).

[20] UN Secretary-General, 'Securing Peace and Development: The Role of the United Nations in Supporting Security Sector Reform', Report of the Secretary-General, UN Doc. A/62/659 – S/2008/39 (23 January 2008).

[21] Ibid.: para. 17.

[22] See Nicola Dahrendorf, 'MONUC and the Relevance of Coherent Mandates: The Case of the DRC', in *UN Integrated Missions: Experience from Burundi, the Democratic Republic of Congo, Haiti and Kosovo*, eds Heiner Hänggi and Vicenza Scherrer (Münster: LIT Verlag, 2008): 91.

[23] Maria Caparini, Philipp Fluri and Ferenc Molnar, eds, *Civil Society and the Security Sector: Concepts and Practices in New Democracies* (Münster: LIT Verlag, 2006): 257.

[24] 'Security Sector Reform', (Washington, D.C.: U.S. Agency for International Development, U.S. Department of Defense, U.S. Department of State, January 2009). This provided lead authority to the assistant secretary of state of the relevant regional bureau; in the Department of Defense to the assistant secretary of defense for global security affairs; and in USAID to the assistant administrator for the Bureau for Democracy, Conflict and Humanitarian Assistance.

[25] See Timothy Donais, ed., *Local Ownership and Security Sector Reform* (Münster: LIT Verlag, 2008) for answers to these questions.

[26] LICUS Overview (web.worldbank.org), quoted by Patrick Doelle and Antoine Gouzée de Harven, 'Security Sector Reform: A Challenging Concept at the Nexus between Security and Development', in *The European Union and Security Sector Reform*, eds David Spence and Philipp Fluri (Geneva: John Harper Publishing, 2008).

[27] In their WDR 2011 background paper, Spear and Harborne considered SSR as part of a political reform intrinsically linked to state (re)formation. The paper also points at the difficulty of reintegration into non-existent economies and the small sums made available for reparations to victims. Joanna Spear and Bernard Harborne, 'Improving Security in Violent Conflict Settings', Security and Justice Thematic Paper, World Development Report 2011 Background Paper (Washington, DC: World Bank, 1 November 2010): 14, available at http://wdr2011.worldbank.org/sites/default/files/pdfs/WDR%20Background%20Paper_Harborne_Spear.pdf?keepThis=true&TB_iframe=true&height=600&width=800.

[28] Articles of Agreement of the International Monetary Fund, adopted at UN Monetary and Financial Conference, Bretton Woods (22 July 1944): Article I Purposes.

[29] Nicole Ball, 'World Bank/IMF: Financial and Programme Support for SSR', in *Intergovernmental Organisations and Security Sector Reform*, ed. David M. Law (Münster: LIT Verlag, 2007): 137–153.

[30] See Vivien Pertusot, 'NATO Partnerships: Shaking Hands or Shaking the System?', IFRI Focus Strategique no. 31 (Paris: IFRI, 2011).

[31] On 9 May 2011 this assembly held its last session in Paris. The treaty was abrogated on 30 June, as the solidarity clauses of the Treaty of Lisbon were considered adequate under current circumstances.

[32] Adopted by the European Council in Brussels, 12 December 2003, under the title 'A Secure Europe in a Better World.' It might be argued that the document was more of a pre-strategic concept, because it did not give precise guidance for force planning. Nevertheless, it was a remarkable attempt to formulate current threats in a way compatible with US and NATO strategy, while insisting on European views for tackling them. See also Willem van Eekelen, *From Words to Deeds: The Continuing Debate on European Security* (Brussels: CEPS, 2006): 181–198.

[33] Ibid.: 12.

[34] General Secretariat, Council of the European Union, 'EU Concept for ESDP Support to Security Sector Reform (SSR)', 12566/4/05 REV 4 (Brussels, 13 October 2005), available at www.initiativeforpeacebuilding.eu/resources/EU_Concept_for_ESDP_support_to_Security_Sector_Reform.pdf. For previous mission concepts, see 'Comprehensive EU Concept for Missions in the Field of Rule of Law in Crisis Management', Council Doc. 9792/03 (26 May 2003); 'EU Concept for Crisis Management Missions in the Field of Civilian Administration', Council Doc. 15311/03 (25 November 2003).

[35] General Secretariat, Council of the European Union, ibid.: para. 1.

[36] Ibid.: para. 12. The paper notes that the Commission was particularly active in the field of reintegration. A concept for DDR was put on the EU agenda during the second half of 2006.

[37] Commission of the European Communities, 'A Concept for European Community Support for Security Sector Reform', SEC (2006) 658, Communication from the Commission to the Council and the European Parliament, COM (2006) 253 final

(Brussels, 24 May 2006), available at http://eur-lex.europa.eu/LexUriServ/site/en/com/2006/com2006_0253en01.pdf.

[38] Ibid.: Section 1 ('Introduction'). Section 5.2 ('Recommendation to strengthen the EC contribution to overall EU support for SSR') advocated a holistic policy and programming dialogue, prioritising support for SSR under the new financial instruments and particularly using the stability instrument for rapid support for critical phases of the reform process.

[39] Ibid.: para. 3.2.

[40] Ibid.

[41] Ibid.: para. 4.3.

[42] Ibid.

[43] Much like the DAC, it also stressed that the norms for transparency and accountability should be the same for the entire public sector. Council of the European Union, 'Press Release', 2736th Council Meeting, General Affairs and External Relations, 9946/06 (Presse 161) (Luxembourg, 12 June 2006), available at www.consilium.europa.eu/ueDocs/cms_Data/docs/pressData/en/gena/90013.pdf: 14.

[44] European Commission, 'A New Response to a Changing Neighbourhood', COM (2011) 303 (Brussels, 25 May 2011).

[45] The earlier Commission document had already defined the EU strategic interests as security and stability. The European Union supported bilateral and regional cooperation, World Trade Organization accession and a strengthened approach towards human rights, rule of law, good governance and democratisation. Michael Emerson and Jos Boonstra (rapporteurs), *Into EurAsia, Monitoring the EU's Central Asia Strategy* (Brussels and Madrid: CEPS/FRIDE, 2010): 128.

[46] E.g., www.makeaidtransparent.com, which promotes the right to know how aid money is spent.

[47] The conference was held in February 2008 in London. See DAC Network on Governance (GovNet), 'Donor Approaches to Governance Assessments: Guiding Principles for Enhanced Impact, Usage and Harmonisation' (Paris: OECD, March 2009), available at www.oecd.org/dataoecd/27/31/42338036.pdf.

[48] See also Chapter 8 in this volume.

[49] On this issue see Nick Witney, *Re-energising Europe's Security and Defence Policy* (London: European Council on Foreign Relations, 2008): 7.

[50] Members are France, Italy, the Netherlands, Portugal, Romania and Spain. Poland and Lithuania are partners and Turkey is an observer.

[51] Further elements are suggested in Alvaro de Vasconcelos, ed., 'A Strategy for EU Foreign Policy', EU ISS Report no. 7 (Condé-sur-Noireau: EU ISS, June 2010).

Chapter 11

Lest We Forget? The Centrality of Development Considerations in Internationally Assisted SSR Processes

Ann M. Fitz-Gerald

Introduction

Security sector reform (SSR) is a concept originally conceived by the international development community, and one which was meant to support the pursuit and promotion of development goals. Since its introduction in 1998 the concept has enjoyed good support from the bilateral and multilateral policy communities, and has developed into one of the few comprehensive programme tools to have outlived over a decade of active debate.

Notwithstanding this debate, and the normative frameworks and policy guidance that have been developed by an active research and policy community, it appears that ground-based SSR does not achieve the level of comprehensive and all-encompassing scope that policy and research recommend. Most importantly, it is questionable to what extent the development community plays a role and influences delivery and implementation of SSR programmes at the 'front end', thereby posing questions as to how SSR directly supports the development agenda.

This chapter is divided into three main sections. The first provides a brief overview tracking developments in the SSR debate from both policy and operational perspectives, and highlights specific issues which contribute to the disconnect between strategic SSR policy and normative frameworks and SSR delivery on the ground. The second section examines both strategic national security and development processes to explore ways in which these exercises can contribute to SSR programmes in a way which advances broader development-related policy goals. The last section is the most substantive of the chapter. It features three country case studies – Uganda, Sierra Leone and East Timor – and investigates issues related to donor objectives, the degree to which national development strategies and plans informed SSR programmes on the ground and the outcomes of the SSR programmes. The chapter concludes with some broad recommendations for further research and policy development.

Developments in the SSR debate

Early debates on SSR which emerged in the late 1990s had their roots in public sector management and governance-related issues, particularly where high and disproportionate levels of military spending were observed. These initial ideas on the need for more comprehensive reforms across a wider sector of actors were advanced by the UK government's Department for International Development (DFID), which, while providing a set of guidelines[1] for DFID governance advisers and managers considering SSR, also reminded the policy community that SSR programmes would only be considered if they contributed to DFID's primary mission: the reduction of poverty. The guidelines gave the first published definition of SSR, which read that the security sector comprised 'all actors responsible for the protection of the state and the communities within it';[2] similar definitions featured in later DFID policy briefs.[3]

The coming together of cross-policy strands involved in SSR work provided sufficient evidence to support the important relationship between security and development. As a result, a number of policy documents were produced on this subject, most notably the 2005 DFID paper entitled *Fighting to Build a Safer World: A Strategy for Security and Development.*[4] However, as different scholars[5] attempted to produce empirical evidence on the underpinnings of this relationship, some doubt was cast on the extent to which 'critical connections' between the two fields could be established.[6] A paper produced by the International Peace Academy in 2005 stated that there were still significant gaps between both policy and research and knowledge and practice; and that preventive (versus reactive) approaches were required to address these gaps.[7]

In the face of this empirical challenge, other scholars[8] have analysed the positive contributions made by institutional and structural developments which occurred as a result of calls for 'joined-up' and 'whole-of-government' approaches[9] to address the relationship between security and development. These trends were important in building cross-government relationships and evolving a less traditional cadre of international development actors who appeared more comfortable about dealing with security actors and institutions. To a certain extent, evidence of in-country cross-government and cross-community relationships has now emerged in the more recently published poverty reduction strategy papers (PRSPs) of low-income countries, which expose the strategic relevance of governance, security and conflict resolution to development.[10]

However, notwithstanding this institutional progress, limitations placed by official development assistance (ODA) on the direct funding of security-related activities have presented some difficulties for support to certain SSR activities. As a way of circumventing some of the more restrictive ODA criteria, some donor countries created new funding mechanisms which stood independent of ODA and could be used to support more comprehensive SSR, and thus a 'security-development nexus'. The UK government's Conflict Prevention Fund and Stabilisation Fund (created in 2008), the Dutch government's Stability Fund and the government of Canada's Global Peace and Security Fund all serve as examples of such funding mechanisms. Despite these mechanisms, having only a small group of international development donors with the wherewithal to support more cross-government, comprehensive SSR programmes has posed challenges for a more 'multinational' approach to SSR to develop on the ground. For a programme area which is so broad, which requires such a wide range of skill sets and which is costly to implement, coordinated, multinational responses become critical for success.

Undoubtedly, the work of the multilateral policy community made an enormous contribution to the continued evolution and practice of SSR. In 2005 the OECD/DAC developed guidelines on security system reform, which were endorsed by the ministers of member states and aid agencies. Notable in the content of the guidelines was an emphasis on extending the dialogue for SSR beyond governments, as well as the vital roles for parliaments and civil society groups to play as advocates, watchdogs and providers of knowledge on policy issues and citizens' needs.[11] The importance of this extended and inclusive dialogue was reinforced by the research findings of the 2011 *World Development Report*, which evidenced an important linkage between the relative success of institutional transformation and the inclusion of a plurality of actors.[12] In addition to successive meetings held by the donor community to discuss a common understanding and practical application of the guidelines, the way in which it shaped and informed national policy and practice also demonstrated the value of this policy advocacy approach. For example, in 2009 the government of Canada developed its own SSR guidelines, largely based on the OECD/DAC publication but with one additional guideline which recognised the linkage between overseas assistance and Canada's domestic security priorities.[13]

Following a two-year consultation period, the OECD/DAC guidelines were taken to a further level in 2007 with the publication of the OECD/DAC *Handbook on Security System Reform.*[14] The purpose of the handbook was to examine the practical challenges in the implementation of SSR and make security and justice delivery more effective. It exists electronically and has been translated into many different languages, which has helped to socialise ideas and lessons from the global SSR experience. The handbook serves as a vehicle to preserve the SSR knowledge and experience built up since the late 1990s, and still features as a reference for international training programmes. Its text has informed the development of wider international policy on SSR in both the European Union and the United Nations, the latter of which released a UN Secretary-General's report in January 2008 entitled 'Securing Peace and Development: The Role of the United Nations in Supporting Security Sector Reform'.[15] The UN's endorsement of SSR gave the concept a serious international profile which further mainstreamed SSR norms and principles across UN member states.

Based on the chronology of SSR strategic conceptual and policy-related developments, and despite some of the limitations posed by conventional principles governing the disbursement of international assistance, it could be argued that the wider community of SSR policy-makers and practitioners is reasonably well served with normative frameworks, guidance and repositories of lessons learned and research to date. Consistent across these approaches are the calls for wide inclusive approaches to SSR which involve a broad range of civil society actors. Throughout this period, these ideas and concepts have been further socialised by a plethora of academic conferences, online debates and the production of edited volumes which address SSR research gaps and new applications. Evidence suggests that the SSR debate will sustain its relevance in broader security and development policy domains in the future, particularly in light of the 2011 *World Development Report*'s suggestion that prioritisation should be given to institutions that provide citizen security, justice and jobs to prevent the recurrence of violence and lay foundations for reform.[16]

The next section explores ways in which SSR has been applied in practice, to understand if development objectives of reducing poverty – and the development-community-driven guidance and role of development in the delivery of SSR – have been prioritised in a way which the original drafters of strategic SSR policy envisioned.

SSR developments on the ground

SSR interventions were originally conceived for all types of transitional societies, including post-authoritarian states and countries in different stages of economic, political and structural transitions. However, to date academic literature indicates that donor-funded SSR programmes have most frequently taken place in countries emerging from conflict, such as Liberia, Sierra Leone, Bosnia, East Timor and Afghanistan. Having said this, the SSR experiences of developing and weak, but not classic 'post-conflict', countries such as Uganda and Ethiopia are also instructive, as are lessons from further afield from activities undertaken in countries like Jamaica and Indonesia.

In most cases SSR is initiated based on a request from the host government for donor-funded assistance. These requests are facilitated in a number of ways: through a direct request to a resident representative of an in-country political mission (such as a defence attaché) with a mandate and funding to support security sector transformation and cooperation efforts of the host state; the result of an evaluation of existing programme/project work of an international agency with a mandate and budget to support further work in this field; as a result of a 'design phase' assessment for the multi-year funding strategy of an international development actor in a particular country of region; or, lastly, as a result of the SSR-related content of a peace agreement.

Resident defence attachés and political and development advisers (governance and conflict advisers included) play a significant role in the facilitation of SSR-related support in a host country. It is often the case that their daily communications and work with different government and community stakeholder groups will lead to a discussion on options for SSR-related project and programme support. For example, requests for support in specific areas of defence management – such as procurement and human resource management – may elicit a response from a resident defence attaché who, on receiving such a request and confirming that these activities could be drawn from his/her own budget for defence assistance, may call on either internal or external support to respond. Irrespective of whether or not funding is drawn from a dedicated defence support budget or from a wider cross-government 'pooled' budget, and in the spirit of 'joined-up' government, the decision of whether or not to support certain activities is often taken as part of a wider consultation with a defence attaché's in-country development and political counterparts. This dialogue supports the sharing of knowledge, ideas and documentation (such as previous reports and assessments) and the pooling of good practice across wider groups and individuals benefiting from such donor support. Similar consultations may be prompted by political representatives in terms of providing support to police forces, or by development representatives in the provision of support to the justice sector or federal capacity-building ministries. In these cases, care must be taken to ensure that what may appear to be externally driven aligns well with local requirements.

Between 2002 and 2010, the use of members of the UK government's Security Sector Development Advisory Team (now called the Security and Justice Team) to respond to requests from resident defence attachés in countries such as Sierra Leone, South Africa, the Democratic Republic of the Congo, Botswana, Jamaica, Malawi, Moldova and many others reflected this approach. Similar approaches have been taken by the US government through its foreign military liaison staff in many Central and East European countries. More recently, and based on its multi-donor funding, the Geneva-based International Security Sector Advisory Team has responded to in-country requirements in a similar way, servicing the advisory and training requirements of a broad group of both developed and developing countries. While delivering services in weak states such as Guinea-Bissau, South Sudan and Albania, the team has also responded to the training needs of donor organisations such as the United Nations and the government of Canada.

Responding to specific in-country requests has proven helpful in many instances. Assistance can often lead to 'quick wins' and the development of improved and lasting relationships between local government leaders and in-country donor representatives. Such assistance can also lead to the demand for wider support, which culminates in programmes being developed from projects. This introduces inevitable coordination issues with other development and political actors servicing the needs of the wider security sector, and often leads to a somewhat 'bottom-up' approach taken to 'strategic SSR'. Notwithstanding the useful and progressive benefits of these initial bottom-up approaches, allowing projects undertaken with traditional security institutions to evolve into programmes without top-down guidance will limit the extent to which SSR interventions support broader development goals in line with both national and international development agendas. This risk can increase when SSR project work takes place in countries which lack well-developed national strategies.

There has also been an increasing recognition of the value of using broader national security consultations and security policy/strategy development exercises to inform more specific and priority areas for further SSR assistance. The utility of these exercises was demonstrated by the development of the security policy framework in Uganda in 2002, the development of a national security policy in Jamaica between 2005 and 2007 and the national security policy consultations in Botswana during 2008 and 2009. In all cases, the perceived need for assistance in one part of the security sector led to a proposal for a broader analysis to be used to inform ways in which support could be best channelled. This approach has proven to be effective in drawing in support from a wider cross-government community, as well as civil society groups and traditional leaders. For example, religious leaders in Jamaica proved invaluable for identifying root causes of gang-related criminal violence; tribal leaders were effective in raising the security concerns of communities in rural locations in Botswana; and academics were effective at drawing on their own research and empirical evidence bases to analyse security concerns in Uganda. These approaches resonate with the principles provided in SSR guidance, such as wide consultations and local ownership; moreover, the outcomes of these processes are often successful at influencing the work of a range of bilateral and multilateral actors working in the country, including wider development interests.[17] Having said this, the 2011 *World Development Report* indicates that much of the assistance for priority development tasks remains slow, in particular where best-fit needs on the ground fall outside the regular donor boxes.[18]

SSR interventions do not always come at the behest of one national actor. International organisations such as the World Bank, the United Nations or the International Monetary Fund (IMF), which often have well-funded multi-year programmes in certain countries, may propose certain SSR activities as a result of reviews and assessments conducted on their activities to date. Indeed, the IMF's concern in 1999 over the increased levels of military expenditure by the government of Uganda led to suggestions for SSR to be considered. SSR assessments carried out by the UN's Office of the Rule of Law and Security Institutions on behalf of existing UN integrated missions may also lead to a number of newly developed SSR-related activities implemented by either UN agencies or national government teams and/or consultants. National and multilateral development actors, such as the World Bank and the UK and Dutch governments, now work with ten-year funding frameworks,[19] which result in support developing and adapting over time based on new challenges and changing strategic environments. In these latter cases, SSR-related activities emerging from multi-year frameworks are often supported by research and analysis undertaken during a SSR programme 'design phase'. One could also make the assumption that these longer-term approaches to SSR take a more strategic approach in line with broader development strategies and international development frameworks (such as the PRSPs), although there is currently no published literature evidencing this link. One challenge to this SSR approach is the difficulty in engaging more conventional security sector actors from the outset and the tendency for the international development community to work with like-minded partners and conventional development 'tools', many of which are not easily transferable to the more traditional security domain. The net result of this parallel engagement is often that the work of security and development communities does not merge together and become mutually supportive and reinforcing. Thus the problem is not only related to poor consideration of development objectives by SSR actors, but also poor SSR considerations of development actors. Both need to develop a better understanding of each other, and the frameworks and *modus operandi* which guide their respective approaches.

Lastly, SSR activities are often linked to the text of international peace agreements and the obligation on the part of one or more parties to the agreement to undertake 'SSR'. As part of the 1997 Conakry Peace Plan to end the conflict in Sierra Leone, the seven-point peace plan devised for the early return of constitutional governance in Sierra Leone included the disarmament, demobilisation and reintegration (DDR) of ex-combatants; the same approach was taken during the Inter-Congolese Dialogue, which saw the 2002 Sun City Peace Accord mandating the UN special envoy to oversee the DDR and resettlement of armed groups.[20] Other examples could also be discussed, such as the Bonn Agreement which called for the rebuilding of Afghanistan's domestic justice system, and the 2005 Comprehensive Peace Agreement in Sudan, calling for the interim formation of integrated units of 21,000 soldiers.

Although many of the DDR and security-related initiatives embodied in peace agreements have implications for, and require support from, development actors, the conditions are not always favourable for broader development programmes to support security reforms, nor are the linkages between security and development requirements spelt out in most peace agreements. In most immediate post-conflict stabilisation and early recovery interventions, long-term development goals are often not considered until a certain degree of institutional capacity to manage and oversee development work is created. In addition, more traditional development donors will be restricted in terms of contributing to post-peace-agreement SSR activities in unstable and high-risk environments.

In summary, a number of different approaches characterise the way in which SSR activities become initiated on the ground. In-country representatives from national political missions, particularly defence representatives, often play a significant role in the development of SSR-related project ideas. While it is often the case that defence-related initiatives signal the need for a more strategic-level national security review (as was the case in Botswana and Uganda), approaches which support the development of national strategic policy frameworks have demonstrated reasonable levels of success at creating inclusive programme which feature a plurality of actors from the outset. In addition, these approaches offer the greatest likelihood of development activities being coordinated with security considerations and 'front loaded' into planning for SSR activities.

The next section further examines these two strategic approaches to broader policy-making in the fields of security and development. Consideration is given to the extent to which both could serve as more effective entry points for strategic SSR.

PRSPs and national security policies compared

In 1999 the IMF renamed the much-discredited Enhanced Structural Adjustment Facility as the Poverty Reduction and Growth Facility.[21] Based on this decision, the PRSP approach was introduced to provide a framework for IMF lending and the World Bank's country assistance strategy, and as a new approach to the challenges of reducing poverty in low-income countries. As well as providing a mechanism for international development donors to align assistance more effectively behind national development strategies, the PRSP acts as a platform for national actors to raise the issue of poverty eradication to the macro-strategic policy level.

One of the central features of the PRSP is a participatory approach and the requirement for civil society organisations (CSOs) to partake in the development of the initial plans as well as subsequent revisions. In this context, PRSP processes have catalysed the formation of civil society networks, such as the Ugandan Debt Network, with some governments having even established minimum standards for CSOs. These participatory approaches have helped encourage dialogue, improve the understanding of poverty and encourage systematic ways of engaging across federal and local authorities in order to develop state-wide perspectives on issues related to poverty and growth. In addition, the requirement for more rigorous and in-depth country-wide data supporting poverty analysis has had a positive effect on the capacity development of national CSOs.

Recognising that full PRSPs would take more than one year to develop, low-income countries were encouraged initially to develop 'interim PRSPs' (I-PRSPs). These initial documents could then benefit from feedback and the findings of other relevant reports and assessments, and subsequently be developed into 'full PRSPs', which would be reviewed and evolved every three years. The IMF retains a current database of all I-PRSPs, full PRSPs and all related review and progress reports.[22]

While the PRSP process served as a positive and progressive step forward from the more controversial issues surrounding structural adjustment, the approach was not without its problems. Based on variable levels of institutional capacity, the quality of papers varied significantly and the completion rates of full PRSPs were much lower than originally anticipated. Countries were encouraged to use existing national and sectoral strategies and not to start from scratch. Despite these and other weaknesses, there was widespread agreement that, based on the input of a plurality of actors, the PRSP process developed a real sense of national ownership, increased the capacity of CSOs, gave prominence in the strategic policy debate to issues related to poverty and encouraged donors to embrace PRSP principles fully, including that they should be country-driven, long term in perspective, results-oriented and comprehensive.

Based on issues related to human security in low-income and weak states, one could argue that many initial PRSP efforts evidenced the interdependencies between security and development in a way which more conventional national security documents did not. The significant participation of CSOs also meant that governance-related issues formed key areas for policy consideration. However, given the secretive nature of the security sector in many of these countries, and the fact that security sector budgets were often dealt with in a way which was distinct from a state's public expenditure management process, a close interface between security and development actors in support of PRSPs was often not achieved. Moreover, the PRSP process remains a concept that was created by traditional international development donors without input from international security institutions. As such, it carries a parlance and a historical background that are deeply rooted in the development institution. These institutional cultural differences can pose further limitations on the interface between security planning and development planning.

On the other hand, a national security policy is a statement or declaration of how a country intends to promote, pursue, defend and protect its national security interests. National security strategies present a methodology for 'operationalising' a national security policy based on current and future strategic environmental trends and available resources. Most normative guidance on strategic planning for national security requires broader implications for economic, social, internal and external security to be identified as a result of a comprehensive strategic environmental analysis.[23] Based on this broad level of analysis supporting strategic policy goals for security, it is reasonable to assume that development concerns central to the human security debate would feature in these approaches. However, state-centric thinking on security which characterised the Cold War and immediate post-Cold War years meant that many of the existing national security strategies, particularly those of the United States and its European allies, focused primarily on military security. These approaches influenced the design and development of the national security strategies of the new Central and East European democracies which were developed as a prerequisite to accede to Euro-Atlantic structures. Having said that, more recently published national security documents, such as the 2010 UK national security strategy and the 2004 Canadian national security policy, evidence wider thinking on issues related to security, development and growth.

Based on the lead role played by defence institutions in SSR interventions in weak and fragile states, the use of a national security framework (policy or strategy) has, in some cases, served as a useful entry point for strategic SSR engagements. Arguably, it has more often been the case that defence-led SSR has been encouraged to revert back to more macro-strategic security policy thinking to understand better the role for defence – and a range of other security actors – in a new strategic environment.

A key challenge to progressive and strategic discussions on national security often relates to the existing understanding of 'security' among the population and the perceptions of the security forces, particularly if such forces have undermined democratic governance and human rights. In these cases, it is sometimes difficult to generate a national dialogue on, and secure widespread participation supporting, a national security process or framework. A comprehensive national security exercise also rests on the assumption that all the different security providers will be accustomed to coordinating together to implement strategic security policy, which is rarely the case.

In summary, both the PRSP and the national security strategy/policy have served as useful entry points for national dialogue and consultation on issues related to security and development. While the PRSP process has been hailed for its engagement and development of CSOs – and the more conventional development-focused ministries of recipient countries – discussions on national security have proven useful in ensuring that initial defence-dominated SSR interventions consider wider security and development issues in defining new roles for the military and other security agencies. Although the PRSP was created by external development actors as a tool for more effective planning for the management of future debt relief, the process has catalysed the production of nationally owned and driven development strategies.

The next section undertakes a more detailed examination of these strategic approaches and the impact each had on the SSR programmes in three different countries. The case studies of Uganda, Sierra Leone and East Timor are used to explore the extent to which strategic development and security processes converged to support SSR activities on the ground. Based on the time periods which characterised each intervention, the different environmental conditions and challenges on the ground and the variable amount of published and 'grey' material supporting each case, a consistent methodology for all case studies could not be followed. However, for each case I attempt to draw some broad conclusions based on an analysis of relevant background material, existing strategic security and development documents, the dynamics of in-country processes and the stakeholders engaged.

Case study 1: Uganda

Even before the concept of SSR had been embraced by the wider donor community, Uganda's strategic approach to development had outpaced any similar effort among its security counterparts. According to Piron and Norton, the roots of the first Poverty Eradication Action Plan (PEAP) were a 'home-grown' government of Uganda initiative and document developed as a result of a 1995 government/World Bank seminar at which concerns were expressed about the lack of systematic consideration of poverty impacts in the Bank's vision of growth in Uganda.[24] The 1997 PEAP gave primacy to the national goal of reducing poverty from 49 per cent in 1997 to 10 per cent in 2017, and served as a political project to achieve national unity.[25] Its strategic pillars included the creation of an environment for enabling sustainable economic growth and transformation, promoting good governance and security, and raising incomes of the poor and increasing their quality of life.[26] Interestingly, despite the fact that the PEAP was initiated by the Ministry of Finance, Planning and Economic Development, this 'home-grown' national development strategy became one of the first ever to reflect the interdependencies of security and development which formed such a key debate in the years that followed.

The publication of the PEAP coincided with calls by the World Bank and the IMF for low-income countries to develop PRSPs, thus the three-year revision of the PEAP in 2000 was used as a mechanism for the development of Uganda's first PRSP, which subsequently became a 'showpiece' for donors. The PRSP consultations-cum-PEAP revisions also gave an opportunity for the recommendations of other parallel studies, such as the 1997 Participatory Poverty Assessment Project (part of the World Bank's country assistance strategy) and the 1998 National Integrity Survey, to be considered; as such, they were supported by over 45 CSOs.[27] For the first time in history, the consolidated impact of these development-related consultations was that the government of Uganda became accepting of CSOs and saw them as a serious stakeholder in the formulation of strategic national policy.

Following the publication of Uganda's PRSP, the IMF and other donors became increasingly concerned over the sustained high levels of defence spending, which in 2002 was 13.2 per cent of GDP (a decrease from 19.6 per cent of GDP in 1994).[28] Based on the UK government's emerging thinking on SSR, and its experience in undertaking strategic defence reviews, DFID funded a defence review process between February 2002 and June 2004.[29] The review featured three phases: a strategic security assessment (to identify both military and non-military security threats and issues), which contributed to the development of the 2002 security policy framework and clarified the security-related responsibilities of different governmental agencies; a defence policy process, which clarified the mission and role of the Ugandan Peoples' Defence Forces and a programme for modernisation; and lastly, a white paper on defence transformation which was approved by Cabinet in March 2004.[30]

Arguably, Uganda was one of very few early SSR experiences prompted by the use of a strategic entry point. The process was also undertaken in a permissive way which kept the defence institutions – unquestionably the key stakeholders of the defence review – playing a lead role, albeit extending the exercise to consider the wider security environment and the broader implications for the full range of security sector actors. The Uganda experience is also an early example of the benefits that a wide consultative process supporting security and defence-related issues can have for future defence planning.

A detailed and well-written description of Uganda's defence review process can be found in a study published by King's College London in 2005, written by one of the lead international consultants involved in the review.[31] For the purposes of this chapter, some of the findings presented in the study provide some interesting insights into strategic processes supporting SSR. First, although the Ugandan experience features government departments such as the police and corrections institutions, it was not successful in including all relevant security actors, such as the Ministry of Justice. Similarly, while the process did secure input from a limited number of academics from Nkumba and Makerere Universities, the participation of CSOs was on the whole quite limited.[32] According to the report, it was thought that this low participation rate limited the degree to which governance issues – rather than capability-focused issues – featured in the discussions.[33] It was also hoped that the strategic security assessment would spur other departments with a security mandate into developing their own policy and plans, which for the most part did not happen.[34] Thus, although there was a broadening and deepening of the national ownership of the review process, the results and impact have been described as uneven.

While the overall outcomes of the defence review in Uganda engaged the interests of the wider development donor community, strategic momentum behind the initial process appears to have waned. Two years after the publication of the security policy framework, there was a divergence of views between the government of Uganda and development partners on the level of resources required to finance transformation and the defence budget as a whole.[35] Even as recently as 2009, Ugandan government officials who were instrumental in supporting the defence review confirmed that the Ministry of Finance, Planning and Economic Development has continued to progress with its plans for development quite independent of the work achieved by the security policy framework.[36] While it appeared that a solid platform of cross-sectoral actors was already in place as a result of the national development agenda, the same level of support and sustainability could not be achieved in support of security reforms. Unfortunately, the PEAP's early identification of the close linkages between the country's security and development agendas did not appear to bridge the two communities at the practical level. These linkages and existing structures could have provided a useful and committed starting base for the analysis of issues related to national security.

Case study 2: Sierra Leone

The international SSR intervention in Sierra Leone has been held up as one of the more successful experiences to date. The most instructive and empirical piece of work covering the comprehensive SSR process and then reviewing it over a period of ten years (1999–2009) was produced by Peter Albrecht and Paul Jackson.[37] Funded by the UK government and further supplemented by a large number of shorter pieces published during the ten-year period, this work tracks the challenges experienced and progress made by the Republic of Sierra Leone government and a range of international actors with regards to the implementation of SSR.

The published literature on the post-conflict intervention in Sierra Leone comments extensively on the degree to which wide consultation was used to support the multifaceted SSR process, which featured a wide cross-section of government authorities as well as civil society. As with most post-conflict settings, one could argue that the consultation process supporting wider SSR was, until after the elections, focused mainly on Freetown and individuals and groups from the political centre. However, efforts were made by 2003 to develop provincial and district security councils and prioritise development of local governance structures in the most rural districts so that all communities could have a voice in security-related matters. Even at the time of writing, efforts to develop local governance capacity/structures and create a pan-national discourse on security and development are still being supported by certain international actors. For example, in 2011 the UN Peacebuilding Fund funded the purchase of vehicles for the use of district authorities and to facilitate regular dialogue across national and local government authorities and the communities they serve.[38]

Led and funded primarily by the UK government, with support from a number of other development partners such as the UNDP and the World Bank, the initial intention of the donor community in Sierra Leone was to create peace and stability in a country whose security and economy had been undermined by poverty, insecurity and violent attacks and human rights abuses on its citizens. Despite the international concern about security and stability, following President Kabbah's request in 1998 for a review of the civilian management of the Sierra Leone armed forces, the origins of SSR were in administration, civil service reform and governance.[39] However, repeated attacks by the Revolutionary United Front and other militia groups on Freetown meant that certain stabilisation measures took precedence over more traditional public sector reforms. This led to the deployment of a West African peacekeeping force (ECOMOG), a DDR programme for former combatants from all armed groups and support to the Republic of Sierra Leone Armed Forces in the form of military training.

These security-related workstreams formed the key components of Sierra Leone's Security Sector Reform Programme (SILSEP), the framework through which the UK government provided support to the Office for National Security (ONS), the Central Intelligence and Security Unit, the Ministry of Defence, the Sierra Leone Police, the Ministry of Internal Affairs and a range of non-security-related institutions with an interest in accountability and enhanced service delivery across the security sector, such as parliament, civil society, media and academia. The purpose of SILSEP was to 'assist in the creation of an enabling environment within the security sector to ensure the successful and sustainable implementation of the Sierra Leone Security Sector Reform Implementation Plan, as articulated within Pillar 1 of the Poverty Reduction Strategy Paper, and the new National Security Policy'.[40]

Other major SSR projects in Sierra Leone included the provision of the Commonwealth Police Task Force, which over time developed into a broader police reform programme. Initial short-term training for the armed forces also transformed into the longer-term International Military Assistance Training Team programme. However, despite efforts to ensure a good level of comprehensive, cross-sectoral engagement, support remained heavily geared towards the operational training of security actors, with very little focus on governance and oversight mechanisms.

In the absence of any 'home-grown' national development plan, an interim PRSP was published in 2002. This interim document was updated and influenced by the work of the broader SSR community, and a full PRSP document was published in 2005 which presented strategic 'pillars' that demonstrated the important and interwoven relationship between security and development.[41] However, these efforts remained 'presentational' only, as the 2005 PRSP document was described as 'overly ambitious and unimplementable'.[42]

Between 2000 and 2005 SSR-related activities focused significantly on developing a well-functioning ONS which could inform a full range of security priorities. Positioned as an executive function of government, the head of the ONS maintained a close and productive relationship with the president which ensured a high degree of buy-in supporting most security-related initiatives. Over time, the ONS prioritised the recruitment, development and training of an effective team that set out to lead on the development of a national security policy. Although not yet officially endorsed by the current government, a draft policy was produced in 2007, but has suffered from a lack of comments – and interest – from other government agencies.[43] At the time of writing, the policy remains unpublished.[44]

In 2007 the ONS also led a donor-funded security sector review. Interestingly, the review uncovered a number of development-related issues that were presenting significant challenges to progressive security reforms. For example, the report called for the reform of the national registration secretariat to professionalise immigration services and various related functions, such as the provision of a national census and the issuance of passports and identity cards.[45] At that time, the absence of reforms in this area was causing serious problems for the law enforcement community. Another factor affecting the development of the security sector was the need for the education system to be more in line with modern requirements and development aspirations. Recognising the emerging opportunities in the area of mineral resources and oil exploration, it was felt that if Sierra Leoneans were unable to manage these resources and opportunities effectively, other external illicit and licit security actors might seek to exploit them.

The outcomes of the nationally led security sector review exercise informed a new PRSP which was officially published in 2008, entitled *An Agenda for Change*. This document departs from the strategic pillars embedded in the earlier PRSP, and outlines four priority areas which will shape Sierra Leone's national growth strategy:

- The provision of a reliable power supply.
- Raising quantity and value-added productivity in agriculture and fisheries.
- Developing a national transportation network.
- Ensuring sustainable human development through the provision of improved social services.[46]

This brief overview of Sierra Leone's security and development experience over ten years exposes a number of interesting facts with regard to strategic entry points for SSR. First, due largely to the short-term stabilisation imperatives, SSR was initiated with a mainstream security focus and a number of activities which concentrated on the operational capacity of the security forces and not governance and institution-building. The mainstream security focus was also reflected in the UK government's SILSEP, which appeared to be devoid of any real development-related targets. While the country's first PRSP document did reflect a good balance between development and security priorities, it was too broad in scope and too ambitious to be implemented. This was evidenced in the PRSP 2004 progress report, which indicated the large extent to which most of the objectives had not been achieved.[47]

Once governance capacity became developed in the ONS, wide and cross-country consultation efforts were undertaken to develop a well-informed direction and policy for the security sector. As consultation broadened, capacity increased and a nationally driven security sector review in 2005 exposed the many development challenges which continued to impair the security sector. The review's findings informed both the revised draft national security policy and a new PRSP. Perhaps one irony is that, unlike its predecessor document, the current national development priorities veer away from security issues and focus on interrelated and interdependent development functions. This indicates that although development activities were not 'front loaded' in the first six years of SSR in Sierra Leone, they proved to be critical to future security sector progress.

Case study 3: East Timor

Following the 24-year occupation by the Indonesia security forces which was ended by the 1999 vote for East Timor's independence, the United Nations took on its first guise as a transitional administration.[48] The mission arrived during a period of widespread violence and destruction in response to the independence vote, and administered the country until formal independence (by way of democratic legislative elections and the approval of a constitution) was achieved in 2002. The transitional administration completed its mandate in 2002 and was replaced by the UN Mission of Support in East Timor, which was mandated to provide assistance in the post-independence period. This mission was replaced by the UN Office in Timor-Leste (UNOTIL) in 2005, which served as a small political mission only. Yoshino Funaki laments the limited impact of this series of UN missions of variable mandates because of departures before institutions and capacity were in place for an effective Timorese takeover.[49]

The 2006 pre- and post-election period was characterised by fierce tension between East Timor's two primary security actors, the military (Falintil-Forças de Defesa de Timor-Leste) and the police (Polícia Nacional de Timor-Leste), with internecine conflict and human rights abuses being committed across the island. Tension also ran internally throughout both organisations. This prompted the United Nations to replace UNOTIL with the UN Integrated Mission in East Timor (UNMIT), tasked to focus on police development and training, reconciliation, institutional governance and capacity-building, support for the 2007 elections and humanitarian relief services. In May 2006 a UN policing mission and the Australian-led International Stabilisation Force were also deployed to help reinstate security, running in parallel with UNMIT. The International Crisis Group (ICG) describes the violence in 2006 as being based on decisions taken on the security sector in the years before and after independence in 2002.[50]

In 2008, following a rebel attack against the East Timor government which seriously wounded the president and killed the prime minister, the government called for a 'state of siege' for several weeks until the surrender of the rebel leader and his followers. These attacks prompted accusations being lodged against external actors,[51] and a growing rift developed between the United Nations and the East Timor government.

No clearly stated 'SSR plans' emerged before 2006, with the creation of a security sector support unit (SSSU) in UNMIT. The SSSU's work continued to support police training and also moved towards providing legislative support for the development of laws governing the security actors, underwritten in a national security policy. Despite such efforts, authors such as Gordon Peake, Edward Rees, Ludovic Hood and Yoshino Funaki have questioned the degree to which any form of 'holistic' and 'comprehensive' SSR has been tackled.[52] Their collective works highlight the dominant focus on police training, with no complementary efforts to tackle institutional governance and civilian oversight. Recommendations set out in a 2008 ICG report entitled 'Timor Leste: Security Sector Reform' further reinforced the fact that SSR should not concentrate on police training and should instead pursue a systematic and inclusive approach, 'combining national ownership and international help'.[53] The report also recommended that UNMIT undertake a comprehensive security sector review.[54]

Although this review did eventually take place, the process and the potential benefit were undermined by three things. First, although both the ICG report and the UN's objectives supporting the review called for a widely consultative process and engagement with civil society and public and private sector interest groups, no such engagement took place. The World Bank confirmed that, until 2007, the ruling party had very little or no engagement with CSOs.[55] By 2008 the CSO community still appeared to exist only in nascent form. Based on prioritisation given to expediency, and the UN's tendency to preserve its network and loyalties with institutional state actors, once embedded in this network it would no doubt experience difficulties in retreating and redeveloping a new stakeholder list. However, as the political landscape changed in 2007, and new political personalities arrived on the scene, a lack of wide and consultative discussions with civil society meant that the new actors had no consensually agreed platform which would provide some continuity.

Second, there have been delays in recruiting experienced and able bodies to staff the SSSU, which contributed to the delay of the 2008 review – and left space to nurture further the rift between the East Timor government and the United Nations. The review enjoyed only nominal national ownership.[56] This exposes the fact that, despite its adoption of a policy on SSR, the United Nations does not have the capacity to implement it. There is also a wealth of evidence to suggest that the United Nations was unable to oversee strategic-level coordination to gear the many and varied wider donor efforts in East Timor into something more comprehensive.[57] Indeed, it appears that the United Nations even failed in the coordination of its own internal mechanisms between, for example, the SSSU and the UN police.

Third, the review lacked any connection with nationally owned priorities, and thus did not reflect the national culture, make-up and traditions of society. The fact that the document was only produced in English also reflects the limited consultation process, and the limited impact its recommendations would have thereafter. Since the publication of the review, with the exception of the delivery of a number of small workshops and seminars which engage broader civil society, only small externally driven activities have ensued.

In many respects, and as many analysts and national authorities have suggested, East Timor 'had nothing' and its requirements were vast. Having said that, the PRSP published in 2002 (entitled 'National Development Plan'), and reviewed in 2005, outlined many of the critical development objectives that would lead to the reduction of poverty and emphasised the critical connection between poverty and conflict. Although the document appears to present a strategic scope which is unachievable for East Timor in the stated timelines, it does offer two important contributions which seem to have been overlooked in the efforts to provide security sector assistance.

The first issue concerns the country-wide consultation which involved over 38,000 East Timorese people from 'every district and every walk of life'. This exercise was undertaken in efforts to develop and achieve a national vision for the new country at the same time as its first constitution was being developed. If the consultation was as comprehensive as it states – and there is no reason to cast doubt on this – then security sector professionals could have benefited from building their plans and programmes from the same qualitative and quantitative data, and from developing the civil society capacity and public dialogue which, in infancy, had supported this work. The 2005 updated PRSP continues to reinforce the importance of a consultative approach and makes reference to engagement with 'civil society, the church, national and international NGOs, and private and public interest groups'.[58]

The document provides clear guidance on what is required to support the strategy's objective in the short and longer terms, referring to effective legal and governance foundations on which to progress cross-sectoral development. More specifically, the strategy reinforces effective administration and governance structures and describes these as 'one of the most critical national priorities', placing emphasis on civil society participation.[59] The UN's focus on police training and engagement with state actors fails to pay heed to the codification of national priorities based on a cross-country consultation which took place just prior to, and in parallel with, ongoing UN assistance to the security sector. Indeed, a closer connection between these security- and development-led efforts may have placed a much earlier priority on the development of a national security policy, and therefore on the administration systems and laws governing the security sector actors.

Despite the criticisms lodged against the PRSP process more generally, the East Timor government's national development plan provided a widely consultative platform sufficient for informing planning in individual sectors. The document was underwritten by both the government and the people. It prioritised the development of a 'national vision' which instruments of power such as the military and law enforcement agencies should seek to protect, defend and pursue. In summary, having even devoted a section to 'Political Development, Foreign Relations, and Defence and Security', it provided a 'good enough' basis to ensure that a better degree of national ownership and a plurality of actors could have entered the security sector assistance process at a much earlier stage.

Conclusion

This chapter reviews a number of issues related to strategic SSR policy and its implementation on the ground. Based on a wealth of academic research on the origins of the SSR debate, and the strategic policy guidance that has developed as a result of an active and progressive policy and research community, the initial section concluded that the international community is relatively 'well served' in terms of SSR frameworks at the strategic policy level.

The chapter then surveyed the various 'entry points' which drive most SSR interventions. A distinction is drawn between types of strategic and technical-level engagements and the suggestion is made that, in practice, normative frameworks are not always followed on the ground. Strategic engagements driven by the international development community have often occurred as a result of different methodologies and consultations based on the requirements of the international development agenda. On the other hand, while discussions supporting national security policies and strategies appear to engage a wider group of security actors and agencies, the extent to which development priorities are considered in these processes remains unclear. Conclusions indicated that, whereas SSR actors have demonstrated a poor consideration of development objectives, based on conventional development tools and international partnerships, development actors have also exhibited a poor understanding of SSR consideration.

A closer examination of both PRSP-driven strategic national development processes and national security frameworks concluded that both mechanisms had potential in terms of ensuring that strategic national policy goals were underpinned by a combined analysis of security and development trends, and thus useful for informing SSR programmes and projects.

The final section of the chapter includes a brief overview of the SSR experiences in Uganda, Sierra Leone and East Timor. In all cases, the impact of both strategic development and security processes was analysed. All experiences demonstrated that the level of consultation and CSO involvement supporting dialogue on national development and poverty eradication was significantly higher than in the dialogue supporting national security discussions. In East Timor there appeared to be no interface between efforts to develop a national development strategy and planning for SSR (in the absence of a national security policy). On the other hand, the Uganda development strategy reflected issues related to both security and conflict resolution, and thus reflected progressive thinking on the security-development nexus within the development community. However, whereas the government of Uganda's efforts to create a development strategy secured the support of the wider national civil society community, dialogue supporting the security policy framework and defence policy was much less successful and thus lost momentum in the years that followed.

The case of Sierra Leone is interesting in exposing the way in which the outcome of the national security sector review informed both the new national development strategy and the redrafted national security policy. However, these linkages were not made until almost ten years after the initial SSR programme was launched. Regardless of the time delay, these mutually supportive events were instructive in demonstrating the support that conventional development programmes can offer to the development of the security sector.

I concede to doing no more than exposing some interesting ideas in terms of potential synergies between existing and new strategic processes focused on development and security. At this time, and particularly in weak and low-income states, more strategic approaches to SSR appear to be driven and influenced by both PRSPs and national security frameworks. Each approach offers benefits in terms of stakeholder engagement, in-depth analysis and the linkage to locally owned processes. However, the case studies explored here demonstrate how development and security reviews have come together to inform each other, and also where both have developed as separate processes. The cases demonstrate that the careful development approach of 'inclusivity' could have helped security reviews and planning to be more 'development-sensitive'. Aligning existing PRSP and national security processes and frameworks to support future SSR processes may go some way to ensuring that the security-development nexus features in national strategic policy, and that development objectives and activities are considered and 'front loaded' at the earliest stages.

Notes

[1] Department for International Development, *Understanding and Supporting Security Sector Reform* (London: DFID Publications, 2000).

[2] Ibid.: 7.

[3] Department for International Development, 'SSR Policy Brief' (London: DFID Publications, 2004).

[4] Department for International Development, *Fighting to Build a Safer World: A Strategy for Security and Development* (London: DFID Publications, 2005).

[5] For example, see Frances Stewart, 'Development and Security', CRISE Working Paper no. 3 (Oxford: CRISE, 2004); Paul Collier, Anke Hoeffler, Lance Elliot, Håvard Hegre, Marta Reynal-Querol and Nicholas Sambanis, *Breaking the Conflict Trap: Civil War and Development Policy*, World Bank Policy Research Report (Oxford: Oxford University Press, 2003).

[6] Necla Tschirgi, Michael S. Lund and Francesco Mancini, eds, *Security and Development: Searching for Critical Connections* (New York: Lynne Rienner, 2010).

[7] Necla Tschirgi, *Security and Development Policies: Untangling the Relationship* (New York: International Peace Academy, 2005).

[8] See David Chandler, 'The Security-Development Nexus and the Rise of "Anti-Foreign" Policy', *Journal of International Relations and Development* 10, no. 4 (2010): 362–386; Ann Fitz-Gerald, 'Addressing the Security-Development Nexus: Implications for Joined-up Government', *Policy Matters* 5, no. 5 (2004), available at www.irpp.org/pm/archive/pmvol5no5.pdf.

[9] The 'whole-of-government approach' was first popularised by the OECD/DAC in 2006. See OECD/DAC, *Whole of Government Approaches to Fragile States* (Paris: OECD, 2006).

[10] For more information see the PRSPs of Sierra Leone, Uganda and Afghanistan, available at www.imf.org/external/np/prsp/prsp.aspx.

[11] OECD/DAC, 'Security System Reform and Governance', OECD/DAC Guidelines and Reference Series (Paris: OECD, 2005).

[12] World Bank, *World Development Report 2011* (Washington, DC: World Bank 2011): 23.

[13] Government of Canada, *Guidelines for Security System Reform* (Ottawa: DFAIT, 2011).

[14] OECD/DAC, *The OECD/DAC Handbook on SSR: Supporting Security and Justice* (Paris: OECD, 2007).

[15] United Nations, 'Securing Peace and Development: The Role of the United Nations in Supporting Security Sector Reform', Report of the Secretary-General, UN Doc. A/62/659–S/2008/39 (New York: United Nations, 2008).

[16] World Bank, note 12 above: 32.

[17] Following the publication of the security policy framework and defence white paper in Uganda, a number of donors that had started to provide the government of Uganda with direct budget support – including the European Commission, the World Bank and the governments of Ireland, the Netherlands and Sweden – were all influenced by the outcomes of these exercises, which were primarily funded and supported by the UK government. For more information see Dylan Hendrickson, ed., *Uganda Defence Review: Learning from Experience* (London: Conflict, Security and Development Group, King's College London, 2005).

[18] Ibid.: 190.

[19] Ibid.: 193.

[20] International Crisis Group, 'Storm Clouds Over Sun City: The Urgent Need to Recast the Congolese Peace Process', ICG Africa Report no. 44 (Brussels/Nairobi: ICG, 14 May 2002).
[21] Zie Gariyo, 'The PRSP Process in Uganda', Discussion Paper no. 5 (Kampala: Uganda Debt Relief Network, 2002): 8.
[22] For more information see www.imf.org.
[23] For example, see Donald Nuechterlein, *America Recommitted: A Superpower Reassesses Its Role in a Turbulent World* (Lexington, KY: University Press of Kentucky, 1991); William Ascher and William Overholt, *Strategic Planning and Forecasting: Political Risk and Economic Opportunity* (New York: John Wiley & Sons, 1983).
[24] Laure-Hélène Piron and Andy Norton, 'Politics and the PRSP Process: Uganda Case Study', Working Paper no. 240 (London: Overseas Development Institute, 2004): 13.
[25] Ibid.: 43.
[26] Government of Uganda, *Poverty Eradication and Assessment Programme* (Kampala: Ministry of Finance, Planning and Economic Development, 1997).
[27] Gariyo, note 21 above: 25.
[28] Piron and Norton, note 24 above: 43.
[29] Hendrickson, note 17 above: 12.
[30] Ibid.
[31] Ibid.
[32] Ibid.: 63.
[33] Ibid.
[34] Ibid.: 45.
[35] Ibid.: 67.
[36] Based on interviews with senior members of the Ministry of Defence and Ministry of Security, June 2009 and September 2010.
[37] Peter Albrecht and Paul Jackson, *Security System Transformation in Sierra Leone, 1997–2007* (Birmingham: University of Birmingham, 2009).
[38] Based on discussions with a representative from Sierra Leone's ONS, 1 June 2011.
[39] Albrecht and Jackson, note 37 above: 28.
[40] Based on discussions with Tom Hamilton-Baillie, who served on the four-member team which evaluated SILSEP in 2007.
[41] Government of Sierra Leone, *Poverty Reduction Strategy Paper: A National Programme for Food Security, Job Creation and Good Governance* (Washington, DC: World Bank, February 2005), available at http://siteresources.worldbank.org/INTPRS1/Resources/Sierra-Leon_PRSP(Feb-2005).pdf.
[42] Based on discussions with a senior member of Sierra Leone's ONS, May 2011.
[43] Hamilton-Baillie, note 40 above.
[44] Based on discussions with a senior member of the ONS.
[45] Albrecht and Jackson, note 37 above: 118–125.
[46] Government of Sierra Leone and World Bank, *Poverty Reduction Strategy Paper: An Agenda for Change* (Washington, DC: World Bank, 2008), available at http://unipsil.unmissions.org/portals/unipsil/media/publications/agenda_for_change.pdf: 19.
[47] Government of Sierra Leone and IMF, *Sierra Leone: Poverty Reduction Strategy Paper – Progress Report* (Washington, DC: IMF, 2008), available at www.imf.org/external/pubs/ft/scr/2008/cr08250.pdf.

[48] At the same time, the immediate security vacuum was filled by an Australian-led peacekeeping force, INTERFET. International Crisis Group, 'Timor Leste: Security Sector Reform', Asia Report no. 143 (17 January 2008), available at www.crisisgroup.org/~/media/Files/asia/south-east-asia/timor-leste/143_timor_leste___security_sector_reform.pdf: 4.

[49] Yoshino Funaki, *The UN and SSR and East Timor: A Widening Credibility Gap* (New York: Centre for International Cooperation, 2009): 2.

[50] International Crisis Group, note 48 above: 4.

[51] Funaki, note 49 above: 3.

[52] See Gordon Peake, 'A Lot of Talk but Not a Lot of Action: The Difficulty in Implementing SSR in East Timor', in *Security Sector Reform in Challenging Environments*, eds Hans Born and Albrecht Schnabel (Münster: LIT Verlag, 2009): 213–238, available at www.dcaf.ch/yb2009; Edward Rees, 'Time to Withdraw from East Timor?', *The Atlantic* (21 December 2010), available at www.theatlantic.com/international/archive/2010/12/time-for-the-un-to-withdraw-from-east-timor/68334/; Ludovic Hood, 'Security Sector Reform in East Timor', *International Peacekeeping* 13, no. 1 (2006); Funaki, note 49 above.

[53] International Crisis Group, note 48 above: 2, 4.

[54] Ibid.: 3.

[55] World Bank, note 12 above: 125.

[56] Ibid.: 234.

[57] Elisabeth Lothe and Gordon Peake, 'Addressing the Cause but Not the Symptoms: Stabilisation and Humanitarian Action in Timor-Leste', *Disasters* 34, supplement s3 (October 2010): S440.

[58] Democratic Republic of East Timor and IMF, 'East Timor: Poverty Reduction Strategy Paper – National Development Plan', IMF Country Report 05/247 (July 2005), available at www.imf.org/external/pubs/ft/scr/2005/cr05247.pdf.

[59] Ibid.: 7.

PART VI

CONCLUSION

Chapter 12

It Takes Two to Tango: Towards Integrated Development and SSR Assistance

Vanessa Farr, Albrecht Schnabel and Marc Krupanski

Introduction

The main objective of this book is to contribute to ongoing discussions on the theoretical and practical relevance of SSR as a building block to facilitate the security-development nexus, with a specific focus on how SSR contributes to development. We examine SSR's ability to foster a positive interrelationship between the security and development communities, and to support their contributions to sustainable human and economic development. The chapters in this book share insights on conceptual debates, gender approaches, regional experiences, lessons from DDR and SSR practice and evolving approaches by international organisations and the broader donor community. The analyses and suggestions presented by the contributors remind both development and security communities that they need to take each other's experiences and concerns into account when planning, implementing and evaluating their own SSR and other reform and assistance activities.

In this concluding chapter we will capture the main arguments presented throughout the book, compare them and translate them into recommendations for the study and practice of security- and development-sensitive SSR planning and implementation. We hope that the lessons learned and arguments presented in this book will prove useful in improving efforts to synchronise security, development and SSR activities in the work of international organisations, national governments, civil society organisations and research communities working on peace, security and development. Those designing and implementing development and SSR activities in transitional societies are particularly encouraged to engage with and contribute to the discussions presented in this volume.

Prevailing gaps in the security-development nexus

Since its introduction in the late 1990s, there has been an abundance of claims made in academic publications and policy statements about SSR's importance for achieving development goals; the claims, rather than solid evidence, have been influenced by and contribute to the belief that the security-development nexus is both common sense and exists. In fact, this book has shown that the point demands greater scrutiny, as there is little clarity about either the empirical nature of the nexus or the role of SSR in embodying and advancing it. In the chapters in this volume, authors differently perceive and approach the question of what the 'security-development nexus' really is. For some, questioning the existence of the nexus is the entry point into their analysis – they then try to work out how this nexus could be proved as a legitimate claim (e.g. Goudsmid et al.; Van Eekelen). Others assume it does exist, but consider it primarily a conceptual and practical approach and look for evidence of practices on the ground, such as DDR programmes, which help reinforce it (e.g. Bryden). Still others view the supposed existence of the nexus as both an approach and an empirical question, but challenge whether or not either security or development practices, as currently implemented, are valid approaches to promote social transformation, the eradication of poverty and an end to all forms of violence (e.g. Hudson; Jackson).

Overall, while most of the authors in this volume conclude that there remains, at least conceptually, some merit to the notion of a security-development nexus, the preliminary findings gathered here show that in practice SSR has not delivered on these idealised and presumed links, while the development field has been reluctant to accept SSR approaches and theories in its own work, deeming them irrelevant or fearing the increased securitisation of development interventions. As will be discussed later in this chapter, the lack of evidence of SSR's development contribution and development actors' relative tendency to disengage from comprehensive SSR programming are mutually damaging. Certainly, for SSR to deliver on its development promises, SSR practitioners need to do a better job of meaningfully integrating both development approaches and actors, while development actors need to open themselves to more significant engagement with SSR processes and accept greater responsibility to guide SSR delivery in order to advance development goals. While the necessity of such cohesion and cooperation has already been outlined in core SSR conceptual documents, such as the UN Secretary-General's 2008 report on SSR,[1] this volume indicates that theory has not yet translated into practice.

The lack of empirical evidence for establishing a development-SSR link

One challenge is that little comparative and empirical research has been amassed to move us beyond the conceptual wisdom that SSR is good for development, or indeed can even be considered a necessary precondition for it to take place. In light of this gap between practice and research, the main purpose of this book was to invite a broad research and practitioner community working on development and/or SSR to reflect on whether security and development, specifically through SSR activities (or 'SSR-type' activities, as in many cases they do not appear to be comprehensive), can be shown to advance economic growth, poverty reduction and human development.

As an overall guiding question for the book we asked if the rhetoric arguing that security and SSR are essential for development would stand up to deep analysis. Do SSR and development initiatives indeed support one another and work in tandem?

Collectively, through reference to a wide array of situations in which SSR programmes are in process or completed, the authors generally acknowledge that an easy and definitive response to the questions posed is not (yet) possible and much more research is needed on this issue. This posed an additional challenge in case study selections, as there are few cases, if any, where a comprehensive SSR programme has been implemented. As a whole, however, even this early attempt to analyse the hoped-for complementarity of the two allows us to conclude that while development approaches have much to teach those undertaking SSR, the far newer discipline has not, on the whole, made enough effort to implement practically the lessons, particularly from development methodologies that might help it realise and implement its idealised claims as a supporter of development aims. The range of analysis presented here shows that what we already know about effective development, especially regarding the implementation of rhetorical commitments and intentions, must be urgently applied for SSR to achieve the broad set of goals claimed for it.

SSR as a vehicle primarily for social change or immediate stabilisation?

Several of the authors (e.g. Schnabel; Hudson; Kunz and Valasek; Jackson; Myrttinen) raise the issue of whether the practically implemented objective or vision of most SSR interventions is compatible with those of development, given that, at field level, they tend to overlook policy directives suggesting they deliver a holistic programme and instead focus primarily on technical approaches. While it is envisioned at the conceptual and policy level as contributing to reform and social transformation, in application SSR has often been reduced to one-sided technical quick-fix exercises (often with a focus on force modernisation or training of armed forces) that may not be the best vehicle for sustained institutional and operational transformation. Moreover, many of these initiatives do not constitute 'real' SSR, yet often are referred to as SSR programmes by both practitioners and researchers alike. The short-term 'band-aid' perspective which views SSR as responsible for kick-starting technical methods to deliver some form of speedy security is, at heart, at odds with transformational development ideals and negates the dominant discourses of SSR which aim, over the long term, to advance fundamental changes in a society. Development aims are profoundly related to vaunted security goals such as the promotion of good governance, citizen-state accountability strategies, the achievement of gender equality and an end to gender-based violence, and justice for all. On this last point, Hudson argues that in order to best address and resolve gender inequities and help create meaningful and lasting social change, other social cleavages that intersect with gender, such as ethnicity, race, class and sexuality, need to be addressed as well.

It is important to emphasise that SSR is not meant to be a quick-fix tool or act merely as a stabiliser, although there is some evidence that it is mis- and under-utilised to these ends. As we learn from several of the case studies discussed here, its 'reform' aspects, which imply a gradual, long-term series of changes, are often flattened and abbreviated for apparent immediate gains. Several authors in this volume warn that this, while seemingly expedient, is an ill-advised choice whose impacts only become visible much later when hoped-for peace dividends fail to be sustained.

Let us concede, for the moment, that an immediate and short-term contribution to stabilisation is a worthy goal, assuming that it is the necessary precursor to evolving, broader structural change in the future. Yet if this is true, why is there so little evidence of proper planning for a transition from the immediate need for a cessation of hostilities to a long-term reform process aligned with national and regional development goals? Is sufficient attention being paid to how this transition can be set up in theory and then made operational? Even if it is true that SSR interventions have to 'start small', the point argued by several authors is that after more than a decade of SSR there is evidence that, although called for in SSR literature, not enough thought is being given to realising a long-term security vision along with the requisite institutions to uphold it and alignment with a broader development agenda.

However, this volume also features contributions by authors (e.g. Goudsmid et al.; Van Eekelen) who approach the question differently. Rather than exploring transformation, a narrower perspective is employed in which SSR is not necessarily viewed as a vehicle for social change, but rather examined as an instrument that allows a stronger and more effective security approach to take hold. Here, the focus is for instance on how best to confront serious crime and other security threats that are seen to impede development, rather than probing whether SSR programmes could or should challenge power inequalities and redress failed governance. Indeed, it appears from the case studies examined that this narrower perspective has been frequently employed in SSR-type activities and, as the authors argue, has demonstrated some success. Although space is made for issues of democratic oversight of security sectors, in this analysis SSR activities are better equipped to confront more 'hard' security concerns that should be addressed through state security institutions. In this way traditional security institutions are made accountable to the populations and effective in development because they are better able to confront immediate security challenges.

Serving both short- and long-term agendas: Calling for a comprehensive SSR approach

There are a number of conditions to be met for SSR to fulfil its development goals. If it is pursued as part of a comprehensive security and development strategy, if it focuses simultaneously on short-term stabilisation and preparing for long-term institutional reform and security governance processes, if it supports this by means of a carefully thought-out vision and implementation plan, then SSR contributes to the immediate provision of stability and security, facilitates humanitarian assistance and enables long-term sustainable development. However, such conditions seem rarely met, perhaps because a synchronised approach to SSR that serves several simultaneous shorter- to longer-term peace-building objectives requires early joint preparation and planning between security, development and humanitarian communities. Given the difficulties of such coordination and the different silos in which humanitarian, development and security actors operate, what seems to happen more often is that SSR and development agendas are pursued in isolation from each other and without close coordination. When they are caught between two possibilities (SSR as a vehicle of social and institutional change, or SSR as a short-term stabilisation tool), SSR interventions tend not to advance either goal very well. As a result, they risk either failing to serve the long-term security and development needs of the population or missing the opportunity created by short-term stability to catalyse the successful onset of broader development activities.

To help facilitate such joint preparation and planning, the chapters as a whole reinforce the need for SSR practitioners to take up development best practices in order to achieve a positive and sustainable impact. In particular, they recommend practices that utilise a socially comprehensive and consultative whole-of-government perspective and coopt multiple civil society stakeholders as inclusively as possible. While these calls for inclusive SSR echo those made in the SSR debate and literature over the past few years, this and preceding DCAF studies show that the message is not getting through on the ground. While conceptually upheld, it is not honoured and practised during the implementation of SSR programmes. At present, the authors show, SSR still focuses far more on state-centred security than the 'people'-centred or human security approach on which it is claimed to be based (Schnabel). Entrenched, traditional military approaches to security, including the institutional violence, sexism and often ethnocentrism that characterise security institutions, prove difficult to shift through a reform process. They are unlikely to happen at all if reform processes are neglected or pursued half-heartedly or incompletely. Despite the 'gospel' of SSR and relevant policy commitment by the donor community,[2] the majority of the authors in this volume conclude that a 'business-as-usual' approach to SSR does not substantially contribute to creating better conditions for sustainable and inclusive development, or facilitate the transformatory social, political and economic processes that are necessary to prevent armed conflicts from recurring.

That said, it is also recognised that there is a vibrant debate within the development field and much diversity of opinion as to best development practices. While the call is made for SSR researchers and practitioners to understand and align their activities better with field-tested development methodologies and objectives, the authors are not arguing that the development field is a homogeneous entity or that everything practised serves as an ideal model. While reflecting on selected positive attributes of the development field, this volume does not intend, or claim, to present a comprehensive analysis and review of development methodologies and debates. Such a mapping and analysis exercise is an important immediate next step, but is beyond the scope of this volume. As previously stated, our emphasis is on development trajectories that offer a socially comprehensive and consultative whole-of-government perspective and coopt multiple civil society stakeholders as inclusively as possible.

Calling for more effective communication of SSR's potential and actual development impact

Additionally, read as a whole, the chapters in this volume demonstrate that the existence of a link is either not clear-cut and easily measurable or estimated, or in some cases that it is clear and called for, but not acted on. Indeed, the size of the security sector's contribution to creating the conditions in which development can flourish appears to be less concretely measurable than the rhetoric about its impact warrants. Conversely, if the impact matched the rhetoric and this could be more effectively shown, partnerships with SSR actors would likely be embraced at an early stage of international assistance, rather than accepted with hesitation or outright resistance. When a number of case studies are examined, whether and how SSR and development find an effective synergy seem as much the result of chance as of design. Indeed, a common finding in the book is that communication between development and SSR implementers is weak. As a result, SSR on the ground rarely derives from or is consciously tailored to advance development goals as articulated in government development plans.

Likewise, however, development planners pay little heed to what is being proposed and delivered on the SSR side. The development community is at least as guilty of missing opportunities to engage in early joint deliberation, planning and design of complementary SSR and development approaches as are representatives of the security and SSR communities. For instance, early community-based and consultative assessments of populations' development and security needs should be conducted jointly, so that the results can inform both development and security interventions. Development actors need to step up and assume responsibility, leadership and ownership. It does not serve anyone to remain passive and simply criticise. Stove-piping of assistance missions between security and development communities is counterproductive to sustainable reforms, change, stability and poverty reduction.

Recognise SSR's positive contribution to transition processes

A further important critique emerging from this book is that SSR, after more than ten years of active debate and some practice and lessons learned, is still intuitively assumed to be a positive activity that will benefit the whole of society and advance development, even – or particularly – in situations where violently destructive norms and serious crime have become entrenched as a result of a protracted war. This is perhaps most obvious in the assumed relationships between beneficiaries and providers of security, between citizens and states. Both to challenge this assumption and to enhance its practical materialisation, several authors (e.g. Schnabel; Hudson; Kunz and Valasek; Myrttinen; Jackson) ask who SSR is actually for, how its main recipients are identified and how their differing needs are determined. Questions are also raised (e.g. Bryden; Myrttinen) about whether those who most often benefit directly from SSR processes, especially through DDR, do in fact go on to contribute to getting development goals back on track. In other words, the authors question whether SSR does in fact live up to its promise of playing a socially transformative role by providing an avenue through which societies can recover from prolonged militarism and new forms of more accountable government can take hold. If empirical and convincing evidence for the existence of both a positive short-term catalytic effect and a long-term transformative role can be established, this will contribute considerably to making a stronger case for investing in collaborative development-SSR planning and implementation.

Common experiences and overall recommendations for change

Bridging silos and divides with improved communication

As is commonly reported and argued throughout this volume, SSR and development actors often do not communicate very well with each other. Asked to identify how, for instance, the most widely used machineries of development planning, such as the UN Development Assistance Framework, relate to planning in the security sector, many of our contributors could find no discernible link at all (see Myrttinen; Kunz and Valasek; Jackson; van Eekelen; Fitz-Gerald). On both sides of the equation, it seems, few conversations and shared visions are informing on-the-ground interventions, creating new difficulties rather than resolving them.

Part of the problem, as argued by Hudson, van Eekelen and Kunz and Valasek, lies in the fact that SSR is a late-coming player to the old and arguably overcrowded development field – a field that has its own problems with inflexibility and gatekeeping, and which pays little attention to traditional security issues, including, most prominently, in the formulation of the Millennium Development Goals. Arriving in an already-established conversation, and faced with entrenched approaches to doing business and even resistance to working on security issues – as these are perceived to compromise development interventions – SSR practitioners have tried to gain legitimacy by attempting to fit into existing discourses about how development and peace-building could act together. This book is no exception. As discussed by Fitz-Gerald and van Eekelen, those responsible for SSR activities also feel political pressure to justify their interventions as 'worth the money'. Indeed, as McDougall and Myrttinen document, SSR actors have routinely been charged with re-routing donor aid that is seen as essential for (re)development after or during conflict to a sector which may seem, on the whole, to benefit very few – and a group of individuals who, as the gender analysts in this volume observe, already wield considerable social and political power.

SSR practitioners show considerable epistemological agency in their wielding of the development discourse – in that they 'know' why development needs a secure environment in which to flourish and feel compelled to explain their interventions' capacity to advance this goal. Nonetheless, they usually overlook national development planning in articulating their own goals. Additionally, as Bryden points out in the context of DDR processes and SSR, there are critical knowledge gaps on both sides that prevent a more significant positive impact.

Development practitioners, by contrast, do not appear to face similar pressure to understand or align with the security sector's influence in the work they attempt to do, even when they work on development in pre-, during- and post-conflict contexts. The lack of such pressure affects their knowledge of the security sector and its role and potential impact on development objectives. While working on this volume we were struck by a sense of mutual ignorance (and perhaps even indifference) about the security and development communities' respective roles and potential – particularly in terms of the potential that joint efforts have for improving the delivery of security and development services to the main 'clients' of both communities, the people themselves.

Moreover, it appears that in this mutual reluctance to examine (or face up to) a perceived or potential positive relationship between development and SSR, development practitioners in crisis settings are greatly influenced by humanitarian actors' responses to the increasingly dangerous environments in which international emergency interventions take place. They, too, would prefer to distance themselves from association with security providers; so they end up overly criticising them from the sidelines. They might hope that denying or remaining unwilling to comment on the ways in which development approaches have become 'securitised' (or more accurately 'militarised', as Schnabel argues) in some post-conflict settings in the past decade will absolve them from responsibility for SSR's shortcomings. However, this attitude blinds them to the obvious fact that problems with SSR are only likely to be rectified through closer cooperation in the planning and implementation of humanitarian, development and security assistance. In fact, the evidence presented in this volume seems to warn that development experts are losing opportunities to influence SSR planning and delivery in ways that make developmental sense. To return to a point made earlier in this chapter, the 'reform' aspects of SSR intrinsically require developmentally sound approaches. As Schnabel proposes, if SSR is pursued in the spirit of its original intentions and in line with its own principles, there is no need for development actors to fear closer ties and engagement with those trying to initiate and implement reforms in the security sector, particularly as SSR has been created to support, not to challenge, development efforts. Thus the development community needs to shed its concerns about perceived securitisation and cooperate early in joint security and development needs assessments and SSR and development programme planning. Patient, incremental, consultative and sustainable changes in national machineries, institutions and individuals, even if these are facilitated by several kinds of quick-impact projects, are necessary, not optional, in both development and SSR.

Synchronising timelines and preventing harm

It may, of course, make some ideological sense for development experts to distance themselves from security sector reformers; but, as this book asked authors to document, there is a much greater affinity between the long-term approaches needed for successful SSR and sustainable development than there is, essentially, between the different objectives that humanitarian and development actors attempt to accomplish (Schnabel). As several authors record, humanitarian work and development work follow (or should follow!) quite different timelines, with the latter successfully taking over from the former's immediate life-saving interventions, via an early recovery approach, in order to build sustainable human and institutional capacities for growth. Development actors therefore have a significant responsibility to work with SSR actors to ensure that SSR interventions, including those that are able to start early, facilitate – or at a minimum 'do no harm' to – the achievement of long-term development goals.

Exploring this point further, many contributors focused their attention on understanding where and how SSR can most usefully fit in the chronology of post-crisis responses. The muddying of timelines between what can be considered strictly humanitarian responses, early recovery and a return to a development approach, and the overlying necessity to make those receiving these interventions as safe as possible are challenges which make everyone's work much more difficult. As is seen in Myrttinen's analysis of the impacts of the tsunami disaster on Aceh's DDR programme, real-time events can overturn the most carefully laid SSR and development plans and set in place an entirely unanticipated threat, such as former combatants converting their combat structures into work-bands as a means to gain access to the most lucrative reconstruction contracts. As he argues, the Acehnese DDR programme was never sufficiently development-focused – or lacked the time – to rehabilitate male ex-combatants. Its interventions were insufficient to stop the still-networked, most powerfully placed ex-combatants from falling back on pre-tsunami social structures and practices to make a living, including by extortion and violence. However, real-time events, often unexpected and thus poorly prepared for, are less of a challenge if structures are already in place that allow for swift, appropriate and synchronised responses by security, humanitarian and development actors. If they constantly analyse a highly fluid context and reflect it in their operational planning and implementation, actors in all these communities will readily be able to judge and act upon their respective comparative advantages in order to develop and follow jointly designed strategies to provide immediate assistance, short- and long-term security and development support.

Participation and exclusion: State-centred versus people-centred approaches

The book offers evidence that SSR tends to do little to lay the bedrock for transformative development and long-term behavioural change. In her gender analysis of SSR, Heidi Hudson argues that it fails to challenge humanitarian interventions and reinforces their stereotyping of women as victims rather than actors in their own right with capacities to define security interventions that will benefit them most. This presents a real danger to both women and men in the aftermath of armed conflict. Her critical analysis of the security-development nexus and the challenges of gender mainstreaming in SSR practices introduces the question of whether, aside from rhetorical statements to the contrary, SSR is in fact able to move beyond its traditional male, security/military audience to speak convincingly to and provide security for all. She focuses on whether SSR really does contribute to women's access to development benefits by ameliorating sexual and gender-based violence and transforming the male-dominated, militarised institutions which often contribute to such violence, and to the social, political and economic subordination of women. Like Rahel Kunz and Kristin Valasek, she observes that worn-out assumptions rather than contemporary consultations shape SSR interventions. In her view, this reinforces neoliberal beliefs about development outcomes and narrows the range of results that could be achieved by SSR. In introducing what becomes collective evidence throughout the book, she suggests that most SSR processes are primarily designed to manage risks, not address root causes of violence and insecurity, unless the latter have in the past emanated from within the security sector and its institutions. She argues that the intended, assumed but often not actual people-centredness of security provision can therefore become a mask for the reiteration of dominant discourses of vulnerability and security. The danger of this happening is greater in cases where SSR efforts are owned by local and national actors whose commitment to advancing true and lasting change is compromised by old habits, allegiances and structures of power and influence.

In addition, through their specific focus on gender, Hudson and Kunz and Valasek argue that SSR practitioners need to utilise positive male identity models – which exist but are overlooked – and emphasise an ethic of care that does not reward dominance and violence. Furthermore, other differences that can create social cleavages, such as ethnicity, race, class and sexuality, need to be addressed so that their potential to destabilise can be ameliorated. Such richly textured analysis can only be accomplished through a methodology that prioritises a participatory and community approach which is well developed in development fieldwork.

Hudson's view is echoed by Myrttinen and Bryden, both of whom also document how structural forms of exclusion are rarely addressed through SSR, which might in many cases contribute to maintaining and even reinforcing the *status quo ante* that contributed or even led to the initial outbreak of armed violence. Through the case studies they offer on DDR in a number of post-conflict locations, Myrttinen and Bryden also show that the power inequalities between civilians and security forces are almost never concretely challenged. The five authors just mentioned all contend that SSR does little to wrest power from those who have misused it, resulting in a widening gap, not a reinforcing nexus, between SSR aims and development goals. This failure to address power may also have detrimental effects on the need to build the fairness and legitimacy of evolving governance structures and institutions as well as their internal and external credibility. Myrttinen, in fact, concludes that post-conflict reform of the armed forces and fighting groups concretely reinforces the political and economic power of former combatants, rewarding what he calls 'their predatory behaviour' and maintaining their separation from the communities into which they are meant to be reintegrated.

This example also shows that all too often what is announced or praised as an SSR programme may in truth have been at best a partial SSR programme – or an SSR-like activity that offers specific reform activities rather than a more comprehensive SSR programme based on solid and consultative assessments, implementation strategies and monitoring mechanisms targeted at the full range of security institutions and civilian state and non-state governance (oversight and management) mechanisms. Closer collaboration with the development community in designing SSR programmes would ultimately point to the inadequacy of quasi-SSR programmes and the need to commit to comprehensive SSR approaches (Schnabel; Kunz and Valasek).

In addition to findings by McDougall, Bryden and Jackson, who focus on the beneficiary question with reference to specific country case studies in which large numbers of combatants were demobilised, evidence is amassed that the people who will be most affected by SSR processes – who are largely civilians, not members of the armed forces or armed opposition groups – are insufficiently considered when SSR is planned, even when they are supposed to be its direct beneficiaries. A number of reasons for this problem are elaborated, many relating to a series of unquestioned assumptions made by SSR planners. Although such criticism has already been present for some time, these authors find that the problem persists. One reason for this is that SSR-type activities are often subsumed into other shortsighted security and institutional programmes. Jackson's chapter, for instance, details the way in which contemporary neoliberal state-building activities have incorporated SSR into other institution-building programmes. This is often to the detriment of SSR's original intent and design, especially when it undermines stated commitments to conduct a national process of self-reflection and self-determination on security priorities in the host country.

Kunz and Valasek agree with Hudson and Myrttinen that gender-blindness and an overemphasis on state-centric definitions of security are significant impediments to the realisation of mutually beneficial SSR and development activities. By prioritising the perspectives of the state and elites and allowing them to define what constitutes security and how to deliver it, reformers reify the security and power of the few, often to the detriment of local communities and non-state networks and bodies. These authors think a basic problem is, despite the copious advice on inclusivity in the SSR literature and guidance material, the fact that SSR planners ignore valuable lessons learned in the development field – and even in the conceptual design of SSR itself – about the benefits of working through inclusive community consultations. In their view, which is shared by Jackson and McDougall, institution- and donor-inspired SSR all too easily facilitates delivery of technocratic management, not politically informed approaches to security. The state-centredness of SSR then emerges as a key problem: it seems the most explicit means by which questions of people's security are overlooked because it facilitates a capital-city, patriarchal, institution-focused and donor-driven approach rather than one which recognises the importance of the localised, informal justice and security institutions that communities actually tend to use and trust. For all the challenges inherent in such local institutions, which are often in themselves sites of local male dominance and violent power, one conclusion from this discussion may be that more effort should be put into building the capacity and accountability of informal justice and security deliverers which operate at the margins of society, such as in urban slums and rural areas where the poorest and least-served populations are likely to live.

Although Fitz-Gerald also questions the state-centredness of much SSR activity, she offers a more optimistic perspective on how consultation and institution-building can articulate together, noting that SSR has become increasingly better conceptualised over the years. She acknowledges that the most helpful SSR experiences have been delivered in situations in which countries in need of security and governance reform take the initiative and directly approach donors. The tone for eventual local and national ownership of externally supported SSR programmes is thus pre-determined by the momentum created through local and national initiatives to engage in SSR. In other words, consultation and buy-in are already built before SSR practitioners begin their work. On the subject of inclusivity, Fitz-Gerald offers examples of how non-traditional actors such as faith communities are making contributions to comprehensive security discussions. However, much like Jackson, Hudson, Schnabel and Kunz and Valasek, she also concedes that, despite many statements of intent to the contrary, on the ground SSR insufficiently relies on (or pays heed to) broad consultation and is compromised by the fact that development and SSR communities are largely disengaged from one another. In her opinion, this indicates a divide between policy and practice, suggesting that SSR may be unable to live up to the comprehensive scope envisaged for it at the policy planning and research levels at which its goals are initially defined. Demonstrated through her case studies, this challenge highlights the need for SSR implementers to partner national security strategies with national development strategies. Such a combination will enhance the active engagement and participation of civil society organisations and bring development concerns into the centrepiece of SSR. Thus, the positive developmental impact of SSR can be better realised.

Connecting SSR to other reform processes

A further recurring theme points to the observation, considered as a whole, that SSR processes do not appear to be connected to broader governance reform. In many instances it is considered to be unduly interventionist to link SSR to democratic political reforms, although making governments more efficient and accountable or creating stronger state-citizen relations may in fact be crucially important for meaningful and sustainable SSR to be pursued. As Jackson discusses, SSR may be seen as a core aspect of reforming a ministry of defence or ministry of the interior, for example, but the reform of these ministries may then be isolated from other governance restructuring processes. Tending towards an ahistorical approach to state-building, he argues, SSR worsens inequalities – a conclusion that is shared by McDougall in his review of East Timor and the Solomon Islands. In his multi-country review, Bryden submits that it is notions of enhanced efficiency in the delivery of hard security, not responsiveness to people's real needs, which drive most of the institutional reform in SSR processes. Fitz-Gerald points to the key role ascribed to governance reform in poverty reduction strategy papers, 'which expose the strategic relevance of governance, security and conflict resolution to development'.

Main messages

A striking feature of the chapters in this book is that all highlight the difficulty of making concrete claims about the supposed security-development nexus and SSR's practical linkages with development – as well as the futility of doing so given how context-specific SSR needs to be. Furthermore, the fact that SSR is so context-specific makes it difficult to draw overarching conclusions that would lead to useful and generalisable recommendations. While those focusing on comparative case studies could come closest to offering lessons learned and proposing measures to help avoid future failures (Hudson; McDougall; Bryden; Myrttinen; Fitz-Gerald), they are cautious about assuming these are universally applicable. Indeed, a solid conclusion of much of the analysis collected here is that it is ineffectual and even counterproductive to take a neoliberal, cut-and-paste approach driven primarily by faith and conviction that SSR will result in positive development impacts. Additionally critical is the need to move beyond technical approaches that fail to meet the contextually and historically specific political engagement required if SSR is to meet the necessary structural reform that supports and facilitates development.

What general messages, then, emerge from this study? Can we, based on the evidence gathered here, posit a return to SSR's original design in which security-related interventions really are shaped by their development roots and the objectives of good security sector governance, and become more flexible, adaptable to the demands of both short- and long-term needs, useful to states without being state-centric and driven by a contextually informed understanding of what individuals and communities, especially those most impoverished and socially marginalised, need and expect?

In our view, this change is possible, and both timely and necessary. Enough lessons have been identified (although not necessarily learned) after more than ten years of conceptual debate and practice for SSR to start making good on its claims to contribute to development.

Generate empirical evidence!

The link between SSR and development needs to be proven, not merely asserted. The same applies to the security-development link. A better understanding of the former may in fact help to develop a better understanding of the latter. Therefore, better empirical evidence needs to be systematically collected to show how SSR and development inter-operate. If we consider SSR to be a potentially effective and powerful instrument for making operational the security-development nexus, and if we regard it as a means to understand and appreciate SSR's role in (and assumed contribution to) sustainable development, we need to monitor and analyse the potential and actual contributions SSR makes to human development and poverty reduction – rather than continuing to assume them.

Monitor progress and failure!

New SSR interventions, as well as ongoing development programming, have to interact more consciously and reliably so that effective monitoring and evaluation systems can be put in place to produce data for analysis. One proposal is that a joint EU military/SSR-civilian/development coordination body might be a useful approach. In addition, more effort, commitment, resources and time must be given to supporting the collection of evidence from which to create appropriate SSR interventions. South-South cooperation in sharing this evidence is another recommendation. The development field has finely honed methodologies for monitoring and evaluation, analysis and community-based ratification of results, as well as good training programmes in place to build the capacities of field researchers to conduct inclusive data collection. These methods can be easily adapted for use in SSR programming. Better communication when formulating SSR and development plans should also be possible by now, especially after several years of field efforts to implement unified methods such as the One UN approach. However, in order to avoid the current fixation with producing outcomes that can be easily monitored and reported on – and for which donors can quickly and visibly claim credit – the imagined or received timelines for successful SSR should be reconsidered.

Invest in and capitalise on consultative processes!

As any development-focused monitoring and evaluation plan will show, the collection and confirmation of data lend themselves well to promoting consultative processes. A common concern expressed in this volume, however, is that little or no consultation takes place before SSR programmes are put in place – or that those community-based discussions that do happen are extractive or facile, taking information from communities, using it to validate (and often unfortunately merely rubber stamp) programme design, but giving little back in terms of responsive security initiatives. Women are still routinely excluded from security-related discussions, while rural communities also suffer from neglect.

Consultations may not be taking place because of a number of problems. Planners of SSR might wish to move quickly and save resources up front. They may imagine that SSR models are easily transferable from one situation to another and require only a little on-the-spot local adaptation, despite plenty of literature and guidelines to the contrary. They may have in mind an end-game that assumes SSR is a process with a specific launch and a specific conclusion – starting at some point in time and stopping at another. They may, indeed, view monitoring and evaluation the same way, forgetting that regular evaluation is needed to establish whether an intervention is delivering on its hoped-for outcomes. They may believe that SSR processes should be tied to other events such as electoral cycles, rather than being a self-iterative practice able to advance and respond to its own achievements. Worst of all, they may try to avoid altogether the messiness of understanding and addressing a community's needs, or might fail to recognise that issues which should have been resolved prior to the consultation were never in fact addressed. The development field can lead SSR practitioners out of these dead-ends, offering a multitude of tools designed to facilitate solid consultation processes, including gender-sensitive ones. These tools can easily be adapted for the purposes of conducting SSR consultations, especially in underserved communities. SSR and development practitioners should also assist each other in their analysis of their findings so that responses are, if not jointly decided, at least mutually comprehensible and reinforcing.

As development practice shows, a consultation process does not mark a time 'before' security reforms begin. Instead, the consultation process in itself becomes a site of knowledge transfer and social transformation through which those usually excluded from security-related discussions can gain real authority and agency on the subject that may give them a new status and authority in the community, making them both safer and more able to realise their rights. Informed citizens can challenge power differences by taking on security sector corruption and violence and playing a watchdog role to ensure that the work done with security personnel has a positive impact. For those marginalised by poverty, sexism and/or racism, their age, location or level of physical ability, gaining access to security discourses is particularly empowering. These are also all features that are inherent in good governance approaches to security governance, and will ensure that the accomplishments of SSR processes will be sustainable and can be continued beyond the duration of a technical reform programme.

In addition, we recommend that community information-gathering is always conducted with a view to creating synergies with the information gathered for long-term development planning. This requires that the different programming timelines of SSR and development – the former in its practice and application all too often favouring quick-impact interventions and the latter valuing long-term approaches – must be reconciled. If SSR truly has an agenda to create sustainable change, then steady transformation must be its vehicle.

In addition, reforms that are perceived to generate little improvement for beneficiary populations are not effective and suffer from a public relations problem: they might be considered useless and thus not worthy of further support by both the donor community and the beneficiaries. Thus integrated SSR and development activities that are generating positive results need to be showcased to both the population and the institutions and organisations that are undergoing reform.

Embark on joint reflection, assessment, planning and implementation!

There is much SSR can learn from development, in both design and implementation. For decades the development community has experimented and come up with better approaches to achieve real change on the ground for beneficiary populations. This does not involve a leap of faith, as people-centred improvements are also at the heart of SSR. This volume presents an invitation to both development and SSR communities to drop wrong assumptions and preconceptions and join a common debate and common mission.

Just as the SSR community needs to engage more seriously with development needs, objectives and approaches, so is there a need for the development community to work more closely with those propagating, planning, designing and implementing SSR programmes.

As the development community faces the challenge of engaging with SSR discussions, objectives and approaches, it needs to resist largely unfounded fears of securitisation – risks which will be even more reduced if joint planning begins at the assessment and planning phase. Mutual benefits can be gained if the respective programmes complement and enhance each other, which is only possible if they are jointly assessed and planned. SSR cannot be sensitive to development objectives if it is not aware of what is required by societal and state actors to achieve them, and able to understand what this implies for the security sector and its reform approaches and priorities. One can only be sensitive to what one fully understands. There is too much to be gained in the long run to be ignorant of the need to collaborate – first in understanding each other's missions (which are in fact very similar in intent if not approach), and subsequently in implementing, maintaining and securing the gains of reform efforts. The earlier and the more intensive their collaboration and joint planning efforts, the more likely both will be in a position to learn from and positively enhance the other.

Future research priorities and concluding thoughts

If we want to learn from past practice and better harmonise security, development and SSR agendas, we need to examine the major difficulties and drawbacks in doing so, analyse how these can be explained and overcome, and identify what remains to be accomplished to ensure that SSR – in theory and practice – is one of the main engines of equitable and sustainable human development. As outlined above, corollary studies from those representing the societies and institutions receiving security and development assistance and support for their reform activities would be an important and welcome follow-up to this study.

Furthermore, we observe that there has been too little dialogue on the issue of the humanitarian-security-development triad (beyond the security-development focus), its chronology and its relationship with SSR. Although not explicitly articulated in most of this book, an enlarged discussion on this three- to four-way relationship seems long overdue.

At present there appears to be a perception, especially apparent at the field level, that the SSR and development communities have incompatible purposes and need to use different means and ends to achieve their goals. The evidence in this book cautions against this view, suggesting that security sector reform programming that complements development initiatives will produce more effective, inclusive and sustainable results. Security and development have travelled parallel roads for more than a decade, crossing paths more often by accident than by design and to the irritation and detriment of each community. This situation can only change if planners, practitioners and donors refrain from making unsubstantiated claims about how security and development reinforce each other, and commit themselves instead to a proper alignment of their goals through better research, stronger communication and a return to their common vision of making the world a safer, better place for all.

Our conclusion and our call to action are the same: embarking on SSR means returning to its development roots. Like good development programming, SSR needs short-, medium- and long-term, top-down, bottom-up and horizontal activities that support, merge and come to fruition in the context of, and in unison with, long-term development objectives and projections.

Notes

[1] United Nations, 'Securing Peace and Development: The Role of the United Nations in Supporting Security Sector Reform', Report of the Secretary-General, UN Doc. A/62/659–S/2008/392 (New York: United Nations, 3 January 2008).

[2] For an elaboration of ideal versus actual SSR approaches, in theory and in practice, see Hans Born and Albrecht Schnabel, eds, *Security Sector Reform in Challenging Environments* (Münster: LIT Verlag, 2009), particularly the chapter by Schnabel on 'Ideal Requirements versus Real Environments in Security Sector Reform': 3–36.

Contributors

Alan Bryden is deputy head of research at the Geneva Centre for the Democratic Control of Armed Forces (DCAF). He recently completed a multi-stakeholder project on demobilisation, disarmament and reintegration (DDR) and security sector reform (SSR), resulting in a new module and operational guide input for the UN Integrated Standards for Disarmament, Demobilization and Reintegration. Ongoing research relates to entry points for security sector reform in francophone West Africa and linkages between SSR and post-conflict peace-building. Prior to joining DCAF he was a civil servant with the UK Ministry of Defence and also worked on secondment to the UK Department for International Development. Alan Bryden holds a PhD from the Department of Peace Studies, University of Bradford.

Vanessa Farr is the social development and gender adviser at UNDP's Programme of Assistance to the Palestinian People, where she focuses on how gender impacts on individual experiences of the intra-Palestinian and Palestinian-Israeli conflicts. She is an expert on gendered experiences of armed conflict, including several aspects of SSR: the DDR of women and men combatants after war, gender and policing, the impacts on men and women of prolific small arms and light weapons (SALW) and women's coalition-building in conflict-torn societies. For the past decade she has worked on gender mainstreaming in weapons collection programmes, written on gender in SSR and DDR processes, undertaken research on the gendered impact of SALW, published on Security Council Resolution 1325 (2000) and its operational implications and co-edited (with Henri Myrttinen and Albrecht Schnabel) a book entitled *Sexed Pistols: The Gendered Impacts of Small Arms and Light Weapons* (UNU Press, 2009). She holds a PhD from York University, Canada.

Ann M. Fitz-Gerald is a reader at Cranfield University's Department of Management and Security and the director of the Centre for Security Sector Management. Her main research interests lie in the field of national security planning and examining the implications for security sector policies and strategies. She is also the course director for Cranfield University's MSc in security sector management, a programme which is delivered in both the UK and Addis Ababa, Ethiopia. She has worked in research and consultancy capacities for a wide range of bilateral and multilateral donor organisations and has advised many national governments on issues related to their national security frameworks. She is presently based at the Royal Military College of Canada in Kingston, Ontario, where she is spending 12 months as the McNaughton-Vanier Visiting Chair. She is currently leading a OECD/INCAF research project entitled 'Global factors increasing the risk of conflict and fragility'.

Tim Goudsmid obtained his master's in advanced international studies, with a concentration on international political economy, from the Diplomatische Akademie Wien. His main research interests include the relationship between crime and (under)development on the local level, governance responses to cross-border crime and the relationship between resource abundance and international competitiveness. He has experience with development on the community level in West Africa and South America (Brazil), and has recently been active for the UN Office on Drugs and Crime in Afghanistan.

Heidi Hudson is professor of international relations and currently academic programme director of the Centre for Africa Studies at the University of the Free State, Bloemfontein, South Africa. Her main areas of teaching include international relations theory, feminist security thinking, Africa's international relations and political strategic planning. Her empirical work on gender has focused on post-conflict African case studies (such as Rwanda and Côte d'Ivoire) and the challenges of peace-building. Her current research interests concentrate on the gender deficits of (neo)liberal peace-building. She has contributed several chapters in books and published numerous articles in, among others, *Security Dialogue*, *Security Studies*, *Agenda* and *African Security Review*. Over the years she has held fellowships in, among others, Sweden, the United States, Thailand and Canada. She is currently co-editor of *International Feminist Journal of Politics*. Heidi Hudson holds a PhD in strategic studies from the University of the Free State.

Paul Jackson is professor of African politics in the College of Social Sciences at the University of Birmingham, UK. He is also head of the International Development Department and a former head of the School of Government at the university. His main research interests lie in the areas of governance and security, with an emphasis on state-building and the boundaries between governance systems. He has published widely in the areas of governance and security, particularly in relation to SSR in Sierra

Leone and beyond. He also has considerable field experience, especially in Africa. He is currently engaged as part of the peace process in Nepal, working with the Maoist and Nepali security forces and the Nepali parliament. He has published two books in the past year: *Conflict, Security and Development: An Introduction* (Routledge, 2010, with Danielle Beswick), and *Reconstructing Security after Conflict: Security Sector Reform in Sierra Leone* (Palgrave Macmillan, 2010, with Peter Albrecht).

Marc Krupanski is a research assistant with DCAF. He previously worked on constitutional and international human rights law in the United States, specifically related to civilian oversight and reform of criminal justice systems. He has also worked extensively on community empowerment and development projects in low-income neighbourhoods in the United States, Mexico, Haiti and Cuba. He holds a degree in history and Latin American studies from New York University and an MA in international history and politics from the Graduate Institute of International and Development Studies, Geneva.

Rahel Kunz is a lecturer at the Institute of Political and International Studies of the University of Lausanne, Switzerland. Her main research interests are gender issues in migration, development and security, the governance of international migration and feminist and post-structural theories. Rahel Kunz has published articles in the *Journal of European Integration*, the *Review of International Political Economy* and *Third World Quarterly*. She has recently completed *The Political Economy of Global Remittances: Gender, Governmentality and Neoliberalism* (Routledge, 2011) and co-edited (with Sandra Lavenex and Marion Panizzon) *Multilayered Migration Governance: The Promise of Partnership* (Routledge, 2011).

Andrea Mancini is an international civil servant working for the UN Office on Drugs and Crime as programme manager for Afghanistan. His main research interests lie in the broad field of development and conflict, with a specific focus on organised crime and drugs. He has fieldwork experience in Latin America (Colombia and Brazil), West Africa and Central Asia, and has authored papers and articles related to human security and violence, as well as surveys on refugees and rural-development-related issues. He holds a degree in international relations from the University of Bologna and a master's in development studies from the London School of Economics. He is member of a non-profit organisation (www.avigam.com) committed to supporting local communities and young photo journalists and film-makers, fielding its missions in developing countries and aiming at exploring marginalised areas.

Derek McDougall is a principal fellow (associate professor) in the School of Social and Political Sciences at the University of Melbourne in Australia. His research interests focus on Asia-Pacific international politics, with particular reference to Australian foreign policy, security regionalism and international interventions (including East Timor and the Solomon Islands). Recent books are *Asia Pacific in World Politics* (Lynne Rienner, 2007) and *Australian Foreign Relations: Entering the 21st Century* (Pearson Education Australia, 2008). He was guest editor for a special issue of *Round Table: The Commonwealth Journal of International Affairs* on "Australia and the Developing World" in August 2011, and has contributed articles in recent years to *Asian Security*, *Australian Journal of International Affairs*, *Australian Journal of Politics and History*, *Commonwealth & Comparative Politics* and *International Journal*. He serves on the editorial board for *Contemporary Security Policy* and the international advisory board for *Round Table*.

Henri Myrttinen is currently carrying out post-doctoral research at the Nordic Institute of Asian Studies in Copenhagen on reintegrating former members of the resistance in Timor-Leste. This is a follow-up to his PhD research on militia, gang, martial and ritual arts group members, which he carried out at the University of KwaZulu-Natal. He has published numerous articles and book chapters on post-conflict developments in both Aceh and Timor-Leste. His most recent publication is a co-edited (with Monika Schlicher and Maria Tschanz) volume on post-independence East Timor, entitled *'Die Freiheit, für die wir kämpfen...': Osttimor nach der Unabhängigkeit ['The Freedom We Are Struggling for...': East Timor after Independence]* (Regiospectra Verlag, 2011).

Albrecht Schnabel is a senior fellow in the Research Division of DCAF. He studied political science and international relations in Germany, the United States and Canada (where he received his PhD in 1995 from Queen's University). He has held teaching and research appointments at Queen's University, the American University in Bulgaria, Central European University, the United Nations University, Aoyama Gakuin University, swisspeace and the University of Bern. His publications have focused on ethnic conflict, refugees, human security, armed non-state actors, SSR, conflict prevention and management, peacekeeping and post-conflict peace-building. His experiences beyond academia include military service in the German armed forces, participation in OSCE election monitoring missions, training and teaching for the UN System Staff College and a term as president of the International Association of Peacekeeping Training Centres. He currently works on SSR's role in sustainable development and peace processes; the agency of women, youth and children in post-conflict peace-building; and evolving non-traditional roles of armed forces.

Kristin Valasek has been active in research and training on gender, peace and security issues for many years. She is a gender and SSR project coordinator at DCAF, where she currently focuses on building the gender capacity of security sector training institutions. She previously managed a project on strengthening the integration of gender into the SSR processes in Liberia and Sierra Leone. Prior to joining DCAF, she coordinated gender, peace and security activities at UN-INSTRAW and worked on gender mainstreaming with the UN Department of Disarmament Affairs. In addition, she is a certified mediator and has grassroots non-governmental organisation experience in the areas of domestic violence, sexual assault and refugee support. She holds a master's in conflict resolution from the University of Bradford.

Willem F. van Eekelen started his career in the Netherlands foreign service with postings in New Delhi, London, Accra, Paris and Brussels/NATO, and in The Hague as director of the Security Department. Elected to the Netherlands parliament in 1977, he served as state secretary for defence, state secretary for European affairs and minister of defence. In 1989 he became secretary-general of the Western European Union and in 1995 was elected to the Netherlands Senate. He participated in the European Constitutional Convention. Currently he is a member of the Netherlands Advisory Commission for European Integration.

Andrés Vanegas Canosa is a lawyer with experience in both the private sector and multilateral organisations. He worked for the Latin America and Caribbean desk of the Division for Operations at UN Office on Drugs and Crime headquarters in Vienna. Prior to this he was a junior international lawyer at Microfinance Foundation BBVA in Madrid, Spain, working in the Legal Affairs Department. Andrés Vanegas Canosa holds a master's in international trade law, specialising in international contracts, and a degree in law, both gained at the University Carlos III de Madrid.

Selected Bibliography

Alkire, Sabina, 'Human Development: Definitions, Critiques, and Related Concepts', Human Development Research Paper 2010/01 (New York: UNDP, June 2010).

Baker, Bruce and Eric Scheye, 'Multi-layered Justice and Security Delivery in Post-conflict and Fragile States', *Conflict, Security and Development* 7, no. 4 (2007): 503–528.

Ball, Nicole 'Promoting Security Sector Reform in Fragile States', PPC Issue Paper 11 (Washington, DC: USAID, 2005).

Ball, Nicole, 'World Bank/IMF: Financial and Programme Support for SSR', in *Intergovernmental Organisations and Security Sector Reform*, ed. David M. Law (Münster: LIT Verlag, 2007): 137–156.

Born, Hans, 'Security Sector Reform in Challenging Environments: Insights from Comparative Analysis', in *Security Sector Reform in Challenging Environments*, eds Hans Born and Albrecht Schnabel (Münster: LIT Verlag, 2009): 241–266.

Born, Hans and Albrecht Schnabel, eds, *Security Sector Reform in Challenging Environments* (Münster: LIT Verlag, 2009).

Boutros-Ghali, Boutros, 'An Agenda for Peace: Preventive Diplomacy, Peacemaking and Peacekeeping', report pursuant to statement adopted by Summit Meeting of the Security Council, 31 January 1992, UN Doc. A/47/277–S/24111 (New York: United Nations, 17 June 1992).

Bryden, Alan, 'From Policy to Practice: Gauging the OECD's Evolving Role in SSR', in *Intergovernmental Organisations and Security Sector Reform*, ed. David Law (Münster: LIT Verlag, 2007): 65–83.

Bryden, Alan, 'Understanding the DDR-SSR Nexus: Building Sustainable Peace in Africa', DCAF Issue Paper (Geneva: DCAF, 2007).

Bryden, Alan and Vincenza Scherrer, eds, *Disarmament, Demobilisation and Reintegration and Security Sector Reform – Lessons from Afghanistan, Burundi, the Central African Republic and the Democratic Republic of the Congo* (Münster: LIT Verlag, forthcoming in 2012).

Bryden, Alan, Timothy Donais and Heiner Hänggi, eds, 'Shaping a Security Governance Agenda in Post-Conflict Peacebuilding', DCAF Policy Paper no. 11 (Geneva: DCAF, 2005).

Brzoska, Michael, 'Development Donors and the Concept of Security Sector Reform', DCAF Occasional Paper no. 4 (Geneva: DCAF, 2003).

Brzoska, Michael, 'Embedding DDR Programmes in Security Sector Reconstruction', in *Security Governance in Post-Conflict Peacebuilding*, eds Alan Bryden and Heiner Hänggi (Münster: LIT Verlag, 2005): 95–113.

Brzoska, Michael and Peter Croll, *Investing in Development: An Investment in Security*, (Geneva: UNIDIR, 2005).

Burton, Cynthia, 'Security Sector Reform: Current Issues and Future Challenges', in *East Timor: Beyond Independence*, eds Damien Kingsbury and Michael Leach (Clayton, Vic.: Monash University Press, 2007): 97–109.

Buzan, Barry, 'New Patterns of Global Security in the Twenty-first Century', *International Affairs* 67, no. 3 (1991): 431–451.

Buzan, Barry, *People, States and Fear: An Agenda for International Security Studies in the Post-Cold War Era*, 2nd edn (Boulder, CO: Lynne Rienner, 1991).

Call, Charles, ed., *Creating Justice and Security After War* (Washington, DC: USIP Press, 2006).

Cawthra, Gavin and Robin Luckham, eds, *Governing Insecurity: Democratic Control of Military and Security Establishments in Transitional Democracies* (New York: Zed Books, 2003).

Collier, Paul, *The Bottom Billion: Why the Poorest Countries Are Failing and What Can Be Done About It* (Oxford: Oxford University Press, 2007).

Collier, Paul, V. L. Elliott, Håvard Hegre, Anke Hoeffler, Marta Reynal-Querol and Nicholas Sambanis, eds, *Breaking the Conflict Trap: Civil War and Development Policy* (Oxford: Oxford University Press, 2003).

Commission on Human Security, *Human Security Now* (New York: Commission on Human Security, 2003).

Coopération Internationale pour le Développement et la Solidarité, 'CIDSE Study on Security and Development', CIDSE Reflection Paper (Brussels: CIDSE, 2006).

DFID, *Fighting Poverty to Build a Safer World: A Strategy for Security and Development* (London: DFID, 2005).

Donais, Timothy, ed., *Local Ownership and Security Sector Reform* (Münster: LIT Verlag, 2008).

Duffield, Mark, *Global Governance and the New Wars: The Merging of Development and Security* (London: Zed Books, 2001).

Duffield, Mark, *Human Security: Linking Development and Security in an Age of Terror* (Bonn: German Development Institute, 2006).

Duffield, Mark, *Development, Security and Unending War: Governing the World of Peoples* (Cambridge: Polity, 2007).

Duffield, Mark, 'The Liberal Way of Development and the Development-Security Impasse: Exploring the Global Life-Chance Divide', *Security Dialogue* 41, no. 1 (2010): 53–76.

Ebo, Adedeji, 'Security Sector Reform as an Instrument of Sub-Regional Transformation in West Africa', in *Reform and Reconstruction of the Security Sector*, eds Alan Bryden and Heiner Hänggi (Münster: LIT Verlag, 2004): 65–92.

Ebo, Adedeji, 'The Challenges and Opportunities of Security Sector Reform in Post-conflict Liberia', DCAF Occasional Paper no. 9 (Geneva: DCAF, 2005).

Ebo, Adedeji, 'The Role of Security Sector Reform in Sustainable Development: Donor Policy Trends and Challenges', *Conflict, Security and Development* 7, no. 1 (2007): 27–60.

European Commission, *A Concept for European Community Support for Security Sector Reform*, communication from the Commission to the Council and the European Parliament, SEC (2006) 658.

Farr, Vanessa, Henri Myrtinnen and Albrecht Schnabel, eds, *Sexed Pistols: The Gendered Impacts of Small Arms and Light Weapons* (Tokyo: United Nations University Press, 2009).

Fischer, Martina and Beatrix Schmelzle, eds, *Building Peace in the Absence of States: Challenging the Discourse on State Failure*, Berghof Dialogue Series no. 8 (Berlin: Berghof Research Center, 2009).

Fitz-Gerald, Ann M., 'Addressing the Security-Development Nexus: Implications for Joined-up Government', *Policy Matters* 5, no. 5 (2004): 1–24.

Gaddis, John Lewis, 'The Long Peace: Elements of Stability in the Postwar International System', *International Security* 10, no. 4 (1986): 99–142.

Garrasi, Donata, Stephanie Kuttner and Per Egil Wam, 'The Security Sector and Poverty Reduction Strategies', Conflict, Crime and Violence Issue Note (Washington, DC: World Bank, 2009).

Hänggi, Heiner, 'Security Sector Reform', in *Post-Conflict Peacebuilding – A Lexicon*, ed. Vincent Chetail (Oxford: Oxford University Press, 2009): 337-349.

Hänggi, Heiner and Vincenza Scherrer, 'Recent Experience of UN Integrated Missions in Security Sector Reform', in *Security Sector Reform and Integrated Missions*, eds Heiner Hänggi and Vincenza Scherrer (Münster: LIT Verlag, 2008): 3–25.

Hendrickson, Dylan and Andrzej Karkoszka, 'Security Sector Reform and Donor Policies', in *Security Sector Reform and Post-Conflict Peacebuilding*, eds Albrecht Schnabel and Hans-Georg Ehrhart (Tokyo: United Nations University Press, 2005): 19–44.

Henry, Marsha, 'Gender, Security and Development', *Conflict, Security and Development* 7, no. 1 (2007): 61–84.

Hentz, James, 'The Southern African Security Order: Regional Economic Integration and Security among Developing States', *Review of International Studies* 35, no. 2 (2009): 189–213.

Hettne, Björn, 'Development and Security: Origins and Future', *Security Dialogue* 41, no. 1 (2010): 31–52.

Hewitt, J. Joseph, Jonathan Wilkenfeld and Ted Robert Gurr, *Peace and Conflict Report 2008* (College Park, MD: Center for International Development and Conflict Management, 2008).

Hickey, Samuel and Giles Mohan, 'Towards Participation as Transformation: Critical Themes and Challenges', in *Participation: From Tyranny to Transformation?*, eds Samuel Hickey and Giles Mohan (London and New York: Zed Books, 2004): 3–24.

Hippler, Jochen, ed., *Nation-Building: A Key Concept for Peaceful Conflict Transformation?* (London: Pluto Press, 2005).

Hoogensen, Gunhild and Kirsti Stuvoy, 'Gender, Resistance and Human Security', *Security Dialogue* 37, no. 2 (2006): 207–228.

Hübschle, Annette, 'Crime and Development – A Contentious Issue', *African Security Review* 14, no. 4 (2005): 1–3.

Human Security Centre, *Human Security Report 2005: War and Peace in the 21st Century* (New York: Oxford University Press, 2005).

Hurwitz, Agnès and Gordon Peake, 'Strengthening the Security-Development Nexus: Assessing International Policy and Practice since the 1990s', in *Security-Development Nexus Program Conference Report* (New York: International Peace Academy, April 2004).

Hutton, Lauren, 'A Bridge too Far? Considering Security Sector Reform in Africa', ISS Paper no. 186 (Pretoria: Institute for Security Studies, 2009).

International Commission on Intervention and State Sovereignty, *The Responsibility to Protect* (Ottawa: International Development Research Center, 2001).

Ismail, Olawale and Abiodun Alao, 'Youths in the Interface of Development and Security', *Conflict, Security and Development* 7, no. 1 (2007): 3–25.

Jensen, Steffen, 'The Security and Development Nexus in Cape Town: War on Gangs, Counterinsurgency and Citizenship', *Security Dialogue* 41, no. 1 (2010): 77–98.

Kabeer, Naila, *Reversed Realities: Gender Hierarchies in Development Thought* (London: Verso, 1994).

Kent, Randolph, 'The Governance of Global Security and Development: Convergence, Divergence and Coherence', *Conflict, Security and Development* 7, no. 1 (2007): 125–165.

Klingebiel, Stephen, ed., *New Interfaces between Security and Development* (Bonn: German Development Institute, 2006).

Krause, Keith and Oliver Jütersonke, 'Peace, Security and Development in Post-conflict Environments', *Security Dialogue* 36, no. 4 (2005): 447–462.

Lemay-Hebert, Nicolas, 'The "Empty-Shell" Approach: The Setup Process of International Administrations in Timor-Leste and Kosovo, Its Consequences and Lessons', *International Studies Perspectives* 12, no. 2 (2011): 190–211.

Luckham, Robin, 'Introduction: Transforming Security and Development in an Unequal World', *IDS Bulletin* 40, no. 2 (2009): 1–10.

McDougall, Derek, 'The Security-Development Nexus: Comparing External Interventions and Development Strategies in East Timor and Solomon Islands', *Asian Security* 6, no. 2 (2010): 170–190.

McFate, Sean, 'Securing the Future: A Primer on Security Sector Reform in Conflict Countries', Special Report no. 209 (Washington, DC: US Institute of Peace, September 2008).

Muggah, Robert, *Listening for a Change! Participatory Evaluations of DDR and Arms Reduction Schemes* (Geneva: UNIDIR, 2005).

Mugumya, Geofrey, *From Exchanging Weapons for Development to Security Sector Reform in Albania: Gaps and Grey Areas in Weapons Collection Programmes Assessed by Local People* (Geneva: UNIDIR, 2005).

Nathan, Laurie, *No Ownership, No Commitment: A Guide to Local Ownership of Security Sector Reform* (Birmingham: GFN-SSR, University of Birmingham, October 2007).

OECD/DAC, 'Conflict, Peace and Development Co-operation on the Threshold of the 21st Century', OECD Policy Statement (Paris: OECD, 5–6 May 1997).

OECD/DAC, 'The DAC Guidelines: Helping Prevent Violent Conflict' (Paris: OECD, 2001).

OECD/DAC, 'Security System Reform and Governance', DAC Guidelines and Reference Series, DAC Reference Document (Paris: OECD, 2005).

OECD/DAC, *OECD/DAC Handbook on Security System Reform: Supporting Security and Justice* (Paris: OECD, 2007).

Scheye, Eric, *Pragmatic Realism in Justice and Security Development: Supporting Improvement in the Performance of Non-state/Local Justice and Security Networks* (The Hague: Clingendael, Netherlands Institute of International Relations, 2009).

Scheye, Eric, *Local Justice and Security Programming in Selected Neighborhoods in Colombia* (The Hague: Clingendael, Netherlands Institute of International Relations, 2011).

Schnabel, Albrecht, 'The Human Security Approach to Direct and Structural Violence', in *SIPRI Yearbook 2008: Armaments, Disarmament and International Security* (Oxford: Oxford University Press, 2008): 87–96.

Schnabel, Albrecht and Hans-Georg Ehrhart, eds, *Security Sector Reform and Post-Conflict Peacebuilding* (Tokyo: United Nations University Press, 2005).

Schnabel, Albrecht and Heinz Krummenacher, 'Towards a Human Security-Based Early Warning and Response System', in *Facing Global Environmental Change: Environmental, Human, Energy, Food, Health and Water Security Concepts*, eds Hans Günter Brauch, Úrsula Oswald Spring, John Grin, Mesjasz Czeslaw, Patricia Kameri-Mbote, Navnita Chadha Behera, Béchir Chourou and Heinz Krummenacher (Berlin, Heidelberg and New York: Springer, 2009): 1253–1264.

Schnabel, Albrecht, Marc Krupanski and Ina Amann, *Military Protection for Humanitarian Assistance Operations – Roles, Experiences, Challenges and Opportunities*, report for Directorate for Security and Defence Policy of Swiss Federal Department of Defence, Civil Protection and Sports (Geneva: DCAF, May 2010).

Schneckener, Ulrich, 'Spoilers or Governance Actors? Engaging Armed Non-state Groups in Areas of Limited Statehood', SFB-Governance Working Paper Series no. 21 (Berlin: DFG Research Center, 2009).

Shaw, Mark, 'Crime as Business, Business as Crime: West African Criminal Networks in Southern Africa', *African Affairs* 101, no. 404 (2002): 291–316.

Short, Clare, 'Security, Development and Conflict Prevention', speech at Royal College of Defence Studies (London, 13 May 1998).

Short, Clare, 'Security Sector Reform and the Elimination of Poverty', speech at Centre for Defence Studies, King's College (London, 9 March 1999).

Smith, Chris, 'Security-sector Reform: Development Breakthrough or Institutional Engineering?', *Conflict, Security & Development* 1, no. 1 (2001): 5–20.

Stern, Maria and Joakim Öjendal, 'Mapping the Security-Development Nexus: Conflict, Complexity, Cacophony, Convergence?', *Security Dialogue* 41, no. 1 (2010): 5–29.

Stewart, Frances, 'Development and Security', CRISE Working Paper no. 3 (Oxford: CRISE, 2004).

Swiss, Liam, 'Security Sector Reform and Development Assistance: Explaining the Diffusion of Policy Priorities among Donor Agencies', *Qualitative Sociology* 34, no. 2 (2011): 371–393.

Tschirgi, Neclâ, *Peacebuilding as the Link between Security and Development: Is the Window of Opportunity Closing?* (New York: International Peace Academy, 2003).

Tschirgi, Neclâ, Michael S. Lund and Francesco Mancini, 'The Security-Development Nexus', in *Security and Development: Searching for Critical Connections*, eds Neclâ Tschirgi, Michael S. Lund and Francesco Mancini (Boulder, CO: Lynne Rienner, 2010): 1–16.

Tschirgi, Neclâ, Michael S. Lund and Francesco Mancini, eds, *Security and Development: Searching for Critical Connections* (Boulder, CO: Lynne Rienner, 2010).

UN Development Programme, *Human Development Report 1994: New Dimensions of Human Security* (New York and Oxford: Oxford University Press, 1994).

UN Development Programme, *Human Development Report 2002* (New York: UNDP, 2002).

UN General Assembly, 'United Nations Millennium Declaration', UN Doc. A/RES/55/2 (New York: United Nations, 18 September 2000).

UN Secretary-General, 'Report of the Secretary-General on the Work of the Organization', Supplement No. 1 (A/54/1) (New York: United Nations, 31 August 1999).

UN Secretary-General, 'Securing Peace and Development: The Role of the United Nations in Supporting Security Sector Reform', Report of the Secretary-General, UN Doc. A/62/659–S/2008/39 (New York: United Nations, 23 January 2008).

Valasek, Kristin, 'Security Sector Reform and Gender', in *Gender and Security Sector Reform Toolkit*, eds Megan Bastick and Kristin Valasek (Geneva: DCAF, OSCE/ODIHR, UN-INSTRAW, 2008): Tool 1.

Vignard, Kerstin, 'Beyond the Peace Dividend – Disarmament, Development and Security', *Disarmament Forum* no. 3 (2003): 5–14.

Visvanathan, Nalini, Lynn Duggan, Nan Wiegersma and Laurie Nisonoff, eds, *The Women, Gender and Development Reader*, 2nd edn (New York: Zed Books, 2011).

Ward, Julius, 'Can Security Sector Reform Alleviate Poverty?', *Polis* 3, no. 2 (2010): 1–10.

World Bank, *Operationalizing the 2011 World Development Report: Conflict, Security, and Development* (Washington, DC: World Bank, 2011).

Wulf, Herbert, 'Security Sector Reform in Developing and Transitional Countries', Dialogue Series no. 2 (Berlin: Berghof Research Center, 2004).

Zoellick, Robert B., 'Fragile States: Securing Development', *Survival* 50, no. 6 (2008/2009): 67–84.

About the Geneva Centre for the Democratic Control of Armed Forces (DCAF)

The Geneva Centre for the Democratic Control of Armed Forces (DCAF) is an international foundation whose mission is to assist the international community in pursuing good governance and reform of the security sector. To this end, the Centre develops and promotes appropriate norms at the international and national levels, determines good practices and relevant policy recommendations for effective governance of the security sector, and provides in-country advisory support and practical assistance programmes to all interested actors.

Detailed information is available at www.dcaf.ch

Geneva Centre for the Democratic Control of Armed Forces (DCAF)
Rue de Chantepoulet 11
PO Box 1360
CH-1211 Geneva 1
Switzerland
Tel: + 41 22 741 77 00
Fax: + 41 22 741 77 05
E-mail: info@dcaf.ch

Judy Smith-Höhn

Rebuilding the Security Sector in Post-Conflict Societies: Perceptions from Urban Liberia and Sierra Leone

In Liberia and Sierra Leone, strategies to reform and reconstruct the security sector have centred on re-establishing the state's monopoly on the use of force. However, little attention is given to the array of non-state actors that often play a major role in how individuals and communities experience security. Rebuilding the Security Sector in Post-Conflict Societies: Perceptions from Urban Liberia and Sierra Leone seek to address this gap by applying a human security approach to security provision across these two contexts. A key point of departure is that in the long run there can be no alternative within post-conflict societies to a locally owned security sector. Operationalising the concept of local ownership means that internationally-supported security sector reform (SSR) activities need to reflect these local realities. As explored within this study, fostering synergies between state and non-state security actors may therefore offer an important avenue to support more sustainable, legitimate SSR efforts.

2010, 256 S., 29,90 €, br., ISBN 978-3-643-80074-9

Alan Bryden; 'Funmi Olonisakin (Eds.)
Security Sector Transformation in Africa
The need for security sector transformation (SST) is prominent in the work of scholars, policy makers and practitioners that focus on the security sector and its governance in Africa. At the heart of this approach is the requirement for comprehensive change in the orientation, values, principles and practices that shape the provision, management and oversight of security on the African continent. The evident obstacles to achieving such far-reaching goals mean that it is particularly important to identify the practical utility of the SST concept in supporting positive behaviour change within different African settings. It is also necessary to clarify the relationship between the concept of security sector transformation and the evolving security sector reform (SSR) discourse. This volume seeks to provide such additional clarity to SST and its relationship to SSR. It includes contributions from a range of acknowledged experts analysing dynamics of security sector transformation at the domestic level as well as 'beyond the state'. The resulting insights are intended to help elaborate an understanding of the challenges to and opportunities for the realisation of an operational security sector transformation agenda in Africa.
2010, 256 S., 29,90 €, br., ISBN 978-3-643-80071-8

Peter Albrecht; Paul Jackson (Eds.)
Security Sector Reform in Sierra Leone 1997 – 2007: Views from the Front Line
Sierra Leone is often cited as a highly influential example of security sector reform (SSR) in practice. In the ten years (1997 – 2007) covered in this volume, Sierra Leone transitioned from open conflict to a process of consolidation and development that culminated in the successful general elections of 2007. SSR is understood as being at the heart of this change. *Security Sector Reform in Sierra Leone 1997 – 2007: Views from the Frontline* seeks to shed new light on this process by giving a voice to stakeholders that were intimately involved in SSR efforts within Sierra Leone. Contributions from both UK and Sierra Leonean personnel provide authentic perspectives that enable us to draw important lessons from a dynamic and evolving relationship between national actors and the wider international community.
2010, 240 S., 29,90 €, br., ISBN 978-3-643-80063-3